Electrical Palestine

Electrical Palestine

CAPITAL AND TECHNOLOGY FROM EMPIRE TO NATION

Fredrik Meiton

UNIVERSITY OF CALIFORNIA PRESS

University of California Press, one of the most distinguished university presses in the United States, enriches lives around the world by advancing scholarship in the humanities, social sciences, and natural sciences. Its activities are supported by the UC Press Foundation and by philanthropic contributions from individuals and institutions. For more information, visit www.ucpress.edu.

University of California Press
Oakland, California

© 2019 by The Regents of the University of California

Library of Congress Cataloging-in-Publication Data

Names: Meiton, Fredrik, author.
Title: Electrical Palestine : capital and technology from empire to nation / Fredrik Meiton.
Description: Oakland, California : University of California Press, [2019] | Includes bibliographical references and index. |
Identifiers: LCCN 2018021244 (print) | LCCN 2018025341 (ebook) | ISBN 9780520968486 (ebook) | ISBN 9780520295889 (cloth : alk. paper) | ISBN 9780520295896 (pbk. : alk. paper)
Subjects: LCSH: Electrification—Palestine—History—20th century. | Electrification—Political aspects—Palestine. | Jewish-Arab relations.
Classification: LCC HD9685.P192 (ebook) | LCC HD9685.P192 M45 2019 (print) | DDC 333.793/2—dc23
LC record available at https://lccn.loc.gov/2018021244

Manufactured in the United States of America

28 27 26 25 24 23 22 21 20 19
10 9 8 7 6 5 4 3 2 1

For my parents

> The capitalistic economy of the present day is an immense cosmos into which the individual is born, and which presents itself to him, at least as an individual, as an unalterable order of things in which he must live.
>
> —MAX WEBER, *The Protestant Ethic and the Spirit of Capitalism*

> Look out, honey, 'cause I'm using technology
> Ain't got time to make no apologies
>
> —THE STOOGES, *Raw Power*

CONTENTS

Lists of Tables and Illustrations ix
Abbreviations and Notes on Sources xi
Acknowledgments xiii

Introduction · The Unalterable Order of Electrical Palestine 1

1 · Expert Revolutionary 21

2 · Contentious Concession 51

3 · The Politics of Thin Circuitries 79

4 · The Radiance of the Jewish National Home 117

5 · Industrialization and Revolt 150

6 · Electrical Jerusalem 188

7 · Statehood and Statelessness 209

Conclusion · Electrical Palestine 218

Notes 225
Bibliography 285
Index 301

TABLES AND ILLUSTRATIONS

TABLES

1. The Palestine Electric Corporation, 1923–1954 *xv*
2. The Jerusalem Electric and Public Service Corporation, 1931–1947 *xvi*
3. PEC Consumption Breakdown, 1925–1954 *xvii*

FIGURES

1. Industrial Zionism: Rutenberg's hydropower station, top left, *Filastin*, November 2, 1932. *3*
2. Rutenberg cartoon, *Punch*, June 7, 1922. *15*
3. Pinhas Rutenberg. *22*
4. Maps of the hydroelectrical scheme at the confluence of the Jordan and Yarmuk Rivers. *54*
5. Jewish electricity workers in Jaffa, ca. 1922. *103*
6. Map of land needed around the Naharayim powerhouse. *131*
7. Naharayim. *147*
8. The National Grid, 1933. *156*
9. Blueprint of Nablus distribution network. *180*

MAPS

1. Borders and power lines at the start of the mandate. *71*
2. Partition proposals and the grid at the end of the mandate. *212*

ABBREVIATIONS AND NOTES ON SOURCES

Note: Unless otherwise specified, the organizations and departments cited are Palestine based. For most correspondence to or from various employees of the Palestine Electric Corporation, only the company is identified, and not the individual names. Individual names are specified only when they could be thought to have some significance. Quotations containing obvious but minor mistakes of spelling or grammar have been corrected.

AG	attorney general
AHC	Arab Higher Committee
CA	Crown Agents
CO	Colonial Office, London, UK
CS	chief secretary
CZA	Central Zionist Archives
DC	district commissioner
DG	district governor
DO	district officer
DPW	Department of Public Works
HC	high commissioner
HMA	Haifa Municipal Archives
hp	horsepower
IECA	Israel Electric Corporation Archives
ISA	Israel State Archives
JEC	Jaffa Electric Company, Ltd.
JEPSC	Jerusalem Electric and Public Service Corporation, Ltd.

JMA	Jerusalem Municipal Archives
kW	kilowatt
kWh	kilowatt-hour
NMLA	Nablus Municipal Library Archives
OETA	Occupied Enemy Territory Administration
Pal Gov	The British Mandatory Government of Palestine
PEC	Palestine Electric Corporation, Ltd.
PICA	Palestine Jewish Colonization Association
PLDC	Palestine Land and Development Company
S/SC	secretary of state for the colonies
TAHA	Tel Aviv–Jaffa Municipal Historical Archives
TNA	The National Archives of the United Kingdom
US/SC	undersecretary of state for the colonies
WHA	Weizmann House Archives
WIA	Weizmann Institute Archives
ZC	Zionist Commission
ZOA	Zionist Organization of America

ACKNOWLEDGMENTS

"Of all the ways of acquiring books," Walter Benjamin says, "writing them oneself is regarded as the most praiseworthy." I certainly hope so, given all the work that went into writing this one. Except, of course, I hardly wrote it alone. I am deeply grateful to the people and institutions that helped me bring this book to completion. I received financial support from the Graduate School of Arts and Science and the Taub Center at New York University, Harvard University's Joint Center for History and Economics, the Institute for New Economic Thinking, Cambridge University's Centre for History and Economics, the Andrew W. Mellon Foundation, the Posen Foundation, and the Palestinian-American Research Center. Thanks also to Niels Hooper, Bradley Depew, and the rest of the team at the University of California Press.

This book draws on sources from many large and small archives in Israel, Palestine, and Britain. Without the generous assistance of many devoted archivists, I would scarcely have been able to write it. I owe a large debt of gratitude to Helena Vilensky at the Israel State Archives and to Haya Ben Jacob, Moshe Feintuch, and Shmuel Shechtman at the Israel Electric Corporation Archives. At the archives of the Nablus Municipal Library I thank the director, Dirar Touqan, and especially the head archivist, Ma'yn Sulayman Abu Ghazal. In Nablus, Ala Abu Dheer, Tamer Khatib, and Naseer Arafat gave of their time and their vast knowledge of their hometown's past and present. Thanks also to Merav Segal at Weizmann House and Mati Beinenson at the Weizmann Institute Archives. Prager Adam at the municipal archives of Haifa was a terrific host, as was Rivka Preshel-Gershon at the Tel Aviv–Jaffa municipal archives. I also owe a debt of gratitude to Ronen Shamir, who graciously shared his work with me while it was still in manuscript form.

I also want to express a heartfelt thank-you to all those who read all or parts of this study and provided helpful feedback: Ahmad Amara, Seth Anziska, Tareq Baconi, Nadim Bawalsa, David Biale, Pat Cooney, Muriam Davis, Steve Epstein, Michael Gordin, Stefanie Graeter, Shay Hazkani, Joseph Hodge, Matthew Kelly, Minna Mathiasson, Joel Mokyr, David Myers, Aaron Norton, Haley Peele, Paul Ramirez, Gil Rubin, Lotta Segerström, Naomi Seidman, Mikael Shainkman, Helga Tawil Souri, Daniel Stolz, Sandy Sufian, Salim Tamari, Alex Winder, and Sunny Yudkoff. I owe an especially large debt of gratitude to Samuel Dolbee, who has the dubious distinction of having read virtually every word I have written in the past decade. Adding saintliness to virtue, he has always insisted—even when it couldn't possibly be true—to have enjoyed the process.

A wonderful group of teachers set me on my path. At Oxford, I learned copiously from Walter Armbrust, Raffaella Del Sarto, Eugene Rogan, and Avi Shlaim. At NYU, I benefited from the thoughtful engagement of Manu Goswami, Myles Jackson, and Andrew Needham. Ronald Zweig was a source of unstinting and enthusiastic support from the first. Knowing he was always in my corner was a great comfort. Karl Appuhn was instrumental in setting me on the path that ultimately resulted in this study, and he has been a source of support and inspiration ever since. To Zachary Lockman I owe the greatest debt of all. I am one in a long line of people who consider it the luck of their scholarly lives to have received Zach's gentle and meticulous guidance. He sets the standard for the scholar-teacher.

At Northwestern, I benefited immensely from the stimulating environment of the Science in Human Culture Program and the Department of History. I am grateful to Hellen Tilley, a fraction of whose vast knowledge of the British Empire and the history of science is faintly echoed in this book. Ken Alder gave generously of his time and considerable intellectual powers. His influence over the past years has been so profound that I can no longer point to anything at all that he contributed. Most recently, I have been fortunate to find a wonderful group of colleagues at the University of New Hampshire. A special thank-you to those faculty who workshopped the book with me shortly before it went to press.

In one important sense I did write the book myself: I decided to ignore some suggestions. For that my fellow travelers should be thanking me. This way, they cannot be blamed for any of the flaws in the finished product.

TABLE 1 The Palestine Electric Corporation, 1923–1954

Year	No. of PEC Consumers Connected	Units Sold by PEC (in kWh)	Gross Revenue (in £P until 1948; then £I)	Length of Transmission and Distribution Network
1923	700	n.d.	n.d.	n.d.
1924	1617	478,824	13,755	n.d.
1925	3326	1,847,225	34,354	n.d.
1926	6,550	2,343,764	66,791	83
1927	7,477	2,527,126	71,315	n.d.
1928	8,582	2,973,701	79,900	n.d.
1929	9,303	3,634,838	90,847	n.d.
1930	10,620	6,168,198	125,582	n.d.
1931	12,029	8,707,917	139,673	633
1932	15,113	11,590,350	145,512	762
1933	21,934	20,136,839	221,128	1,051
1934	35,397	34,385,515	363,900	1,435
1935	53,246	50,362,193	488,443	1,759
1936	66,537	65,495,957	582,765	1,952
1937	75,805	71,265,889	611,051	2,198
1938	80,384	72,253,610	611,364	2,378
1939	85,526	84,077,141	675,375	n.d.
1940	86,190	98,873,482	689,932	n.d.
1941	83,337	103,031,247	846,503	2,497
1942	88,288	123,105,000	1,055,964	2,539
1943	85,546	149,540,000	1,285,630	2,603
1944	90,319	173,636,900	1,519,106	2,649
1945	94,910	199,125,538	1,804,870	2,688
1946	107,474	233,058,000	2,127,451	2,806
1947	124,320	281,341,000	2,557,716	3,074
1948	128,563	246,247,651	2,663,751	3,202
1949	150,769	315,007,523	4,008,000	3,394
1950	181,273	443,000,000	4,833,000	3,668
1951	209,062	532,000,000	n.d.	4,012
1952	231,722	638,000,000	n.d.	4,317
1953	265,631	722,600,000	23,418,449	4,814
1954	297,926	901,000,000	28,718,666	5,322

TABLE 2 The Jerusalem Electric and Public Service Corporation, 1931–1947

Year	No. of Consumers	Units Sold (in kWh)
1931	n.d.	839,699
1932	n.d.	1,053,000
1933	n.d.	1,390,000
1934	n.d.	2,186,000
1935	n.d.	3,308,000
1936	n.d.	4,516,000
1937	n.d.	5,724,321
1938	n.d.	6,549,497
1939	18,672	7,398,000
1940	19,954	7,514,294
1941	20,799	8,661,982
1942	21,612	10,383,000
1943	22,572	12,084,000
1944	24,102	14,173,733
1945	26,000	16,768,994
1946	n.d.	18,603,000
1947	n.d.	22,117,000

TABLE 3 PEC Consumption Breakdown, 1925–1954

Year	Units Sold by PEC (in kWh)	Irrigation	Industry	Other Purposes
1925	1,847,225	0	n.d.	n.d.
1926	2,343,764	0	1,427,000	n.d.
1927	2,527,126	n.d.	1,509,180	n.d.
1928	2,973,701	n.d.	1,870,886	n.d.
1929	3,634,838	n.d.	2,426,953	n.d.
1930	6,168,198	1,727,132	2,190,464	n.d.
1931	8,707,917	3,075,748	3,239,258	3,132,000
1932	11,590,350	4,399,533	4,058,629	4,532,000
1933	20,136,839	9,029,436	6,575,526	10,248,000
1934	34,385,515	14,281,892	9,855,466	17,061,000
1935	50,362,193	16,134,366	17,166,602	22,664,000
1936	65,495,957	24,122,151	18,710,245	25,617,000
1937	71,265,889	25,334,888	20,314,114	26,571,149
1938	72,253,610	25,502,139	20,180,322	30,468,000
1939	84,077,141	28,505,454	25,104,000	29,948,000
1940	98,873,482	28,234,000	35,692,000	36,868,000
1941	103,031,247	33,298,000	32,865,000	44,197,000
1942	123,105,000	37,411,000	41,497,000	54,272,000
1943	149,540,000	45,767,000	49,501,000	67,272,000
1944	173,636,900	49,965,000	56,401,000	78,871,000
1945	199,125,538	53,954,000	66,301,000	94,298,000
1946	233,058,000	64,228,000	74,532,000	122,149,000
1947	281,341,000	n.d.	86,109,000	119,000,000
1948	246,247,651	57,000,000	70,000,000	154,000,000
1949	315,007,523	66,000,000	95,000,000	221,000,000
1950	443,000,000	85,000,000	137,000,000	266,000,000
1951	532,000,000	113,000,000	153,000,000	327,000,000
1952	638,000,000	140,000,000	171,000,000	352,000,000
1953	722,600,000	172,000,000	199,000,000	438,000,000
1954	901,000,000	201,000,000	262,000,000	n.d.

Introduction

THE UNALTERABLE ORDER OF ELECTRICAL PALESTINE

> Of course, most of us don't know anybody who knows how any of it works. It's background stuff—*infra* means "below," and a good bit of this "below-structure" actually runs below ground, out of sight, or above our heads in skinny little wires we don't notice anymore; what isn't out of sight remains out of mind—until it fails.
>
> —SCOTT HULER, *On The Grid*[1]

ON A DAY IN LATE MAY 1923, more than a thousand people descended on the Arab town of Tulkarm, a community of four thousand inhabitants perched at the top of the fragmented limestone promontory that connects Palestine's hilly interior to its eastern Mediterranean coastal plain. For centuries, Tulkarm had served as a staging post for armies of conquest and, in more peaceful times, for regional trade. But the agitated multitudes flocking to the town on this late-spring day in 1923 were not looking to truck, barter, and exchange. Nor was the gathering in Tulkarm the only one taking place in the spring and summer of 1923: in Jaffa, Haifa, Jerusalem, Nablus—all over Arab Palestine—people were turning out in unprecedented numbers to discuss the same emergent threat. They came to discuss the electrification of Palestine.[2]

Four years earlier, in 1919, the renowned Jewish engineer Pinhas Rutenberg had turned up in Palestine with designs for a countrywide power system and promises of capital influx and industrial modernity. He arrived in a land of wretched poverty. The Great War, which had ended only the year before, had cut a wide path of disease and starvation through the population, while the Ottoman war machine, before its ouster by British forces, had decimated the land and the livelihoods that depended on it.[3] Rutenberg's proposal offered a way out of this dire state.

Yet as Rutenberg a few years later prepared to throw the switch at Palestine's first powerhouse, in June 1923, the area convulsed with protest. Three days before the meeting in Tulkarm, a large crowd of Palestinians moved through the streets of Jaffa, chanting: "Rutenberg's lampposts are the gallows of our nation!"[4] A few weeks after that, the Sixth Arab Palestinian Congress convened in Haifa and adopted a resolution condemning "the erection of poles and extension of wire" and called for a countrywide boycott of the works, making Rutenberg the only Zionist mentioned by name in that or any other official document of the Palestinian congresses.[5]

But if Rutenberg was offering a solution to a pressing problem, why did so many Palestinians react with existential alarm? Never before had a substantial cash injection been more urgently needed. The answer is that while few denied that Rutenberg's proposal was poised to solve one problem, as far as the Palestinians were concerned it stood to compound another. They looked at Rutenberg's designs—the ring of high-tension wire delineating a new border; the projected load centers portending urban industry; the medium- and low-tension lines progressively filling in the interior spaces—and sensed not just a power system but also the base plate of a future Jewish state.

Proponents of Rutenberg's plans, for their part, dismissed such concerns with reference to a distinction that to this day remains part of what the cultural theorist Stuart Hall called "the horizon of the taken-for-granted."[6] They insisted that Rutenberg's plans were strictly technical. As such, they were "outside of politics" and therefore "should be considered strictly 'neutral.'" The sentiment was commonplace among supporters of Rutenberg's project, whether Jewish or Arab, throughout the interwar period. These exact words, however, belong to the secretary of the Palestine Electric Corporation and came in response to a number of instances in which "irresponsible people" had "blown up or damaged" parts of the grid. The secretary expressed regret that "some newspapers give publicity to these facts as acts of warfare" when in fact they constituted mere "theft and robbery." The letter's clear depoliticizing agenda appears all the more remarkable when we note that it was addressed to the mayor of Nablus, a stronghold of Palestinian nationalism, in March 1948—in other words, in the midst of what Israelis know as "the War of Independence" (*milchemet ha'atzma'ut*) and Palestinians as "the

FIGURE 1. Industrial Zionism: Rutenberg's hydropower station, top left, *Filastin*, November 2, 1932. The fifteenth anniversary of the Balfour Declaration. "Balfour and the woes inflicted on Palestine by his fateful declaration."

Catastrophe" (*al-nakba*).[7] Less than two months later, Prime Minister David Ben-Gurion would declare the independence of the State of Israel from the main hall of Beit Dizengoff in Tel Aviv.

In contrast to the apolitical pose of its promoters, electrification was a central bone of contention between Arabs and Jews throughout the period of British rule, and the terms of the debate persistently turned on the question of whether the technical was political. Unlike in 1923, however, by 1948 Rutenberg's abstract vision had materialized in the form of a dense skein of wires crisscrossing the length and width of Palestine, delivering light, power, and heat to industry, agriculture, public spaces, and private homes. Of the quarter million kilowatt-hours sold in the year before the 1948 War, more than 90 percent was consumed by Jews. This book, therefore, tells a story of how a particular relationship between technology and politics was made in Palestine in the period of British rule, and then tracks its consequences. It is a story with global echoes, one that, through an account of Palestine's electrification, seeks to offer a new perspective on the making and substance of modern political power.

Electricity is central to all the things we associate with the modern age: the accelerating rate and increasingly uneven distribution of economic growth, the making of heterogeneous systems through deepening and multiplying connections between diverse (technical and nontechnical) elements, the exponential increase in our reliance on nonsomatic (chiefly fossil-fueled) motive force and the consequent decoupling of human work and play from earthly rhythms. Electricity is also the technology that allows us to package these developments as spectacles.

So integral is electricity to modern everyday life that we literally cannot imagine our world without it. Electricity's vocabulary and grammar condition how we think and talk about ourselves and our societies: we are shocked by the unexpected, wired from stress, and electrified by excitement; controversial issues are charged, while too much work requires us to recharge. Or most of us, anyway. The human dynamos among us can, of course, power through endlessly. Despite the ubiquity of such metaphors, we rarely register that their origin is in fact technological, just as we take for granted the conveniences that electricity provides. The expressions are what George Orwell called dying images—that "huge dump of worn-out metaphors" whose ubiquitous usage has uncoupled them from the underlying dynamic that gave the expressions force in the first place. To Orwell, dying metaphors contain a political danger. The imprecision of the unmoored images reflects and encourages a "reduced state of consciousness," priming us for unthinking political conformity.[8]

Orwell was no doubt correct. In fact, his insight can be extended beyond the linguistic domain. It is not just in how we talk that electricity has lost much of its original valence, making its politics hard to see. The hollowness of the metaphors corresponds to a larger truth: that building power systems is always rich in politics and fraught with controversy, yet once built, their contested nature disappears from view, as they are sublimated into an apolitical language of technics.[9] Despite the fierce political contention that surrounds the Arab-Israeli conflict, the electrification of Palestine is no exception. Indeed, who still remembers that the first cause of Palestinian national mass mobilization was the construction of an electric grid in Jaffa? That the Zionist movement identified electrification as one of the chief vehicles of Jewish state building? Or that British officials saw the electrification of Palestine as the linchpin of a new era of global peace and prosperity? To recover some of what has been lost, this book considers the power relations that inhere in the technical processes and material objects involved in generating and distributing electrical power. By paying attention to the processes by which the grid took on its seemingly natural form, we gain access to the previously hidden realm of politics and social formation, and so recover the significance that hides behind the dying metaphor of electricity.

More specifically, this book follows the efforts to build a countrywide electric grid in Palestine, an endeavor that corresponded to the years of British rule from 1917 to 1948. In 1921, the British granted a charter for an electricity monopoly to the Russian Zionist and engineer Pinhas Rutenberg, who had prepared a proposal over the course of a year following his arrival in the country in 1919. Three decades later, the power system of the Palestine Electric Corporation, the firm Rutenberg founded, covered the entire country. Atop this technological complex, this *electrical Palestine*, stood an independent Jewish state, whose economy, society, and politics reflected the electrical logic that had helped call it into being.

In the course of the roughly thirty years that this study spans, Palestine underwent a radical transformation: from a vaguely defined area of some six hundred thousand inhabitants in 1917 into a distinct territory with a population of some two million and all the fundamentals of modern statehood by 1948. The economy was eleven times larger, and per capita incomes had more than tripled, although, as we will see, that growth was distributed rather unevenly. The character and composition of the population had also changed: from a population made up of three religious groups—Muslims (80 percent), Christians (10 percent), and Jews (10 percent)—into two ethnically distinct societies, one Arab (68 percent) and one Jewish (32 percent). While the population of Palestine as a whole grew by a factor of three, the number of Jews grew by a factor of ten. Moreover, at the start of the period, roughly three-quarters of the population were rural dwellers, towns were small, and there was little by way of industry.[10] By 1948, however, industry produced a value double that of agriculture; and one-third of Arabs and three-quarters of Jews lived in cities and large towns that were interlinked by dense networks of roads, rails, and wire.[11]

The elevenfold economic growth of Palestine exceeded that of all other Middle Eastern economies. But the economic growth of the Jewish sector, also known as the Yishuv, was globally unique. To economic historians, the Jewish growth miracle in interwar Palestine is a staple truism. The Yishuv's annual growth rate of 4.8 percent, we learn from the historians' tables, made it one of the fastest-growing economies in the world and, with the exception of Venezuela's oil-fueled boom, the world's fastest-growing non-Western economy.[12]

But the countrywide electrification project was also instrumental in launching the accounting practices by which Palestine's transformation was and still is being measured, in the form of the statistics just cited. To comprehend Palestine "in numbers"—to quote the title of the economist Yehoshua Ziman's book from the time—became a pressing imperative for the Zionist movement.[13] The reality conjured by these statistics was not neutral. It enacted the "calculative rationality" that, according to Max Weber, constitutes the beating heart of modern capitalism. The figures announced the sound working of Schumpeter's market mechanism, which rendered the complex social reality of Palestine politically legible as an ethnonational binary, a territory reducible to "Jewish" and "Arab" things. Countless statistical tables published in the years of the British mandate listed "Arab" agriculture, industry, education, life expectancy, and so on in one column, with corresponding "Jewish" figures running alongside. One index, for example, published by the Jewish Agency in 1945, showed that one-third of "Jewish" energy needs were satisfied by electricity, whereas the corresponding "Arab" figure was 7.5 percent.[14] Statistics, therefore, not only reified the simplification of the population into the Jewish-Arab binary we have come to take for granted; they also served as an important means by which the Zionist movement created a distinction between the way the Yishuv produced, consumed, and performed modernity—with the modern bookkeeping to prove it—and the Palestinian Arab community, which, in the words of one Jewish Agency publication, possessed "a mentality that does not always view with favour the exact and numerical approach to reality."[15] Going forward, we will have reason to complicate some of the figures just cited, the assumptions on which they were based, and the politics of their production and dissemination. Like the supposition of an unbridgeable divide between "the technical" and "the political," parsing Palestine into "Jewish" and "Arab" domains was itself a form of politics, whose workings are central to the story at hand.

The central argument of this book is that the story of Palestine's transformation is largely a story of the precipitous and uneven development of its infrastructures, and that its ethnonational conflict is largely a story of diverging economies coevolving with those technologies. Indeed, the Jewish State of Israel, founded on May 14, 1948, was arguably *infrastructural* before it was anything else. From the perspective of fixed capital, the precipitous growth rate of the Jewish infrastructural state outpaced the already significant growth differential in population or capital between the Jewish and Arab sectors.[16] This was not accidental but the outcome of a deliberate effort to erect the material predicates of sovereignty, which proved hugely important to the outcome of 1948. Among the many grand infrastructural projects undertaken in the late Ottoman and mandate periods that are detailed in these pages, electrification was especially important to Palestine's socioeconomic transformation, as well as to the reconfiguration of the area as a modern and Jewish national space. Like no other undertaking, infrastructural or otherwise, the power system made every inch of the territory the target of a single, centralized undertaking, with far-reaching conceptual and material consequences. Like electrification projects elsewhere, the work of the Palestine Electric Corporation integrated local environments into a systemic network of institutions and culture that mimicked and foreshadowed the characteristics of the nation-state that would emerge in 1948.[17] The national culture that the grid participated in producing in Palestine encompassed only one ethnic minority, the Jews, while to the extent that the grid touched Palestinian Arab national culture, it did so largely in terms of opposition and exclusion. The nature of the forces driving the transformation of Palestine from a vaguely defined imperial backwater into a precisely bounded modern state therefore contained the seed of another radical break, made manifest in 1948 in the form of Jewish statehood and Palestinian statelessness.

To contemporaries there was nothing surprising about the fact that a large technological system was essential to the circulation and accumulation of political power in Palestine. One British official commented that the person in charge of electrification would become "the absolute dictator of Palestine's fortunes."[18] And when it turned out that that person would be Pinhas Rutenberg, a committed Zionist, there were many who considered the implications to be obvious and far-reaching. "We are giving to a Jewish organisation a grip over the whole economic life of Palestine," wrote the head of the Middle East Department of the British Colonial Office a few days before granting the concession to Rutenberg in 1921.[19] The Palestinian Arabs were no less convinced that the stakes involved were high. In a petition to the British Parliament, the Palestinian Arab Executive claimed that "the Zionists, through Mr. Rutenberg, are aiming at getting a stranglehold on the economics of Palestine, and once these are in their hands they become virtual masters of the country."[20] As we have already seen, it was not just the political elite that understood the matter in those terms. And no group was more attuned to the political power of electrical power than the Zionists. In a letter from 1935, the Jewish philanthropist James de Rothschild looked back at the power system, in which he had been an early investor, and what it had accomplished: "The purpose of this undertaking was to become—as it in reality did become—an important instrument for the Jewish people in Palestine."[21]

It may seem self-evident, in other words, that electrification, with its vast, sprawling infrastructure and ubiquitous fields of use, was destined to insert itself at every level of private and public life, and so be bound up with the economic and political transformations that Palestine underwent during the thirty years of British mandatory rule. Yet the history of Palestine's electrification and its political, social, and economic ramifications have only rarely been studied, and even when they have, then only partially. This reflects a wider propensity among historians to ignore such linkages as those between technology and politics, despite the great significance attributed to them by people at the time, preferring instead to treat the realm of technology as separate from the realm of human relations. This suggests that the notion of there being an unbridgeable gap between technology and politics, a notion on which much, though by no means all, of the present story turns, is alive and well in contemporary society, including much of contemporary social theory. More often than not, historians overlook the question of how societies evolve in conversation with their technologies, despite the interventions of numerous technology scholars over the past half-century.[22] Yet, in the most straightforward sense of the term, an electric grid is inescapably a social construction, that is, a product of human interaction.[23] This study therefore joins a longstanding effort by technology scholars to demonstrate that ignoring the mutual influences running between technology and politics amounts to a form of question-begging that obscures the way technological objects function as both causes and effects of social change. Like all large technological systems—sewage networks, railroads, telegraphs, and so on—electric grids are cultural artifacts.[24] Mistaking a political object for a natural one elides and thus perpetuates, even intensifies, the politics built into it, as it naturalizes an aspect of political claims-making.

By being embedded in a larger sociotechnical network, the power system shaped and was shaped by evolving political agendas, economic activities, and social visions on a multitude of scales, from the global to the imperial, regional, and local. Putting electrification at the center of the story of Palestine's transformation therefore makes new connections visible, with far-reaching implications for how that story should be understood.

For one thing, it becomes clear that the history of empire matters a great deal more to the history of Palestine than scholars have acknowledged, and in ways previously unexplored. The tendency among scholars has been to treat British policy in Palestine in isolation from Britain's imperial project. But in fact, as this book shows, Britain's attitude toward Zionism and the Arab population was merely another provincial articulation of its empire-wide concern with non-Western development. It was this concern, and the critical role of technology within it, that caused the British to persist in their support for the Jewish national home policy, insisting, despite mounting evidence to the contrary, that Jewish "industry and capital" would facilitate material and moral progress for all inhabitants. At its core, this imperial vision was one of economic development, and one that, given the limited resources that the British state was willing to devote to its imperial project, was necessarily dependent on attracting private capital. The present story, therefore, turns on a technological vision that was profoundly about capital and about capitalism. This is the second insight that electrification makes legible. To call the reader's attention to the social networks that capitalism and technology sustain and are sustained by in turn, this study introduces the term *technocapitalism*. Its purpose is to bring out the deep yet underappreciated connection between the two primary driving forces of this story—technology and capitalism—specifically, to explore how they were mutually sustained by means of the same discursive and material practices, and how in the imperial context of modern Palestine they produced distinctive national movements and, ultimately, territorial partition.

As the case of Palestine demonstrates, capitalism and technology are closely interrelated and share many characteristics.[25] Technological and capitalist reason both rely on self-reinforcing ideas, discourses, and practices that put an ever-growing distance between themselves and alternative systems. Specifically, Zionism's territorial claim was based, to a far greater extent than is recognized in the existing scholarship, on the promise of organizing an economically viable territory in the context of global trade, and of doing so by means of infrastructural technologies. Its advocates justified their claim to Palestine through their promise to transform the territory into an area of modern production and consumption, and crucially also into a viable node in the global flow known as "free trade."[26] That capitalist proposition was underpinned by a belief, on the one hand, in science's ability to stake out the most efficient way forward, and, on the other, in the endless powers of technology to transform apparently backward lands into productive and dynamic participants in global trade.[27] As we will see, this aspiration was expressed on the ground through the application of specific technologies chosen for their supposed ability to engender "free trade," and whose precise properties were instrumental in shaping the endeavor as it evolved, in both expected and unexpected ways.

The material element of the story is as important as it is usually overlooked. Several scholars of nationalism have pointed to the close relationship between capitalism, industrialization, and nationalism, and so have more recent scholars of capitalism.[28] Most famously, Benedict Anderson identified print capitalism as a critical element in the emergence of national communities in Western Europe. He also emphasized the importance of censuses, maps, and museums, which, he argued, produced a "totalizing classificatory grid," on the strength of which "tightly bounded territorial units" could be established. That, as this book shows, is true not merely if we understand "grid" in the metaphorical sense in which Anderson uses it.[29] The electric grid bounded and structured modern Palestine through its symbolical power but also, and perhaps more important, through its material properties. It imposed its electrical logic on the political entity it helped create.

This was so even for the unelectrified. It is a well-known and much-lamented fact among historians of Palestine (and Israel) that there is far more historical documentation on the Israeli side than on the Palestinian side, another manifestation of the power differential that results from statehood and statelessness. This book draws heavily on the records of the Israel Electric Corporation, the successor to Rutenberg's Palestine Electric Corporation. The company documents offer a unique view into the economic and political life of Palestine for both Jews and Arabs, though for reasons explored in these pages the relationship of the company to each was by no means symmetrical. To further compensate for the lopsidedness of the sources, this book draws on the holdings of various local archives, including those in Nabuls, Tel Aviv–Jaffa, and Haifa, and much printed Arabic material. In this, I depart from the approach taken by Ronen Shamir in *Current Flow*, the other extant scholarly work on electrification in Palestine. To begin, Shamir employs a more focused lens; his book deals mainly with Tel Aviv-Jaffa in the 1920s, whereas the present book takes a broader view of the whole of Palestine and the whole of the period of British rule. Moreover, Shamir, who did not consult any sources in Arabic, offers little insight into Palestinian perspectives on electrification, and explicitly denies any Palestinian agency in the making of the power system, even as he insists on claiming agency for the grid itself.[30] For Shamir, electrification, in a process he characterizes as governed "primarily by commercial considerations and the technological imperatives of machines," enacted an "episteme of separatism," which in the course of the

1920s and 1930s caused the Jewish community in Palestine to disengage from the Arab population, politically, socially, and economically.³¹ Thus, even though he initially dismisses distinctions between technology, economics, and politics as "only shorthand," his narrative effectively reaffirms those distinctions. Possibly, it is his reliance on British and Zionist documents that blinds him to the close practical links between technology, politics, and economics, as the British and the Zionists were the ones with a stake in denying such links. The problems inherent in Shamir's approach become especially apparent in his discussion of the 1948 War, as his approach effectively makes commercial and technological logics, which he treats as effectively distinct from other logics, responsible for the flight and expulsion of some 750,000 Palestinians. In the course of researching this book, I have come to believe that Shamir's focus on separatism is misplaced. Like the categories of the "political," "economic," and "technological," ethnonational separatism is better understood as an ideological effect of a de facto relational dynamic. Therefore, I depart from Shamir both methodologically and empirically. In what follows, I seek to show that commercial considerations are never separate from politics, machines have no independent "imperatives," and, most important, Palestinians were an essential part of the network of forces that created Palestine's electric grid.

Palestinian opposition to Rutenberg assumed a central role in the overall struggle against Zionism less than a year after his arrival in the country. Electrification therefore became central to producing Palestine as an object of nationalist contention. The struggle against the first powerhouse, in Jaffa, coincided with a reorientation of Palestinian Arab politics from Greater Syria to a nationalism centered on Palestine. For the Palestinians, then, electrification came to participate in the making of a new inside and outside, constituting Palestine, conceptually and materially, as an object of national politics. The tactics that the nationalist movement adopted, moreover, began from a technological fact, namely, the young electric grid's vulnerability to sabotage, which the Palestinians used to gain purchase for their political demands. Rutenberg countered by switching electricity-generating technologies, from the vulnerable sprawl of waterpower to a contained and thus easier-to-protect fuel-powered station. He then endeavored to expand and thicken the grid ahead of commercial demand, in order to further reduce its physical vulnerability. But most important, Rutenberg engaged in *boundary-work*; that is, he endeavored to align his project with a "free-market" rationale and emphasized the technological exigency that supposedly governed the grid's development, the better to deny the political quality of his work. In so doing, he managed to characterize Palestinian opposition as politically motivated, in contrast to his own scientific posture. In practice, then, the strategies that Rutenberg adopted, including the design of the grid and the location of the power stations, responded to circumstances generated by the interaction of the system's technological properties and the oppositional politics of the Palestinians. Palestinian nationalism continued to evolve together with the grid throughout the mandatory period. Palestinians experienced continuous friction as a result of their desire to "be modern," on the one hand, and their rejection of "Zionist current," on the other. The internal struggle over electrification through the 1930s remade the political fault lines within the Palestinian community in ways that would bear heavily on the great anticolonial revolt of 1936–39, and Palestinians' drastic change of fortunes during the 1948 War.

MATERIAL MEDIATION

The standard historical account of the emergence of modern Palestine in the period following World War I begins with a written text, namely, the 1917 Balfour Declaration. Virtually all histories produced in the past quarter century highlight that declaration's fateful distinction between Jews, to whom it granted national rights, and "non-Jewish inhabitants," who were accorded only "civil and religious rights." This despite the fact that Jews at the time made up only some 10 percent of the population, and the overwhelming majority of the remainder consisted of Palestinian Arabs.[32] The author of the definitive account of the Balfour Declaration, James Renton, writes that it "became the basis for the British Mandate for Palestine, which, in turn, enabled the birth of the Jewish state almost thirty years later" and "led Palestine into one of the most bitter conflicts in modern history."[33] The standard narrative then continues to track this diplomatic history of texts and declarations: the white papers of 1922, 1930, and 1939 and the emergence of the idea of partition, born of the 1937 Peel Commission and given international legal sanction with the UN partition resolution of 1947.[34] In these familiar accounts, each such diplomatic moment is narrated as a response to anti-Zionist or anticolonial violence, such as the Jaffa Riots of 1921, the Wailing Wall Riots of 1929, and the Great Arab Revolt of 1936–39. The result, we are to understand, is that the political course set by the Balfour

Declaration culminated three decades later in the Palestinians' political and physical dispossession in the 1948 War and the creation of the State of Israel.

To explain this pattern of diplomacy and the apparent pro-Zionist bias of British policy, scholars have cited attachment to the Bible and messianic Christianity, imperial realpolitik, and/or the influence of Jewish and Arab racial stereotypes.[35] In all these accounts, the significance of the text itself is undisputed, and so is the *in*significance of all other people and things. Indeed, according to the prominent historian Anita Shapira, "The Balfour Declaration belongs to an era in which a handful of statesmen in smoke-filled rooms decided on the fates of peoples and states and how to divide up declining empires, with no participation by the media or the masses."[36] For the Palestinians, the result of the declaration was an "iron cage," in Rashid Khalidi's famous image, constraining the Palestinian political action for the duration of the mandate and beyond.[37]

This book does not dispute that the history of mandatory Palestine is to a large degree a history of Palestinian dispossession, or that this process was expressed and enforced on the levels of diplomacy and jurisprudence. It does, however, challenge the standard account's assumption that the Balfour Declaration in itself overdetermined the history of Palestine, such that political power flowed directly from the writ of the document. This book also challenges the common claim that British policy muddled through on a tide of incompetence and prejudice, or that, as Tom Segev has claimed, British policy makers somehow were so enraptured by "the mystical power of 'the Jews'" that it "overrode reality."[38] This book retells the story of evolving power relations in Palestine, as the function not of chance, prejudice, or written proclamations, but of the material enactment, over the course of the 1920s and early 1930s, of a Zionist-dominated technocapitalist order centered on a bounded Palestinian territory and economy. Palestine was not transformed directly by the words of the Balfour Declaration, nor by subsequent proclamations, or the civilizing rhetoric of the Permanent Mandates Commission of the League of Nations. Instead, these texts depended for their implementation on material and discursive vehicles of a far more contingent and often rather provincial sort. And those vehicles, as we will see, ended up determining the outcome of 1948 as much as, if not more than, the ideas as they were originally expressed through paper and ink. In other words, this book focuses on the process of material mediation that translated and transformed the Balfour Declaration's ideas into reality. This mediation involved the dispersal of agency, beyond a seminal text and its high-powered authors, over a range of human and nonhuman actors.[39]

Modern Palestine—and the Jewish state that emerged from it in 1948—was forged as people, goods, information, and capital moved through the space in patterns largely determined by its infrastructures. The electrification scheme in particular was essential in setting the territorial scale of modern Palestine, pulling local communities together by virtue of being stakeholders in the grid's growth. The concession that the British granted Rutenberg involved a countrywide monopoly, a requirement, as he successfully argued, of the capital-intensive nature of the enterprise. Thus, even before the borders of Palestine were determined, a nascent *electrical Palestine* was conceptually fixed in terms of an exclusive right to electrify the "Palestine" of the concession text, whatever the precise geographic delimitation would turn out to be. In the event, the electrification venture grew to a vast scale whose technical requirements demanded certain borders, and implied a particular economic future for the land, involving large-scale industry and global capital.

Once completed, the electric grid constituted the first material manifestation of what until then had been a mostly abstract claim for Jewish sovereignty in Palestine. It set Palestine up as a site capable of hosting a modern Jewish national home, complete with a (Jewish) national industry, economy, and culture. By the same token, the Palestinians' struggle against electrification amounted to a concrete campaign to prevent de facto Jewish sovereignty over the land, conducted all over Palestine against a network that seemed to threaten local control over every inch of the territory equally. In short, the power system was essential in shaping out Palestine within the larger agendas of technocapitalist colonial development and Jewish nationalism. The system, for its part, was possible only because of its central role in the Zionist gambit to organize a viable political and economic national entity within that technocapitalist framework. Thus, Zionism, Palestinian nationalism, and the electric system enabled and produced each other, as well as modern Palestine. As a result, the conflict between Arabs and Jews inscribed itself on the grid, as the grid in turn inscribed itself on the conflict.

The final product was an entity I call *electrical Palestine*: a shared lifeworld composed of a set of tightly integrated components, conceptual and material, drawn together in continual violation of the received domains of social theory—those of economy, science, culture, and so on. We might consider using concepts like paradigm, habitus, or episteme. Or if electrical Palestine were soccer, it would be the pitch, the sidelines, the goal posts, the referee,

THE HALF-PROMISED LAND.

FIGURE 2. Rutenberg cartoon, *Punch*, June 7, 1922. Reproduced with permission of Punch Ltd., www.punch.co.uk.

the rules, the players, the ball—the entire "complex of men and things" that make soccer distinct from, say, tennis or fly fishing, and whose rules privilege certain attributes over others, creating certain strong path dependencies.[40]

THE MACHINE IN THE MIDDLE

To understand this history, it is necessary to understand the technology at the center of this book. Electric power systems, like all networks, depend on being viably located within a matrix of other networks. It is this network dynamic that makes power systems so instructive to study. They are, moreover, site-specific technologies that connect people with each other and their environment, and yet they depend significantly on technology transfers from far away. They are structured by universal technological laws, while also subject to local contingency.

Technological exigency suggests a certain way of organizing power systems and, by extension, of organizing the societies they serve. For starters, electricity cannot be economically stored and is produced at the same moment it is consumed. Output, therefore, needs to be both high and even over time, something that is usually accomplished through diversification of production (horizontal integration) and of consumption (private, commercial, industrial, etc.). For instance, one power utility in 1920s Germany benefited from servicing different religious communities with different holidays, since it diversified the timing of demand on the system. In the early history of electrification in America, traction companies often built amusement parks at the ends of their lines to increase load diversity by encouraging off-peak electricity use.[41] Furthermore, building power systems is the most capital-intensive enterprise in history. Only steam railways during that industry's formative years approached the sums required for the initial stages of electrification. The huge capital demands of power systems create the need for multinational business interests with the financial muscle to absorb short- to medium-term losses for the sake of long-term profitability.[42] The capital intensity and exponential nature of electrification's economies of scale encourage the creation of legally protected monopolies, similar to those of railways and water utilities. To see the largest possible returns on the outsize investment, electrical utilities often operate subsidiary ventures that depend on access to cheap and abundant power, what economists call vertical integration.[43] Finally, the introduction of large-scale power systems intensifies labor divisions, by shifting demand on the workforce to skilled and semi-skilled labor, both in the electrical industry itself and in the wider industry that electrification spawns. It also intensifies socioeconomic divisions more broadly in that it, like all increases in technological complexity and their attendant increases in production costs, raises the economic threshold for ownership of the productive means.

These structuring conditions have far-reaching implications for society as they interact with local conditions in shaping social, political, legal, and economic orders. But for all the significance of these structuring factors—the "natural" inclination of power systems, as it were—none of the technological exigencies of power systems is deterministic. For every rule, there are myriad exceptions, motivated by context-specific considerations that often fall outside conventional definitions of the technological realm. This is why electric grids, besides being technical systems, are also cultural artifacts.[44] Moreover, because of the site-specific nature of the technology, operating in the context of a highly sensitive ecosystem of international, national, and local governments, businesses, and consuming publics, it serves as a useful bellwether of historical change. There is, in short, an intimate link between electric grids and social fabrics. And the communication, moreover, is two-way: power systems do not simply form according to some presocial technological exigency, and then shape society; nor are power systems simply products of social influence. Technology is always already social, just as society is always already technological.

This, however, hardly prevents people from making grand claims on assumptions of either technological or social determinism, depending on interests or worldview. In fact, the subtle yet vitally important power play inherent in these claims is one of the most significant elements of the history of technology. For instance, it is often in the interest of operators of large technological systems—because of the accelerator effects associated with scale economies and diversification—to see that they expand; and the particular properties of a given technology predispose its promoters to pursue expansion along certain lines. This, however, does not in itself mean that power companies *should* or *must* be monopolistic large-scale enterprises. What is efficient on the level of the system might not be efficient on the level of society as a whole. Yet it is a central feature of the history of large systems that what is good for the system has uncritically been promoted to a holistic prescription for an entire social order.[45]

This insight is critical to understanding the full significance of electrification in Palestine. The story unfolded in what historians of electricity call the first age of systems electrification. The shibboleth among proponents of large-scale electrification of the time was "rationalization," a term that implied a strictly rational search for "the optimal combination of economic gains with a minimum input of economic resources, including capital and labor."[46] In fact, however, the metrics used in these seemingly technical calculations were based more on the perceived virtues of certain technologies than on their actual efficiency. For instance, when Britain put its own National Grid into commission in the 1930s, overcoming decades of resistance by local power companies and their local political patrons, it did not in fact revolutionize the patterns of British industrial development, although that was usually how it was described at the time. What it did accomplish was to make electricity much cheaper in many parts of the country, thereby significantly increasing consumption of that particular commodity.[47] No surprises there: after all, maximizing output efficiency is what large systems and scale economies do. In other words, technology is surrounded by a great deal of confusion with respect to means and ends. Arguments for a certain technological solution are often justified by means of a circular logic according to which the increased output justifies expansion of the system. In Palestine, the unspoken assumption was that maximizing system efficiency unproblematically equated to a science-based, apolitical model for the organization of society along "modern" lines. To such a model, political objections were inadmissible, category errors that served only to indict those voicing them.

In addition, the electrification of Palestine unfolded at a time when economists and politicians increasingly came to turn to per capita consumption of electricity as a catchall metric of scientific and industrial development. Engineers and government officials then took those figures to index the state of the economy overall, which in turn came to serve as a proxy for civilizational standing. The power of electricity consumption as a civilizational metric derived in part from its status as a science-based technology (assumed to operate independently of softer values, like ideology or culture), and in part from the fact that it was easily quantifiable, and thus easily rendered precise, giving it an aura of objective truth.[48]

The widespread intellectual slippage between technological exigency and social order is not a static feature of human society but a function of a historical conjuncture of the early twentieth century, and especially the emergence of modern large technological systems. In fact, it was precisely in order to name this new kind of interconnected system, combining multiple concepts and objects into a sociotechnical whole, that the use of the word *technology*, in the 1930s, took on the broad meaning it has today. The new concept signified not just a *means* of social and political progress; technology itself was seen as an embodiment of the essence of progress. Yet despite the unprecedented heterogeneity of these systems—which drew on both theoretical and applied science, various forms of expertise, formal institutions like the law, as well as informal social relations—they still tended to be conceptualized in terms of their mechanical element: the machine remained in the middle, the steam engine in the case of the railroad, the generators and wire in the case of power systems. Despite their rapidly growing diversity, then, "technology" was reified as something concrete, mechanical, and objective, and its social relations, its constructed nature and the politics flowing from it, were hidden from view.[49]

Yet while large technological systems transcend borders in practice—stitching together things like mechanics, expertise, production, profit, and social values—in the modern era they also engendered an entirely new discursive edifice of distinct domains, of economics, science, politics, culture, and so on, which served as the foundation of much of contemporary social theory, as well as the common sense that guides us in everyday life. Stuart Hall, as mentioned earlier, calls it "the horizon of the taken-for-granted." Max Weber, in the context of the psychological hold of the capitalist spirit, talks about a seeming "unalterable order of things." Arguably, it is in this paradoxical story of the abundant spread of technology throughout twentieth-century society, and its simultaneous cordoning off as a separate realm, that we begin to see why the historical mainstream has been reluctant to consider the technological elements of the stories we tell. By contrast, the goal of much of Science and Technology Studies, with its explicit goal of scrutinizing the pieties of the "moderns," has been to remind us that there is much more to technological systems than their mechanical elements. "B-52s do not fly," as Bruno Latour reminds us, "the U. S. Air Force flies."[50]

The central and complex role of large technological systems in the modern world matters a great deal to the history of electrification in Palestine and to why that history is so important. The transformation of the area took place at a moment when technocapitalists all over the world were hard at work paving the way for commerce and civilization, combing the globe for untapped potential that was to be identified and unlocked by the tools that science and technology provided. The power system in Palestine derived its political power from its massive presence on the land, from the way it embodied rationality and progress, and from its status as key to unlocking Palestine's hidden potential. At the same time, its influence was both enhanced and shaped by its quasi-invisibility as political power. This served its promoters well, and they, like technocapitalists elsewhere, often pursued their objectives by playing to the commonsensical understanding that the grid was "outside of politics." From the start, Pinhas Rutenberg and the Palestine Electric Corporation expended considerable effort denying that there was anything political about their activities, and insisting that their power system was conceived of, built, and maintained in strict accordance with scientific reason. These efforts involved an array of tools and discursive strategies, mobilizing the power of objectivity, efficacy, precision, reliability, authenticity, predictability, sincerity, desirability, and tradition.[51] Through their success, they created new sources of power and legitimacy. If this sounds familiar, it is because we are living in a world organized and directed by entrepreneurs like Pinhas Rutenberg. If Palestine is electrical, so too is the world.

But this kind of power is never total. Entities amenable to being claimed for science are equally vulnerable to being pulled back into the profane realm of politics. Some Palestinians never lost sight of the political ramifications of electrification, despite the efforts of the British and the power company to deny them. For example, Palestinian commentators at the time excoriated those who took "Rutenberg's *Zionist* current" as "traitors to their city and homeland."[52] Moreover, during their protests in the early 1920s against Jewish national aspirations in Palestine and then again during the revolt of 1936–39, Palestinian Arabs targeted the grid as one of the forces threatening to strip them of their homeland. To many Palestinian Arabs, electricity had not only technological properties but also political ones: whatever else electricity was, in the context of Rutenberg's monopoly network, it was also Zionist. For the Palestinians, therefore, contesting the social and political order also involved contesting a technoscientific order and the civilizational assumptions that underwrote it.[53]

ONE

Expert Revolutionary

> We war with rude Nature; and, by our resistless engines, come off always victorious, and loaded with spoils.
> —THOMAS CARLYLE,
> Critical and Miscellaneous Essays *(1896)*[1]

THE MOTIVE FORCE BEHIND ELECTRICAL Palestine was Pinhas Rutenberg. To many, he appeared as the Platonic embodiment of the iron-willed builder. Right angles on a squat frame topped by a shock of charcoal hair, gimlet gray eyes set narrowly behind round wire-rimmed glasses. According to his biographer Eli Shaltiel, Rutenberg's legacy stands as the man who "single-handedly effected a mighty technological revolution that changed Palestine beyond recognition."[2] Winston Churchill once described him as "a man of exceptional ability and personal force."[3] Rutenberg was unapologetically authoritarian, and many remarked on it. One American Zionist complained to another that Rutenberg had "so little experience in the democratic world of give and take" that he was "all but unmanageable." Yet, he confided, "I cannot help admiring his courage, vision and limitless energy."[4] Chaim Weizmann, the long-standing president of the Zionist Organization and first president of Israel, looked at Rutenberg and saw a "tremendous turbine harnessed to a single great purpose."[5] In short, so intimately linked was Rutenberg with his power system that to observers he *was* electric power. In both Hebrew and Arabic, his name was soon used interchangeably with the charged particles of his grand venture. "Electricity has two names in the Holy Land," *Popular Mechanics* told its readers in 1930. "One is 'hashmal,' mentioned in Ezekiel... and the second name is 'rutenberg.'"[6] In Arabic, *rutenberg* was a word no less pregnant with meaning. In the mid-1930s, for instance, the Palestinian Arab newspaper *al-Difa'* would charge those who wanted "to introduce Rutenberg in Nablus" with having "surrendered to imperialism and Zionism."[7]

FIGURE 3. Pinhas Rutenberg. Courtesy of the Israel Electric Corporation Archives.

Rutenberg was born in 1879 in the small town of Romani in the Poltava District of the Russian Empire. His family belonged to the town's large segment of well-heeled Jewish merchants. As a young child, he attended a Jewish primary school for traditional religious learning (*heder*). His aptitude for math convinced his parents to send him to a secular gymnasium, from which he graduated to the prestigious St. Petersburg Polytechnic Institute in the capital. This set the young Rutenberg apart from the great majority of Eastern European Jews, who were barred from higher education on account of the discriminatory *numerus clausus* in force on Jewish admissions since 1887.[8]

The curriculum of the St. Petersburg Polytechnic Institute was composed of a mix of theoretical and applied sciences on a model that had begun to dominate technological thinking across the industrialized and industrializing worlds. The institute was founded in 1902 at the prompting of Dmitri Mendeleev—he of the periodic table—and constituted an early attempt to deal with the Russian Empire's failure to sustain the economic growth of the 1890s. Russia needed a skilled workforce with practical knowledge of economics, statistics, technology, and "scientific" management. The school's great strength was the curriculum's innovative blend of various technical and theoretical skills, always with an eye to practical application. Economics and engineering each had a division of its own—a first for economics, which had traditionally been adjuncted to law faculties. The institute soon emerged as the flagship institution in a countrywide network of polytechnical institutes.

The divisions mixed not only with each other but also with the humanities. The institute's students were trained to think in terms of large integrative systems, seamlessly linking economic, technical, and social issues, steered by small groups of trained experts. The combination of practical economics with mathematical and statistical training ensured that the students graduated to a world of abundant opportunity. They were in high demand in both government and business, and later became the makers of the central planning that would develop into the hallmark of Soviet economic policy, though it clearly had pre-Communist as well as non-Russian precedents. The graduates of the institute moved nimbly across the borders of academic disciplines, as well as those of nations. They spoke foreign languages—the most common being German and English—and read journals from America, Britain, Germany, France, and elsewhere, and worked all over the European continent and beyond.[9]

Whether in Russia or abroad, electricity formed an important part of their work, and the institute's old-boy network has been characterized as a "kind of electrician's mafia."[10] The great importance of electrification can also be gleaned from the popular culture of the time. The cult of "Ilich's lightbulbs," referring to Lenin's given name, soon evolved into a mass movement, so much so that in the 1920s Elektrifikatsiia became a popular girls' name.[11] But most of all, electricity's importance shone through in Bolshevik statecraft. In Russia, as elsewhere, electrification was considered the linchpin of industrial advancement after World War I. In February 1920, the Council of People's Commissars created the State Commission for the Electrification of Russia (GOELRO) plan. It was the first economic plan of Bolshevik rule and became the prototype for the subsequent Five-Year Plans. The GOELRO plan—whose importance was underscored by its informal sobriquet, "the second party program"—was the cornerstone of the project to industrialize the country. Developed at the same time as Rutenberg's venture in Palestine, the plan envisioned building thirty regional power plants, including ten large hydroelectric stations.[12] Although Rutenberg had fled Russia by then, the training he received at the Polytechnic Institute profoundly influenced his work and taught him the importance of moving fluidly between the worlds of technology, economics, and politics.

This should give us enough background to begin sketching who Rutenberg was to electrical Palestine. First and foremost, he was a highly skilled engineer of heterogeneous systems. He thought holistically, in terms of overall system health and growth, even before all the necessary components were in place. He did not invent or innovate, and he cared little for the cutting edge. Instead, he worked with battle-tested tools and devices, whether technologically, by running the industry-standard, three-phase alternating current through his wires, or by promoting his project by speaking to such colonial hobbyhorses as the profligate native and the white man's burden, and racial stereotypes, such as Jews' financial acumen, Ottoman degeneracy, and Arab backwardness. His overall vision was firm but also vague. In this, Rutenberg belonged to an international tribe of entrepreneurial engineers who constructed and managed complex, integrated, centrally controlled technological systems, by applying grease where grease was needed, whether the nature of the friction was technical, social, economic, or political. He was, in a word, a *systems entrepreneur*. Success depended on situating his system so as to make it viable across the different domains with which it came into contact. As a systems entrepreneur, Rutenberg faced the challenge of transferring and adapting international technologies of proven effectiveness to local conditions, which involved communicating about one's project so that it seemed not just desirable but indispensable to whoever was listening at the moment, and so that the project appeared as the solution to the most pressing problems of the day, whatever those problems may be. This, as Rutenberg well knew, was crucially about presenting a solution so attractive that it helped define the problem to be solved.[13]

Revolution was what first brought Rutenberg to Palestine. A decade on, it was revolution again that pulled him away, and then, finally, what made him return and settle in Palestine for good. As a student in St. Petersburg, he adopted the left-wing politics that was endemic to the faculty and students at the Polytechnical Institute, and he joined the Socialist-Revolutionary

Party.¹⁴ After graduation he was hired as a junior engineer at the Putilov metalworking factory, where the workers were renowned equally for their high technical skill and their vibrant left-radical politics.¹⁵ Rutenberg took part in the abortive revolution of 1905 and soon afterward was forced to go into exile. The imposition, however, followed not from his involvement in the revolution itself but from his role in its denouement. When it came to light that Father Gapon, the man who had led the workers' march on the Winter Palace, sparking the revolution, was a double agent for the czar's secret police, it was Rutenberg who had him taken out to the countryside to be exectued.¹⁶

As rumors of the deed began to spread, he was forced to leave Russia. He settled in Italy, where he devoted himself to large-scale irrigation projects. The reputation he garnered there would stand him in good stead with the British in Palestine a decade and a half later, as Italy was a leader in the field of irrigation and dam construction.¹⁷ These were also the years when Rutenberg began to take an interest in Zionism, and shortly before the outbreak of World War I he paid his first visit to Palestine. His initial intention was to explore the possibilities for electrification and irrigation, but he was soon pulled in a different direction. Together with Vladimir Jabotinsky, the future leader of the right-wing Revisionist strand of Zionism, he began organizing a Jewish self-defense force. In 1915, Rutenberg traveled with Jabotinsky to America on a fund-raising trip. While there, he worked closely with the Zionist socialist Ber Borochov on promoting a Jewish Legion, and he became active in the American Jewish Congress.¹⁸

The same year, he published a Zionist pamphlet under the pen name Pinhas Ben-Ami, titled *The Revival of the Jewish People in Their Land*.¹⁹ By that point, many Zionist manifestos had already been written, including Theodor Herzl's *The Jewish State*, which upon its appearance in 1896 had launched the movement known as political Zionism. Rutenberg, like Herzl, was an assimilated Jew, whose preferred language throughout his life was Russian (for Herzl, it was German). Like Herzl, Rutenberg had never read Leo Pinsker, the first modern Zionist thinker and author of *Autoemancipation*. Nor, for that matter, had Rutenberg read Herzl. Despite his Hebrew pen name, Rutenberg wrote his Zionist pamphlet in Russian and had it translated into Yiddish; only much later did it appear in Hebrew.

Although by all accounts Rutenberg was deeply involved in Zionist politics by the end of the war, he dropped everything and returned to Russia when Alexander Kerensky offered him the post of chief of police in St. Petersburg (then Petrograd) following the so-called bourgeois-democratic revolution in February 1917. In his role as chief of police, and later as the city's deputy governor, he took aggressive action against the Bolsheviks, hunting them with all the powers of his office. After the Bolshevik takeover in October, he was temporarily imprisoned. He was released six months later and went into exile again, this time for good. With his career as a socialist revolutionary over, he redirected his productive energies once and for all to Zionism and the electrification of Palestine.[20]

He first went to Paris and the ongoing World War I peace conference, where his name had preceded him, and began his campaign to obtain a concession to electrify Palestine.[21] He went from Paris to Palestine, arriving in the fall of 1919. That winter, he and his associates began survey work in northeastern Palestine, the area around Beisan, south of the Sea of Galilee. Together with his team, he measured distances, velocities, solidities, frequencies, depths, and altitudes. They represented the space through photographs, sketches, cross-sectional models, diagrams, and charts.[22] In so doing, Rutenberg was staking a new kind of territorial claim on behalf of his system and of Zionism. It was a quintessentially modern claim that laid the basis for what Edward Said once described as the Zionists' "policy of detail."[23]

One member of Degania A, a kibbutz located in the area in which Rutenberg began his survey mission, tells the story of his encounter with the iron-willed engineer. Rutenberg appeared one day without warning, wearing a long coat, boots, a big Russian hat, and a pistol on his hip, and demanded the use of a horse. When he returned later that night, he had the blackboard from the child's ward brought into the dining hall and proceeded to lecture the settlers about his plans. After he finished, a weathered kibbutznik stood and asked: "Who are you going to sell all these horsepowers to? The Arabs of Samakh?" The room laughed uproariously as Rutenberg, his face turning a deep red, shot back, "Even Abbadiyeh [a small Bedouin village below Degania] will buy electricity!" This threw the room into another fit of raucous laughter, and Rutenberg stormed out.[24]

Throughout the years of the mandate, Rutenberg was often praised by external observers for his survey work. In Parliament, Churchill, the colonial secretary, defended the electrification concession by observing that Rutenberg "produced plans, diagrams, estimates, all worked out in the utmost detail."[25] On the basis of such data collection and the politically powerful construction of precision it allowed, Rutenberg and other Zionist actors would transform Palestine into a modern territory and also, by virtue of the identity of the surveyors, into a Jewish national space.[26] The data collection would both facilitate greater control over the natural environment and play into a modern notion of governance based on centralized legibility, which, as Said observed already in 1979, would decisively tip the scales in favor of Zionism.

Before Rutenberg launched his undertaking, access to electricity in Palestine had been very limited. In Jerusalem, the first city in Palestine to be lighted by electricity, there was a small number of privately owned generators. Some smaller towns and villages had generators for lighting. Nazareth, for instance, operated an old direct-current dynamo of 115 volts, powered by an even older car engine, and a small-scale distribution system consisting of 1.5 miles of iron wire.[27] In Tel Aviv, there were two direct-current generators, one by the township's municipal building, powering a small network of streetlights, and the other powering an electrical pump of twenty horsepower at one of the townlet's largest wells.[28]

As early as February 1919, the Zionist Commission and the Zionist Organization were involved in a study of the waters of the Auja River (Yarkon in Hebrew) for the purpose of electricity generation.[29] Such initiatives were enthusiastically received by the Jewish settlers in Palestine. About six months before Rutenberg arrived in the country, a group representing the Jewish inhabitants of Tel Aviv and Jaffa wrote to the Zionist Commission requesting that it immediately take the necessary steps to "obtain a concession for using water power of the river Audja for the purpose of irrigation and for the establishment of a central Power Station," an undertaking, the letter claimed, "which is so important for the agricultural and industrial progress of our district."[30] In early spring, Chaim Weizmann sent a letter to Louis Mallet at the British Foreign Office, in which he stressed the importance of immediately tending to the welfare of the inhabitants of Palestine. He handed Mallet a list of development projects that should be begun "without delay"; it included electrification using the waters of the Auja.[31]

CENTRALIZED LARGE TECHNOLOGICAL SYSTEMS

In their eagerness to secure a steady supply of electricity, the Zionists were working from a universal script. Centrally controlled large-scale systems dominated technological thinking in both colony and metropole at the time. In Western societies, the 1920s was the decade of large-scale electrical transmission. Large power systems were often created by interconnecting several smaller grids, coupled with the construction of large, new generating plants (often hydroelectric) as the principal power source.[32] More than any other large technology, the scale of electric supply networks came to be regarded as metrics of civilizational advancement at large. Lenin, for instance, famously defined communism as "Soviet power plus electrification of the whole country," and claimed that electricity would help to "raise the level of culture in the countryside and to overcome, even in the most remote corners of the land, backwardness, ignorance, poverty, disease, and barbarism."[33] In the United States, John Carmody, head of the New Deal's Rural Electrification Administration, told Congress in 1937, "We believe in the economic wisdom of bringing farm families out of the dark into the light, out of stark drudgery into normal effort, out of a past of unnecessary denial into a present of reasonable convenience."[34] Besides exemplifying the intimate coupling of technology and civilizational uplift, these quotes also illustrate the striking similarities between technological thinking in capitalist and socialist societies.

In the introduction, we saw a colonial official predict that Rutenberg's plan would "do more than anything else to pacify Palestine, facilitate immigration, and develop the country."[35] His was not a lone voice; the record abounds with similar statements by statesmen, journalists, Zionists, military men, and consulting engineers. One report by the consulting engineers Sir Charles Metcalfe & Selves, commissioned by the British industrialist Lord Mond (later raised to the peerage as Baron Melchett) stated: "The generation and distribution of hydro-electric power is one of the first essentials to the development and settlement of Palestine." The report also characterized "power, light and heat" as "fundamental necessities of modern civilisation."[36] In a German newspaper, the prominent engineer I. Archavsky asserted in reference to Palestine, "The installation of cheap electrical power will entirely revolutionise the agricultural economy of the country."[37] Thus, the question at the heart of this chapter is: How do we make sense of this unbridled technological utopianism, and the way it bestowed such tremendous, yet largely invisible, power on one man and his power system? The answer, as we will see in what follows, is that Palestinian electrification took place in an imperial context marked by technological utopianism, and specifically by a vision of centralized, large-scale technologies as engines of "moral and material" progress.

Palestine policy was configured in a colonial matrix of development, capitalism, and technoscience, whose origin goes back to the Enlightenment view of nature as a separate entity, governed by regular and therefore knowable laws. Since what is knowable is amenable to modification, the Enlightenment understanding of nature created the idea that both nature and society could—and should—be improved upon by refining their organization according to the rational principles already inherent in them.[38] This program began at home, in Western Europe, but its advocates soon set their sights farther afield. A second generation of thinkers, including Alexis de Tocqueville and John Stuart Mill, ushered in the Enlightenment's "turn to empire" in the 1830s. The decade saw the emergence of a new rationale underwriting European rule of non-European peoples and lands.[39] The new colonial program of improvement was linked to an increased civilizational confidence, resulting from the economic and technological advances of the scientific and industrial revolutions, elevating "men of science" to a new status, in the imperial historian Richard Drayton's words, as "masters of nature's secrets."[40] Development was thus part of a larger scientific program, guided by supposedly objective laws of nature and striving toward goals that were also given by nature.[41]

In Britain at the turn of the last century, this policy came to be known as *constructive* imperialism, after the policy advocated by the statesman Joseph Chamberlain. As mayor of Birmingham in the 1880s, Chamberlain undertook various large public works projects. His views developed further during his tenure as foreign secretary from 1895 to 1903, and visits to the Americas and Egypt, where he witnessed firsthand Lord Cromer's "enlightened" despotism. Chamberlain aspired to supplant the laissez-faire system of the nineteenth century with a more carefully managed form of imperialism. His vision was founded on a close conceptual link between science, the state, and development, similar to the mercantilism of William Pitt the Younger and Sir Joseph Banks a century earlier. Chamberlain, however, went much further in his emphasis on state planning and technocratic expertise. In so doing, he paved the way for what the colonial historian Joseph Hodge describes as "the triumph of the expert," whereby colonial authority shifted away from the local official who "knew his natives" to the technocrat who "knew his science."[42]

In the view of Chamberlain and his many influential followers, advances in science and technology did not merely make possible the development of the world's natural resources; the advances mandated such activity as a matter of moral duty.[43] In British Africa, constructive imperialism manifested itself as active intervention on behalf of white settler communities. According to this logic, strengthening the economic activities of settlers would enhance their role as engines of development and models for emulation by the natives.

In a speech in the late 1890s, Chamberlain explained, "I consider many of our Colonies as being in the condition of undeveloped estates, and estates which never can be developed without Imperial assistance."[44]

It was as part of his ambition to mobilize white settler groups to serve as vanguards of development in the colonial world that Chamberlain first became acquainted with Zionism. During his tenure as colonial secretary, he was approached by Theodor Herzl, the founder of the Zionist Organization, who asked him to sanction the establishment of a Zionist settlement in Cyprus. Chamberlain was receptive to the idea but suggested they find a location where there was not already a white settler population.[45] The conversation between Herzl and Chamberlain resulted in Britain's formal offer to the Zionist Organization in 1903 of a Jewish autonomous zone in East Africa. The so-called Uganda Offer was ultimately rejected at the seventh Zionist Congress in 1905, due to the refusal of most Eastern European Zionists to consider establishing a Jewish state anywhere but in Palestine.[46] The episode nonetheless demonstrates the affinity of Zionism and "constructive" settler colonialism in the minds of Zionist activists and British officials alike.

While the "scientific" thinking that underwrote the concept of colonial development has received a fair amount of attention from scholars, the fact that it was equally shaped by a more concrete technological vision has not. In contrast to the developmental vision that took hold after World War II, which focused as much on political development as on economic and social issues, the dominant concern during the first age of development was to bring about material progress, from which moral progress was thought to follow spontaneously.[47] Starting in the mid-nineteenth century, development efforts were focused on providing what was believed to be the material prerequisites of economic growth: waterways, irrigation systems, roads, bridges, ports, and railways. By the turn of the last century, telegraph and telephone lines, airports, and electric grids had been added to the list. The Liberal MP F. C. Linfield, one of the members of the East African Commission of 1925, argued, "We have within the Empire millions of acres of fertile land and enormous mineral resources only awaiting development by the provision of roads, bridges, and railways."[48] At the annual meeting of the American Society of Civil Engineers in 1922, the former president Charles Marx pointed out that "the creation of wealth" was "the essential factor in the development of civilization." Indeed, so securely was this notion rooted in our common sense, Marx continued, that engineers had lost the ability to "marshal the facts in support" of it. He thus offered: "The creation of wealth implies a surplus, and it is only when there is a surplus of wealth—a quantity greater than is needed to meet the physical needs of peoples from day to day—that there can be had that leisure for the development of the thinking power of man on which depends both his material well-being and spiritual growth."[49] In short, by the interwar period, if not before, "civilization" was an active working concept of politicians and civil engineers alike.

The link perceived between "moral and material progress," where the former would follow from the latter, was thus firmly contained within the horizon of the taken-for-granted by the mid-nineteenth century. Indeed, to many it seemed as if there was no limit to what technology could accomplish. Marquess Dalhousie, governor-general of India from 1848 to 1856, characterized the railway, telegraph, and postal system that he had been responsible for introducing to the subcontinent as the "three great engines of *social* improvement."[50] During the Scramble for Africa that began in the 1870s, engineers, characterized by S.J. Thompson as the "true missionaries of progress and enlightenment," were among the first to arrive in the new European possessions.[51] Their task was to unlock the vast potential that the continent was believed to hold in store.[52] In each new place they went, they designed wildly optimistic schemes for economic development: a railway in the British Lagos Colony was projected to trigger an economic boom for the entire Niger region; the Aswan Dam would turn the desert of Upper and Middle Egypt into the largest cotton field in the world; the Uganda Railway would open up the entire eastern part of Africa to the civilizing influences of science and commerce; the road from Dar-es-Salaam to Lake Nyasa was taken as a sign by the British consul in Mozambique that "the march of civilization has commenced" in East Africa.[53] "There is no civilizer like the railway," proclaimed H.H. Johnston, a colonial official and one of the most influential spokesmen for the British colonial party in Africa in 1889.[54] Indeed, in British India and British Africa alike, railway construction proceeded at an extraordinary pace; by the turn of the last century, India had the fourth-largest railway in the world, and by 1907 the length of the British-built railway lines in Africa exceeded ten thousand miles.[55]

The British were not alone in embarking on this mission, based on the belief that technology held the key to progress for indigenous peoples, for civilization, and for the global economy all at once. In 1882, for instance, a special commission of the Indies Society stated that the tramways built by the Dutch in Indonesia were "the most powerful incentive to labor, exchange in values, and civilization."[56] By the end of the nineteenth century, French, German,

Dutch, and other imperialists were scavenging the earth for untapped economic potential, and found it all over Indochina, Africa, the Caribbean, and the Middle East, ushering in a period of more than two decades during which European empire grew by an area the size of France each year.[57]

The emphasis on facilitating movement through infrastructure followed from a commitment to the ideal—if never the reality—of the free flow of people, things, capital, and information on the global market, which was believed to be the key to expediting both local development and global peace and prosperity.[58] "We develop new territory as Trustees of Civilization," Foreign Secretary Chamberlain exclaimed in a speech in the late 1890s, "for the Commerce of the World."[59] Indeed, as Jason W. Moore has recently observed, the Enlightenment ambition to quantify, map, and ultimately control nature evolved together with the emergence of capitalism. The two implied each other from the start.[60]

Equally important, however, the "free" flow had to be closely managed so as to restrict the mobility of undesirable elements, whether human, microbial, or inanimate.[61] Enter, therefore, centralization, the second leg of interwar development policy. While large technological systems would spur economic development, technological *centralization* would solve the issue of control. Technological utopianism coupled with the exigencies of colonial rule made large centralized schemes terrifically attractive. On the one hand, large-scale projects constituted a spectacular display of British superiority, creating the image of "a technological empire" capable of mastering the forces of nature.[62] On the other hand, technological centralization facilitated an unprecedented level of control, enabling small numbers of officials to govern vast territories. In India, a country that at the end of British rule counted four hundred million souls, was never governed by more than fifteen hundred British officials; in Nigeria a few hundred officials administered a population of tens of millions.[63] This logic was also on display in the construction of the Aswan Dam in British-ruled Egypt in the 1860s. The dam concentrated knowledge and control of the river's flow to a single point, the dam, whereas previously knowledge and control of its flows had been distributed evenly along the length of the basin and therefore across a multitude of actors. In this way, a new source of political power was called into being.[64]

Indeed, resorting to technological centralization in order to consolidate political power is a globally recurring feature of modern history. When, in 1920, the new Bolshevik leaders of Russia (what in December 1922 would become the Soviet Union) deliberated over the question of how to organize the electricity supply, they opted for the most ambitious of the schemes considered: a centralized system, directed from Moscow, of high-powered regional generating stations across the country. In part, the project was justified on technological grounds, with reference to the same "economic rationalization" in evidence all over the Western world. But, more important, it was also motivated by the degree of central control that it accorded to the party. In effect, then, technological centralization was a facet of the larger Bolshevik agenda of centralization, the issue that famously split the Russian socialists into Mensheviks and Bolsheviks around the turn of the last century. Conversely, in cases where the state was not involved in electrification, it tended to restrict the scale of the venture to avoid the consolidation of technological power in the hands of a single nonstate actor. Such was the governing logic of the British power supply before the National Grid, and indeed the czarist state in Russia before the Revolution.[65]

In the colonial context, moreover, large-scale projects helped maintain (and in some cases create) local political hierarchies, constituting another way in which technological centralization engendered concentration of political power. In British India, a portion of the revenue from the large systems was doled out to local rulers, enabling them to distribute the dividends in turn as patronage. This allowed the British to maintain a power structure that gave them control over the masses through local middlemen. As with the waters of the Nile, smaller systems would have contributed to wider wealth distribution, and thus diffusion of political power.[66]

Centralized large technological systems, therefore, served as both the means by which European colonial powers were able to exercise control over vast non-Western areas and what justified their doing so. The mission was a "trust of civilization" because it would raise the living standards of indigenous peoples, while releasing the untapped potential of their lands for the benefit of global commerce.[67] Understanding the logic behind such systems is crucial to understanding why Rutenberg's scheme appealed so to British officials, why it was ascribed such tremendous importance, and how it came to have such a powerful influence on the development of mandatory Palestine.

DUAL MANDATE

Yet for all that, hardly anyone advocated a development program of unchecked economic growth. The belief in the power of technology and unrestricted markets to effect material and moral progress had, throughout its history, been tracked closely by a countervailing concern, namely, that modernity had come to the West at great cost.[68] This concern was built into the concept of overseas development from the start, in the form of the corollary goal of ameliorating some of the chaos and upheaval that were thought inevitably to attend progress.[69]

The challenge was commonly expressed as a balancing act between "progress" and "well-being," where progress was understood as the forward motion of economic growth and technological innovation, which, if left unchecked, was a chaotic and potentially destructive force.[70] The balancing act was apparent in myriad non-Western development schemes and the subject of numerous conversations among British imperialists of the early twentieth century, such as those involved in the African Survey of 1929–39, the Round Table movement, as well as Alfred Milner and the members of his so-called kindergarten.[71] It received its most famous treatment in Frederick Lugard's essay *The Dual Mandate in British Tropical Africa* (1922).[72] The "dual mandate" of the title referred to Britain's twofold obligation to develop the "undeveloped estates" of Britain's African possessions without compromising the well-being of indigenous populations. In the book, Lugard argued that the two "mandates" were in fact compatible and that European activities in Africa could be mutually beneficial to settlers and natives alike.[73]

The effort to develop the non-Western world entered a partly new phase with the creation of the mandates system of the League of Nations following World War I. The Great War brought about the greatest territorial reorganization the world has ever seen, as four empires collapsed in the postwar shuffle. While the Russian and German empires were brought down from inside in February 1917 and November 1918, respectively, Austria-Hungary crumbled in the fall of 1918 after being roundly defeated on the battlefield. For the Ottoman Empire, the war ended with military defeat and the Armistice of Mudros in October 1918.

The collapse of the Central Powers confronted European statesmen with the question of what to do with the vast territories, some 2.3 million square miles in- and outside of Europe, formerly under these empires' control. The European territories became subject to a program of nation building through the twin strategies of "unmixing the peoples" and minority rights.[74] The non-Western world, by contrast, was deemed not yet amenable to such politics.

Instead, the empires' former holdings in Asia, Africa, and the Middle East became targets of a revamped civilizing mission. The vehicle of this mission was the mandates system and its international oversight body, the Permanent Mandates Commission of the League of Nations. The mandates system and the League were born together of the same document, Articles 1 through 26 of the Treaty of Versailles, collectively known as the Covenant of the League of Nations, signed by the Allied powers and Germany on June 28, 1919.[75] Article 22 concerned the establishment of a tutelary program for the purpose of facilitating the "progressive development" of the non-Western areas and populations of which the Central powers had been divested. Under the mandates system, European rule would continue until the mandated territories could "stand by themselves under the strenuous conditions of the modern world."[76] Lugard, who had sought the position as Britain's representative on the commission when it was formed, got his wish in 1922, when the original representative, William Ormsby-Gore, left to take up a position in government.

Article 22 also introduced a classification of mandated territories as A, B, or C, according to their degree of civilization. The areas formerly under Ottoman control that became subject to the mandate system—Palestine, Mesopotamia, and Greater Syria[77]—made up the most civilizationally advanced class, Class A. According to the League covenant, Class A mandates had "reached a stage of development where their existence as independent nations can be provisionally recognized subject to the rendering of administrative advice and assistance by a mandatory until such time as they are able to stand alone."[78] From April 19 to 26, 1920, the Allied Supreme Command met at San Remo on the Italian Riviera and agreed that Britain would assume the role of mandatory with respect to Palestine (split into Palestine and Transjordan on September 16, 1922) and Mesopotamia (whose name was formally changed to Iraq on August 23, 1921), while the mandate for Greater Syria (today's Syria and Lebanon) was conferred on France. At the time, however, these designations were territorial approximations without clearly defined borders. The San Remo Agreement did not contain any specifications with regard to borders, but left it up to Britain and France to hammer out the specifics in subsequent bilateral negotiations.[79]

The Palestine Mandate, moreover, was set apart from the others by virtue of what came to be known as Britain's "Jewish national home policy," first laid out in the Balfour Declaration of 1917. In this letter from Foreign Secretary Arthur Balfour to Lord Rothschild, Britain declared its support for the establishment of "a national home for the Jewish people" in Palestine. The Balfour Declaration, which was later incorporated into Palestine's mandatory charter, contained a since much-debated qualification in the form of a pledge to respect the "civil and religious rights" of Palestine's "non-Jewish communities," a group then constituting more than 90 percent of the land's inhabitants.[80] To most later observers it has seemed obvious that this split responsibility—often referred to as a "twofold duty" or "dual mandate"[81]—of ensuring the well-being of the majority, while also promoting the establishment of a national home for the small Jewish minority, constituted a stark and obvious contradiction.

But, in fact, neither the British government nor the Permanent Mandates Commission saw it that way initially. Indeed, it is no coincidence that the phrase "dual mandate" was ubiquitous in the context of both the Palestine Mandate and late British imperialism. Works discussing the concept in the Palestinian context do not mention this lexical overlap; nor do they give any reason to think that its significance extends beyond the nominal. In fact, however, the duality at work in Palestine—between supporting Zionism and ensuring Palestinian well-being—was a local articulation of the empire-wide balancing act between effecting progress and protecting indigenous groups from the social ills that development was thought to bring if gone unchecked.

In terms familiar to any settler-colonial context, European Jewish immigration to Palestine was seen as an engine of development, and thus represented the progress side of the equation. By investing their industriousness, skill, and capital in Palestine, European Jews would develop the country for the benefit of all its inhabitants. Although commonly overlooked today, this was widely understood and discussed at the time. From the conference where the Middle Eastern mandates were allocated, the *London Times* reported that "the practical consequence of the decision at San Remo will be that Jewish energy and capital will begin to flow towards Palestine to be devoted to the development of the country and to the benefit of all its inhabitants."[82] In 1922, Colonial Secretary Winston Churchill stated in Parliament, "By their industry, their brains, their science, their money [Jewish immigrants to Palestine] must create new sources of wealth . . . [to] benefit and enrich the entire country."[83]

This, of course, had been the central argument of political Zionism from the start, as noted earlier in the context of Britain's "Uganda Offer." Already in 1899, only two years after the first Zionist Congress and the founding of the Zionist Organization, Herzl assured the prominent Palestinian politician Yusuf Diya' al-Khalidi, "In allowing immigration to a number of Jews bringing their intelligence, their financial acumen and their means of enterprise to the country, no one can doubt that the well-being of the entire country would be the happy result."[84] Other Zionist representatives were even more explicit about the value of the movement to the British Empire. Chaim Weizmann, the long-standing president of the Zionist Organization and first president of Israel, consistently stressed the point when making the rounds in London on behalf of Zionism during World War I. He told C. P. Scott, the editor in chief of the *Manchester Guardian* and an early supporter of Zionism, that the Zionists would serve as Britain's "man on the ground" in Palestine. Within a few decades, he assured Scott, "we could have . . . a million Jews out there, perhaps more; they would develop the country, bring back civilization to it and form a very effective guard for the Suez Canal."[85] In another indication of the close perceived link between material and moral progress, as well as the natural fit of Zionism within this worldview, Weizmann referred to his movement's contribution to Palestine as "moral industrial progress."[86]

Despite an abundance of voices around the time of the establishment of the mandate for Palestine claiming that the dual obligations were in fact complementary, later scholars have not taken the claim seriously. Statements supporting Zionism on developmentalist grounds have commonly been read as so many fig leaves, and scant significance has been attached to the precise content of the claim. But as Arthur Balfour once observed regarding the declaration that bore his name, the idea of "planting a minority of outsiders upon a majority population, without consulting it, was not calculated to horrify men who worked with Cecil Rhodes or promoted European settlement in Kenya."[87] In fact, there is good reason to take seriously the words of Herzl, Churchill, Weizmann, and innumerable other British politicians, officials, military men, technical consultants, and journalists: support for Jewish immigration was understood as a developmental policy, not contradictory to general Palestinian well-being but, provided that the right balance was struck, conducive to it.

Indeed, the main actors behind Britain's Jewish national home policy were all committed "constructive" imperialists in the mold of Joseph Chamberlain. Their support for Zionism flowed from the belief that the Jews would bring economic progress to Palestine, and that economic progress would automatically engender social advancement.[88] Leopold Amery, for instance, an influential British Conservative Party politician and journalist, wrote in a letter to a friend shortly before helping to draft the Balfour Declaration that "the

Jews alone can build up a strong civilization in Palestine." William Ormsby-Gore, another prominent new imperialist strongly influenced by Joseph Chamberlain, wrote in 1919 during his time on the government's Advisory Committee to the Palestine Office that "we have got to get men and women from around the world for the development, peaceful settlement and redemption of Palestine as a sort of punctum from which progress and development can take place and, looking around the world, we can get it nowhere but from the Zionist Organization."[89] Similar ideas were expressed by all the major figures behind the Balfour Declaration and the national home policy, including Lord Cecil, Alfred Milner, and Jan Smuts.[90]

From Palestine, Herbert Samuel, the newly appointed high commissioner, wrote a series of reports arguing that the Zionist "development works contemplated would benefit Jew and Arab alike, and would tend to obviate any chance of trouble" because of the "moral effect" they would have "upon the non-Jewish population."[91] His first proposals to this effect date back to 1915, five years before his appointment as high commissioner. In a memorandum titled "The Future of Palestine," he put forward the idea that British support for Jewish settlement in Palestine would prove advantageous to the economic development of the area and to the British Empire.[92] This was not simply cloaked Zionism; as the famous political theorist Harold Laski observed about Samuel: "When he speaks of Free Trade with the accent of religious fervor, he is completely sincere."[93] General Louis J. Bols, who ran the military government that preceded Samuel's civilian administration, wrote to his commanding officer, General Edmund Allenby, in December 1919, proposing that Britain grant a "big loan" to Palestine to facilitate its development. "If this is done," he assured Allenby, "I can promise you a country of milk and honey in ten years, and I can promise you will not be bothered by anti-Zion difficulties."[94] As in East Africa, the most important fact of British policy in Palestine was Britain's needs as a major manufacturing and capital-exporting country.[95] Fortunately for empire, governing wisdom held that there was no conflict between those self-interested imperatives and the lofty mission to civilize the world.[96] On the contrary, global progress and imperial profit were two sides of the same happy coin.

Although the connection between the Jewish national home policy and British imperialism is almost entirely overlooked today, it was clearly on the minds of many European statesmen at the time. When the Permanent

Mandates Commission called for the creation of a basis for "an understanding between Jews and Arabs," or when the British expressed hope that ethnic strife would soon subside despite all evidence to the contrary, it was not because they were ignorant of the widespread opposition among the Palestinian Arabs toward Jewish immigration.[97] Nor did they willfully ignore it, as some have claimed, because of their Zionist sympathies. Rather, as the 1937 Peel Commission Report on the Arab Revolt stated, the Mandate Charter "implied the belief that the obligations thus undertaken towards the Arabs and the Jews respectively would prove in the course of time to be mutually compatible owing to the conciliatory effect on the Palestinian Arabs of the material prosperity which Jewish immigration would bring to Palestine as a whole."[98] Indeed, so powerful was this logic on the technical and political minds of the time that the former president of the American Society of Civil Engineers held Palestine up as a model to be emulated in other trouble spots around the world. "If hydro-electric power development can successfully wipe out hatred of race and creed, as claimed by Mr. Rutenberg," he wrote, "might it not be desirable to push its development wherever possible in other centers of strife?"[99]

The clearest indication of the importance of imperial policy to Palestine are the striking parallels with other areas under British influence at the time, especially in East Africa. Like British officials' attitude toward the Arab population of Palestine, East African colonial development was guided by the assumption that the native population was not mature enough to cope on its own with the challenges of modern industrial society and thus required a European vanguard.[100] In *The Control of the Tropics*, Benjamin Kidd argued that Africa would "never be developed by the natives themselves." Therefore, they had no right to "prevent the utilization of the immense natural resources which they have in charge."[101] Jan Smuts, a South African statesman, similarly argued that "without large-scale permanent European settlement on this continent the African mass will not be moved, the sporadic attempts at civilization will pass."[102] In keeping with the dual mandate doctrine, there was no contradiction, in Smuts's view, between the interests of Africans and the interests of the European settlers. "From the native point of view . . . just as much as from the white . . . point of view, nay even more from the native point of view, the policy of African settlement is imperatively necessary."[103] The key to successfully developing southern Africa for the benefit of all its inhabitants, he informed an assembly of German settlers in Windhoek, was "whites and capital."[104]

A few decades later, British colonial officials would echo the same sentiment in the context of Palestine. In a parliamentary debate over Rutenberg's electrification project in 1922, Winston Churchill, as colonial secretary, claimed that "left to themselves, the Arabs of Palestine would not in a thousand years have taken effective steps towards the irrigation and electrification of Palestine."[105] They would have been content to let the Jordan River "continue to flow unbridled and unharnessed into the Dead Sea." Palestine and the Palestinians, therefore, needed the Jews: "Give [the Jews] the ordinary scientific apparatus they require, and they will return you a plentiful reward for every pound that is invested in them."[106] Examples abound of how thinking on Palestine and Zionism paralleled "constructive" imperialism's settler-colonial program in Africa, in which white settler communities would serve, in Frederick Lugard's words, as "useful object-lessons in improved methods."[107] Even the well-documented tension between local British officials in Palestine, who tended to be skeptical of Zionism, and metropolitan officials, who regarded Zionism favorably, corresponded to similar tensions elsewhere between local and metropolitan officials, where the former were more concerned with preserving "the native way of life," while the latter prioritized hard-nosed reform and modernization.[108]

For all that, however, the Yishuv was not a typical settler colony. It enjoyed few of the explicit privileges of most settler communities elsewhere, such as Kenya, Angola, Mozambique, and South Africa. In all those places, the colonial overlords actively and systematically manipulated markets in the settlers' favor, by granting exclusive rights to grow and market certain crops and helping to bring the produce to market through transport subsidies and the like. While it is true that the British mandatory in Palestine adopted a range of economic policies that were designed to benefit the Jewish community, including Rutenberg's electricity monopoly, they were of a lesser scale than those enjoyed by most other white settler communities.[109]

Also as in East Africa, it was widely acknowledged that the development that Jewish immigration to Palestine was expected to generate was also a source of potential harm to the well-being of the indigenous Arab population, at least in the short term. The effort to resolve this issue in the Palestinian context again followed an imperial precedent. In a move familiar to scholars of science and technology, the question of the extent to which the British should support the Jewish national home was sublimated into an economic calculus, rendered in the Palestinian context as "absorptive capacity." Set down in a white paper from 1922, Jewish immigration was capped at "the economic capacity of the country at the time to absorb new arrivals." This called into being a new calculative apparatus, supported by a host of new technopolitical concepts. The most important of these was the "lot viable," a unit of land defined by the government as "sufficient for the subsistence of a farmer and his family... to maintain a decent standard of living." Naturally, each of the calculations making up the aggregate metric of "absorptive capacity" became a site of contestation between the various interest groups.[110] In this way, an issue that would seem clearly political (of promoting Jewish state building in a majority-Arab territory) was displaced onto a technical one (the question of how many Jewish immigrants the land was able to support economically).[111] In other words, science was not only what guided the progress side of the equation, with its calls for centralized large technological systems that would generate material and moral progress; it also provided the means for striking the right balance between progress and well-being.

The sharp distinction between economics and politics was of course artificial. "For," as Weizmann explained, "'absorptive capacity' does not grow wild on the rocks and dunes of Palestine; it must be created, and its creation calls for effort, enthusiasm, imagination—and capital."[112] As we will see in later chapters, calculating absorptive capacity or making economic projections of any kind was itself inescapably political, as there was no neutral way of crunching the numbers, or indeed no neutral numbers. Yet such economism would dominate Palestinian policy at least until the Great Arab Revolt of 1936–39, which the British took as definitive proof that Arab-Jewish relations were not just an economic issue, or solvable by purely economic means. By the second half of the 1930s, the British finally had to reckon with the fact that for the Palestinians, just as for the Zionists, the question of Palestine's future was fundamentally a political one. Figures related to absorptive capacity, megawatts generated, or gross national product were not—and had never been—an answer to the basic questions posed by the conflicting claims of Zionism and Palestinian nationalism.

TECHNOCAPITALISM IN PALESTINE

In the second half of the nineteenth century, after the Ottoman Empire reasserted control of the area in 1841, the end of the Crimean War in 1856, and the Ottoman land reform of 1858 (opening up most land in the empire to private ownership), Western capital began to flow into Palestine. Ottoman reform initiatives and the influx of European capital in Palestine spurred a conceptual transformation of the area from a vague landscape of religious promise to a precisely defined territory of economic potential. This kind of economic thinking, as we have seen, lay at the heart of late imperial developmentalism and was both structured and fueled by the promise of a particular technological vision. The emergence of a precisely bounded modern Palestine was a function of the deeply spatial and scalar quality of this economic vision. This view runs counter to the bulk of scholarly and popular accounts, which have tended to locate the emergence of modern Palestine in the agreements and arrangements concerning the area's future after World War I and the fall of the Ottoman Empire. Arguably, there is a not so subtle teleology at work in such accounts, for the way they treat the territory as an always-already present entity, which the wartime agreements formalized but did not create.

It is true, of course, that as a territorial designation "Palestine" is ancient. As Edward Said has pointed out, "If one were to read geographers, historians, philosophers, poets who wrote in Arabic from the ninth century on, one would find innumerable references to Palestine."[113] In fact, the term dates back at least to Roman times and has been used on and off since then to designate various portions of the eastern Mediterranean coastal plain between Sidon and the Sinai, plus hinterlands. Elements of a local Palestinian identity were in evidence among Arabs as early as during the Mamluk era.[114] Yet by the time of British mandatory rule, Palestine's borders had never been precisely defined, much less politically unified. During the late Ottoman Empire, the term was commonplace but vague, and governed under a fluctuating set of administrative districts.[115] The European powers, including the British, also had only a vague sense of Palestine's borders, even after the increased interest in the Holy Land in the latter half of the nineteenth century. The 1910–11 edition of *Encyclopaedia Britannica* states that Palestine is "a geographical name of rather loose application."[116] The entry does go on to attempt a definition, but as Bernard Lewis has noted, it "differs in several important respects from that laid down for the British mandate only a few years later."[117] The 1919 *Handbook of Syria (including Palestine)*, produced by the British occupation force, states, "In modern usage Palestine has no precise meaning. No definition can be got merely by discussing or attempting to follow the limits of the territory of the ancient Hebrews."[118] Nor did the Zionists have a better sense of Palestine's exact boundaries.[119] Through the end of World War I, the most common geographic reference in Europe and among Jews was the biblical designation "from Dan to Beersheba." This was vague indeed: besides giving only a northern and southern point, there was no consensus at the time regarding the true location of Dan, and the Beersheba of the Bible was not the same as the relatively recently founded Ottoman market town of the same name.[120]

Although historians have read modern Palestine back into the many agreements made during World War I, in fact none of them, including the Husayn-McMahon Correspondence and the Sykes-Picot Agreement, contain any explicit references to a territory called "Palestine," much less identified it as a desirable geographic concept around which to create a political entity. Arguably, the borders proposed in these and other wartime agreements and arrangements are recognizable as Palestine only from the perspective of a future in which Palestine does indeed exist. The only one among the wartime declarations and agreements that explicitly mentioned Palestine was the Balfour Declaration of 1917.[121] But it should also be noted that the British pledge of support for "the establishment *in* Palestine of a national home for the Jewish people" did not presume that Palestine itself (whose borders were again not specified) would be constituted as a discrete political unit, though that is how it has been read by later observers.[122] Finally, many of the postwar proposals concerning the Middle East, including the proposal of the 1919 King-Crane Commission and that of Amir Faysal of the Syrian Arab government, rejected the idea of a separately administered Palestine outright.[123]

Neither can it be said that Palestine was created in an effort by the British to safeguard the Suez Canal. To be sure, the need to protect the canal was a vital consideration, calling for some arrangement of territorial control, but this imperial concern did not provide a sufficient rationale for the establishment of a British-ruled Palestinian territory.[124] Imperial rationale is even less helpful as an explanation for the creation of the precise boundaries of that territory. In other words, the existing explanations for the making of modern Palestine provide important and surely necessary factors, but neither on their own nor together are they sufficient. The salient presence of "historic" Palestine in the historical consciousness of our own time has arguably prevented us from seeing that a more precise guiding logic is needed. That guiding logic, I suggest, was technocapitalism.

Palestine in the late Ottoman period became more closely integrated into the wider region by means of the first steamship lines between Palestine and other Mediterranean ports, and the Hijaz Railway, whose construction began in 1900. In 1904, the spur to Haifa from Dar'a in present-day Syria was completed, connecting the grain producers of the Hawran with the Mediterranean coast for further transport to Europe, with the effect of quadrupling wheat exports. Furthermore, the land was drawn more tightly together through telegraph lines (the first one, between Jaffa and Jerusalem, was erected in 1865), roads (the first wagon road between Jaffa and Jerusalem was built in 1867), and railroad systems (the Jaffa-to-Jerusalem line was completed in 1892, reportedly doubling commerce over the next two decades).

The Sublime Porte managed this development from afar, mainly by having the Ottoman Ministry of Public Works put out various works projects for public tender. While the capital for such projects was usually European in origin, preference was frequently given to local contractors. The public works ministry granted concessions for, among other things, the development of the Tiberias hot springs for commercial purposes by two Beiruti notables, Dr. Fakhouri and Abdel Nour. Another pair of Beiruti notables, Michel Sursock and Mohammed Omar Bayhum, were granted a concession to develop the Huleh marshes and later transferred the concession to an Ottoman company, the Syro-Ottoman Agricultural Company. In 1912, Platon G. Franghia received a concession to develop the port in Jaffa. In 1914, another concession was granted to the company Midian Limited for developing the mineral and petroleum resources of southern Palestine and northern Arabia. The same year another concession was granted whose true significance would only become apparent several decades later. The bid belonged to the Ottoman national Euripides Mavrommatis and concerned the development of a new tramway, an electric grid, sewage, and waterworks.[125]

The rapid intensification of economic activity in Palestine attracted not only European businessmen, missionaries, and religious tourists but also economically distressed European Jews, mainly from the Pale of Settlement. The Jewish population of the area had been dwindling since antiquity, and although Jewish migration had turned to a net positive by 1700, mainly as a result of various messianic movements, the real turning point coincided with the influx of European capital. From 1840 to 1880, the Jewish population doubled, reaching some 24,000. This was only a small fraction of world Jewry (0.5 percent), and a small minority of Palestine's population, which counted some 450,000 souls at the time.[126]

The growing economic attractiveness of Palestine is manifest in the writings of several European Jewish intellectuals of the time. In essays such as "It Is Time to Plant" (1875–77) and "Let Us Search Our Ways" (1881), Peretz Smolenskin pioneered the idea of collective Jewish migration to Palestine as a solution to Jews' economic woes in Eastern Europe. For Smolenskin, and the many Jews who heeded his call and migrated to Palestine, its attraction was as a land of economic, rather than religious or political, promise.[127] In other words, while in the fullness of time Zionism would evolve into a nationalist movement on the model of the Italian risorgimento and various Eastern European nationalisms, the movement's intellectual roots lay in an economic development project for Jews by Jews. Even as its adherents increasingly came to espouse a form of romantic nationalism, Zionism never lost its commitment to developmentalism, ideologically, rhetorically or, most especially, practically. The seventh Zionist Congress in 1905 adopted a resolution enumerating the four key activities of Zionist settlement in Palestine: "exploration," "promotion of agriculture and industry," "cultural and economic improvement," and "acquisition of concessions."[128] Economic development and commercial concessions would remain key tools of Zionist state building until the establishment of the state in 1948.[129]

From 1880 to 1914, fifty thousand more Jews arrived in the country, in the so-called first and second aliyot (sing. aliyah). Their arrival was mirrored by a significant indigenous economic and population growth: on the eve of World War I, Jews made up about 7 percent of the population total of 750,000.[130] For the first few decades of Zionist activity in Palestine, progress was slow and hesitant, largely due to their small numbers and the Ottoman Empire's suspicion toward the new arrivals.[131] The first survey of northern Palestine's river regime was undertaken in 1904 by the Eastern European Jewish immigrant Nahum Vilbush. The outbreak of World War I and the Ottoman Empire's entrance on Germany's side raised the prospect of new rulers in Palestine, which sparked a flurry of Zionist activity in Palestine and London. When the British ousted the Ottomans from power in Palestine in 1917–18, the Zionist Organization and Zionist Commission immediately set about collecting all necessary information on concessions and landholdings the more effectively to lobby the authorities for commercial contracts of their own.[132] They also demanded that all concessions be suspended for the time being.[133] Ironically, considering the charge commonly leveled against the new arrivals, the Zionists' chief justification for suspending foreign concessions was their fear of a colonial dynamic of resource extraction that would not serve the country's development.

The Zionists were not the only ones in the early aftermath of the war vying for commercial concessions from the new rulers of Palestine. As the British settled into their new role, they were inundated with appeals. One group of supplicants consisted of those who had obtained concessions under the Ottomans, such as Euripides Mavrommatis for electrification and irrigation, Ismail Bey-Haseni and Salim Ajub's concession for the exploitation of petroleum in the south of Hebron near Kornov (later sold to Standard Oil), and the Dead Sea concession granted to an Arab company in 1911.[134] Palestinian Arab businessmen and municipalities also sprung into action, applying for various concessions. Jaffa, Haifa, Ramallah, Tulkarm and Nablus all submitted electrification bids.[135] There was also some interest from foreign companies.[136] The local tenders were modest in scope, involving limited street lighting schemes by means of fuel-powered generators and localized grids. Given the British preference for large-scale projects, however, these schemes held limited appeal.[137] The official reason for rejecting the applications was that no commercial concessions could be put up for public tender, let alone granted, pending the ratification of the mandate by the League of Nations and the conclusion of peace negotiations with the Ottoman Empire.[138]

Around the same time, we begin to see the earliest applications of a technocapitalist logic toward the geographic delimitation of Palestine. In 1918, David Ben-Gurion and Yitzhak Ben-Zvi, later the first prime minister and second president of Israel, respectively, coauthored the book *Eretz Yisra'el: Past and Present*. Written in Yiddish and published in New York, the book provided a meticulous historical account of "the Land of Israel" from ancient times. Yet, they argued: "If we want to determine the borders of Eretz Israel today, especially if we see it not only as the past domain of the Jews but as the future Jewish homeland, we cannot consider the ideal boundaries that are promised to us according to the tradition, and we cannot be fixed to historic borders that have changed many times and that have evolved by chance." Instead, they proposed that Palestine's borders be drawn according to "all of the Land of Israel's natural-physical signals" of the present, and its "cultural, economic and ethnographic conditions."[139] Palestine, according to Ben-Gurion and Ben-Zvi, was "a land without a people" poised for redemption by European Jews, "a people without a land," and could sustain as many as ten million people.[140] The book was an immediate best seller, with twenty-five thousand copies sold in the first three years after publication.[141]

The same shift was also manifest in the pages of *Palestine*, the British Palestine Committee's newspaper and *The Round Table: A Quarterly Review of the Politics of the British Empire*, a journal under the sway of the new imperialist statesman Alfred Milner and described by Jacob Norris as "the official mouthpiece of the new imperialist movement."[142] During the war, Milner was a key member of Prime Minister Lloyd George's War Cabinet, and in 1919 he was appointed colonial secretary. This "apostle of empire" was also one of the central forces behind the Balfour Declaration.[143] In 1918, an article in *The Round Table* argued that "Palestine has ... an economic future; and in making the most of its economic possibilities the Jews will not merely lay a secure foundation for their own national life, but will enrich the world by the addition of one more to the number of productive territories."[144]

Such notions, as we have already seen, were ubiquitous among government and colonial officials at the time. In his autobiography, Leopold Amery writes that his interest in Zionism "was at first largely strategical," but that "it was not long before I realised what Jewish energy in every field of thought and action might mean for the regeneration of the whole of that Middle Eastern region."[145] Both quotes are also indicative of the close interlinking of commerce and civilization that was integral to "constructive" imperialism: rendered productive by the Zionists, Palestine was believed to have the potential to serve as an engine of development for the whole region, and be a boon to the global economy as a whole. The biblical formula "from Dan to Beersheba," meanwhile, was explicitly set aside.[146]

In the fall of 1918, in preparation for the Paris Peace Conference, the Zionist Organization began studying the question of Palestine's borders more closely. Tellingly, it did not turn to Bible scholars or prominent Jewish historians for expert consultations. It turned to the agricultural engineer and citriculturalist Shmuel Tolkowky and the famous Palestinian agronomist Aaron Aronsohn, discoverer of wild emmer. Aronsohn in particular exercised a dominant influence on the proposal that the Zionist delegation presented at Paris in February 1919. It was based exclusively on economic considerations that, in the words of the Zionist proposal, would ensure Palestine's ability to "sustain a large and bustling community, that can better carry the burden of a modern government." The proposal, which was the most detailed and carefully prepared vision for the area by far, emphasized certain "methods of economic development," such as "drainage, irrigation, roads, railways, harbours and public works of all kinds," requiring "modern scientific methods" and a population that was "energetic, intelligent, devoted to the country, and backed by large financial resources that are indispensable for development."

In fact, every section of the border was justified in terms of technocapitalist development: the northern border was drawn to include sufficient sources of water for electrification and irrigation; the eastern plains of the Jordan

River should bound Palestine to the east, according to the proposal, since they were potentially cultivable. Although not strictly relevant to the border issue, the proposal also mentioned the planned deep-sea port in Haifa.[147] In another sense, however, the port was deeply relevant. The Zionist border proposal constituted a complete technocapitalist package: natural resources, the technological tools and know-how to exploit them, and the means to deliver them to the global market. In other words, technologies of local-to-global movement traced out the borders of Palestine.

The Zionist vision for modern Palestine was based on a spatial reconfiguration that refashioned the area according to a capitalist scalar logic. A number of local markets were agglomerated into a national economy that in turn would feed into a global market of production, circulation, and consumption of capital and commodities.[148] In 1929, Sir Hugo Hirst, the founder and chairman of the General Electric Company (GEC) in Britain, was interviewed regarding Rutenberg's work in Palestine, in which the GEC had recently become a substantial investor. Hirst stated that the electrification work had made possible "a bright economic future for Palestine, with the newly-started Haifa Harbour as centre, with the oil pipes from Mosul, and with roads as well as other means of communication connecting Palestine to all the centres of the East."[149]

For the same reasons that the Balfour Declaration lacked the capacity to determine material realities, however, the Zionist proposal at Paris did not have the power to dictate the borders of modern Palestine. To both the north and the east, the Zionists' proposal was significantly more expansive than the borders that were ultimately adopted. But the Zionist proposal in Paris both reflected and furthered the shift away from a vision for Palestine centered on the past to a forward-looking, technocapitalist rationale.

This, in turn, would pave the way for Pinhas Rutenberg's electrification project. Although not explicitly mentioned in the proposal, electrification was the topic of many conversations in early 1919. Rutenberg, who happened to be in America during World War I and would sometimes join Ben-Zvi and Ben-Gurion in the New York Public Library during their work on *Eretz Yisra'el* was engaged as an expert at Paris and gave testimony that the "development of electricity and an irrigation network in Palestine was a precondition for every plan for industrial and agricultural development."[150] As we will see in chapter 2, once the shift to technocapitalism had taken place, Rutenberg's proposals became critical to the precise border delimitations and the overall reconfiguration of Palestine as a capitalist and Jewish national space. To be sure, this is not to say that the past ceased to matter. At least on the rhetorical level, it remained the most important overall justification for the Zionist project. Rather, the shift occurred on the level of designing the material reality of the Jewish national home. It was the vision for its future, not its past, that gave Palestine its precise shape and character.[151]

In the immediate aftermath of World War I, numerous schemes were proposed for Palestine, all of which exhibited the powerful influences of the concepts and ideas outlined here, not least the utopianism that seemed to saturate so much of the thinking around non-Western development. One proposal envisioned digging a canal through Palestine, connecting the Red Sea to the Mediterranean with terminal points at Haifa and Aqaba, as a way of relieving the Suez Canal. Upon learning of this particular proposal, the British press extolled "the enormous value the new undertaking will represent to civilization" and predicted that the canal would create the conditions to set up "the industrial centre of the world.... It is hoped that before long Palestine will become the scene of the busiest workshop in the world."[152] Around the same time, a French engineer designed a plan involving a canal from Haifa to the Dead Sea, where power would be generated from the fifteen-hundred-foot level drop between the two bodies of water.[153] In 1919, the Norwegian engineer Alfred Hiorth proposed a scheme for the irrigation and electrification of all of Palestine. Hiorth's proposal called for a tunnel to be dug from the Mediterranean, passing beneath Jerusalem, to the Dead Sea, utilizing the fall of thirteen hundred feet, to produce 100,000 horsepower of electricity. In an illustration of the impurity typical of all modern visions, Hiorth stated that he had designed his plan after intensive Bible study and that it was modeled on the visions of Ezekiel 47 and Zachariah 14. Hiorth's plan received attention all over Europe, and the Zionist Organization, which studied it in detail, regarded it favorably.[154]

CONCLUSION

In his autobiography, Chaim Weizmann noted Rutenberg's "lack of Jewish background," claiming he was first and foremost a socialist. If the Bolsheviks had not chased him out of Russia, Weizmann speculated, Rutenberg might not have concerned himself with Jewish life at all.[155] But presenting it as one or the other misses what was most important about Rutenberg. Whether as a socialist, Zionist, or engineer, his brief was revolution. It was the word most commonly used to describe him, the word he himself used to describe his work.

In presenting himself this way, Rutenberg was channeling the zeitgeist. For a few decades around the turn of the last century, it seemed as if every canal, bridge, port, railway, and telegraph line possessed the power to revolutionize the world, as much as any Jacobin or Bolshevik. The hopes attached to Palestine's electric grid were no exception. And therein lay a new form of political power, which Rutenberg, the expert and revolutionary, would marshal toward the creation of a Jewish state. It was not simply Zionism, colonialism, or capitalism that enabled him to do so. None of these forces would have been sufficient on its own. Certainly, *this* Palestine—electrical Palestine—would not have emerged were it not for the precise impact of the complex of mutually influential forces outlined in this chapter. To be sure, they included Zionism, colonialism, and capitalism but also a particular technological vision, as well as the prevailing influence of technological and material properties.

To say that Zionism was not the only or even most important idea that animated Rutenberg's work, or informed British support for Jewish colonization, or determined the borders of Palestine is not to depoliticize this history and its contemporary reverberations. It is to point to the diversity of elements that politics comprise. Nor is the claim that huge powers were vested in one man and his power system intended as a return to a "Great Men" version of history. As a highly proficient systems entrepreneur, Rutenberg was as much a product of this history as was electrical Palestine itself. Most important, we now begin to see why modern Palestine was imbued with a logic that was attached to a particular technology, that of the electric grid, and thus why modern Palestine emerged as an entity that, at its core, both figuratively and literally, was electrical.

TWO

Contentious Concession

In all matters relating to Palestine, we stand under the shadow of the Balfour Declaration. The Rutenberg concession has always been regarded as the most practical example of the policy of setting up a National Home for the Jews. It is so regarded by the Zionists themselves. We are always trying to divert the attention of the Zionists from political to industrial activities, and preaching to them from the text that their best chance of reconciling the Arabs to the Zionist policy is to show them the practical advantages accruing to the country from Zionist enterprise. For these reasons we have supported and encouraged Mr Rutenberg's projects and I submit that we must continue to support and encourage them, so far as circumstances permit.

JOHN SHUCKBURGH,
Middle East Department Head (1922)[1]

WITHIN A YEAR OF ARRIVING in the country in the fall of 1919, Rutenberg had put together a first draft of his electrification scheme, supported by a wealth of maps, charts, and graphs. Next, he needed to convince three groups of its value: the British governments in Jerusalem and London, investors, and the Zionist movement. Naturally, the three were connected; success on one front would help him on the other two. Rutenberg found his first loyal supporter among the ranks of British officialdom in the summer of 1920 when he brought his proposal to High Commissioner Herbert Samuel. In July 1920, his first month on the job, Samuel reported to London: "I have discussed the project in detail with Mr. Rutenberg, and I am deeply impressed by the possibilities that would be opened out for Agricultural and Industrial development in Palestine should it be successfully executed. . . . I am so conscious of the high importance to Palestine of his project and of its being commenced at the earliest possible moment, that I am anxious to render every assistance to Mr. Rutenberg in his present journey and mission."[2] So much for securing political support. The most plausible sources of funding, meanwhile, were Jewish businessmen in Europe and America. Attracting

Jewish investors and ensuring that the project be Jewish owned was important politically to Rutenberg and even more so to the Zionist Organization, on whose support he also depended. Furthermore, Jewish investors were easier to attract, given their greater willingness to invest in projects that would benefit Jews in Palestine. Even Jewish businessmen who were agnostic or even opposed to Zionism as a political program were often sympathetic to Jewish economic development in Palestine, and thus more inclined to invest in commercial ventures there over and above what a strict business calculus would favor.

August 1920 was a good month for Rutenberg. Mutual acquaintances facilitated a meeting between Rutenberg and the British industrialist and Liberal MP Alfred Mond. Rutenberg explained his scheme to Mond, who passed it on to his own engineering consultants, Sir Charles Metcalfe & Selves. Their assessment of the scheme was uniformly laudatory. Encouraged by their endorsement, Mond committed to five thousand shares and forwarded the consulting engineers' report to the British government.[3] The same month, Rutenberg was able to convince the Executive of the Zionist Organization of the merits of his plan. Through the Jewish National Fund, the Executive pledged £100,000, or one-fifth of the £500,000 that Rutenberg then estimated the scheme would require. Julius Simon and Nehamie de Lieme of the Zionist Executive in London wrote to High Commissioner Samuel informing him of the Zionist movement's financial commitment, stressing the importance "to maintain its national character as an undertaking providing an essential service to the community."[4] Maintaining the support of both groups would require a constant dialectic of mutual reinforcement.

Later in the fall of 1920, Rutenberg met with Samuel and several department heads at Government House in Jerusalem. Rutenberg went over the technical details of the plan and told them about the funds that he had secured so far. He indicated that his project would benefit the British mandatory both directly as a key means of economic development, and indirectly as a way of attracting Jewish capital to Palestine. Plenty of wealthy European and American Jews were looking to contribute toward the "upbuilding" of Palestine and the Jewish national home, Rutenberg assured them. They would be particularly attracted by a project of such obvious utility and technical sophistication. The government representatives were impressed and agreed not to consider any scheme that conflicted with Rutenberg's for the duration of their negotiations. Samuel concluded the meeting by expressing his gratitude to Rutenberg for his services to Palestine and the Jewish people.[5]

THE PROPOSAL

On December 8, 1920, Rutenberg formally submitted his report, a glossy bilingual publication—Hebrew on the right, English on the left—of more than sixty pages, complete with lavish color illustrations. It stated that to develop Palestine, "large numbers" of Jewish immigrants had to be admitted to Palestine. Sustaining them through the initial phase required immediately launching "industrial undertakings and public works." Only later, when the immigrants had gotten used to the country and acquired some experience with physical labor, could they go into agricultural colonization. Such reasoning flew in the face of mainstream Zionism, whose ideological core was formed around the Tolstoyan notion that Jews needed to return to the land as farmers. As we will see throughout this study, there is good reason to doubt whether the rhetorical stress on agriculture within the Zionist movement, as well as in most histories of Zionism, accurately reflected what really mattered to the development of the Yishuv.

In any event, to Rutenberg industry, not agriculture, was the first necessary step toward building a Jewish state in Palestine. Industry, in turn, required cheap power. The only natural source of such power in Palestine, in his estimation, was water. Furthermore, developing Palestine's water resources for the purposes of generating electricity would have the added advantage of simultaneously preparing the ground for irrigation.[6] Rutenberg envisaged drawing on the motive force of the entire riparian system of the Jordan basin, from its tributaries in the north (the Hasbani, Leddan, and Banias Rivers) to its efflux into the Dead Sea, including Lake Huleh, the Sea of Galilee, the lower part of the Yarmuk Valley, Jordan's tributaries below the Sea of Galilee, and the Litani River basin.[7]

The first stage of Rutenberg's scheme called for the construction of a power station at the confluence of the Jordan and Yarmuk Rivers, eight miles south of the Sea of Galilee. By harnessing the motive force of the 164-foot level drop between the lake and the confluence, Rutenberg estimated that the station could generate 100,000 horsepower. That represented ten times the estimated present demand, according to his own figures. As we have seen, the grand—arguably utopian—scale followed a colonial precedent and a widespread preference among engineers and policy makers in Western societies. Equally, Rutenberg's scheme resonated with the concurrent energy system developments in the Soviet Union, which should not surprise us, since it was his classmates from the St. Petersburg

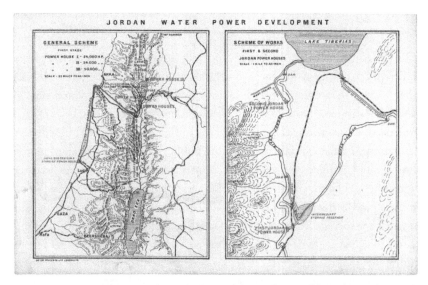

FIGURE 4. Maps of the hydroelectrical scheme at the confluence of the Jordan and Yarmuk Rivers, south of the Sea of Galilee (left), and of the high-tension electric distribution mains (right). Courtesy of the Israel Electric Corporation Archives.

Polytechnical Institute who presided over them. He was also not the first to home in on the Jordan-Yarmuk confluence. Sixteen years earlier, Nahum Vilbush had identified the spot as one of the most promising for a waterpower installation.[8]

The technical specifications of the scheme adhered to the emerging Western standard for a large-scale transmission network. In the technical literature, it is sometimes referred to as the *essential grid* model: a number of relatively short transmission sections tied together at major load centers (points of high demand), whence low-tension lines grow outward in step with demand. Under Rutenberg's plan, polyphase (or three-phase) alternating current would be distributed throughout the country by a large high-tension circuit. It would run due west from the power station to Haifa, and from there fall off south along the coast to Tel Aviv; then it would make a right angle inland and cut across the fertile lowlands of the coastal plain and ascend the Jerusalem hills, before returning north along the Jordan Valley, coming full circle. On the way, it would pass by most of the major population centers. Beisan, Haifa, Jaffa, Jerusalem, and Jericho were slotted to receive electricity in the first stage. Later, lines would extend southward, from Jerusalem to Beer Sheva, and from Jaffa to Gaza.[9]

The idea of harnessing the falling waters of the Jordan River was also in line with contemporary technological best practice, as indicated at the first World Power Conference in 1924, where waterpower held pride of place throughout the proceedings.[10] Two years earlier, Charles Marx, the former president of the American Society of Civil Engineers, claimed that "all engineers are so firmly convinced of the important role which hydro-electric power is playing and will continue to play as a factor in human development, that they do not take time to marshal the facts in support of this statement." Marx went on to extol waterpower's "supreme importance for the civilization, not only of this country, but also of other countries where only a beginning of its use is taking place."[11]

The choice of power source, the choice of location for the power plant (at the Jordan-Yarmuk confluence), and the design of the electric grid were all products of what the environmental historian Brett Walker has referred to as "hybrid causation." That is, it brought together factors from realms normally treated as separate, such as ecology, topography, economy, hydraulics, hydrology, meteorology, Bedouin tribal relations, epidemiology, and the availability of survey data.[12] In fact, these factors were so intimately intertwined that no obvious entry point presents itself. Let us start with the Bedouin. As it was becoming increasingly clear that Britain would not honor its promise to support Amir Faysal as ruler of Greater Syria (*bilad al-sham*), the area grew increasingly restive. Through the negotiations in Paris during the winter of 1920, and the subsequent establishment of the Middle Eastern mandates at San Remo in April, France was officially conferred the mandate for Syria. The tension that resulted extended to the Bedouin tribes of Beisan, Samakh, and Hawran.[13] Rumors had begun to circulate that Amir Faysal's government in Damascus had declared war on the French, and that tribes were mobilizing to prevent France from establishing control of the area. Either out of a desire to join Faysal's war effort—the reports of which, as it soon turned out, were incorrect—or out of a desire to fight Zionism, some tribes prepared for battle. Meanwhile, other tribes, acting on behalf of the British and, ironically, of Faysal, prepared for counterinsurgency. There are conflicting accounts of how the hostilities began, and no systematic study of the issue exists. Some at the time attributed the outbreak to the arrest of Amir Bashir, a prominent tribal leader, early in the fall of 1920. Others mention the arrest of Amir Muhammad al-Zaynati, another leader of the Ghazawiyya tribe in Beisan.[14]

As a result of the restiveness, Rutenberg had to depend on a British military escort as he began his survey of the Jordan basin in late 1919. Nevertheless, he was forced to halt his work in early 1920 because of what he characterized as "the present state of war in the region of Beisan."¹⁵ On April 22, 1920, some five hundred Bedouin from the Ghazawiyya tribe engaged British gendarmes in a pitched battle, which ended only after both sides had incurred heavy casualties and the arrival of a large contingent of British troops, along with bomber planes, driving the Bedouin back into the hills. The following day, an even larger Bedouin party, numbering about two thousand attacked Jewish colonies near Samakh. This attack was repelled by the British, who called in several airplanes and tanks. The British predicted that "matters will get worse instead of better."¹⁶ In the short time before Rutenberg was forced to abort the mission, he managed to survey the southern shore of the Sea of Galilee, and the Jordan Valley from the Sea of Galilee to Melhamia, that is, only a fraction of the area for which Rutenberg's scheme required extensive topographical and other data. Even with respect to the area he did manage to cover, his data were incomplete.

As a result of the unsafe conditions, the spotty survey data, and the water economy of northern Palestine, only one location appeared suitable for immediate construction. That was the confluence of the Jordan and Yarmuk Rivers, a few miles south of the Sea of Galilee. It had a number of advantages. It was part of the area with which Rutenberg had firsthand experience. The area also had a significant asset in that the Sea of Galilee offered a natural water storage reservoir. This not only would make it possible to even out the significant annual variability of water levels but also solved the problem of incomplete survey data. If, for instance, the quantity of water available from the Yarmuk turned out to be less than estimated, it could easily be compensated "to the desired amount from the vast yearly reservoir of Lake Tiberias." Finally, the land surrounding the confluence happened to be the spot in the area most densely populated by Jewish settlers. In addition to making land acquisitions easier, the presence of Jews would afford a certain level of physical protection to the workers and the works. "In the present political conditions in Palestine," Rutenberg stated in the report, "this region is the only one where the works can be carried out in safety." According to Rutenberg himself, all these things combined to make a site just below the Jordan-Yarmuk confluence, known in Arabic as Jisr al-Mujami', the only site on which it was possible to begin construction of a powerhouse "immediately."¹⁷

The design of the grid, meanwhile, was complicated by the fact that present demand was far below the amount of energy that the powerhouse stood to generate. This created the challenge of deciding where the power should go in anticipation of demand, instead of in response to it. In part, Rutenberg designed the grid on the basis of an assessment of future growth. It seemed safe to assume, for instance, that consumption in Jaffa and Tel Aviv would grow rapidly, and that consequently they would be important load centers. But far more important in a country of negligible electricity consumption, Rutenberg's predictions ended up creating demand through supply. Together with his associates he designated certain areas as "points of intense immigration and colonization"; supplying electricity to those places was based not simply on an estimate of future demand but on a desire for the construction of energy-consuming activities, such as industry, at those locations.[18] In a report from 1923, prepared by Rutenberg, he elaborated on his view of the relationship between electricity and development: "In general it is to be noted that the presence of an electric power station in a town raises the value of the town. This is reflected for instance in the rise of the value of land, houses, etc. Electricity modernises labour work by erection of electric cranes, spills, etc. enables the building of tramways and facilitates the development of industries. Electricity in short is a material factor for the growth of the economic life and prosperity of a town."[19]

Once construction began on the power plant toward the end of the decade, spokesmen for the Palestine Electric Corporation would often boast to the press that "not the least interesting feature of the enterprise is that it reverses the usual sequence of supply awaiting demand."[20] This is not as unusual as it may sound, and Palestine was not the only place where these kinds of considerations figured into the planning of electrical engineers.[21] But Palestine does appear to be an exceptional case as far as the extent to which the grid functioned as a tool of socioeconomic engineering. The foremost example in this regard is Haifa, a town that at the start of the mandate had no industry to speak of. With a vision of making Haifa into a major industrial and international trading hub, Rutenberg's plan called for sending 1.6 million kilowatt-hours there for industrial purposes. Indeed, largely as a result of this commitment, Haifa's industry grew rapidly, and so did its population: from just over 20,000 in the early 1920s to over 145,000 in 1946. Perhaps more significant, the proportion of Jews in the town doubled, from one-quarter to about half.[22] Indeed, in the most intense period of Palestine's industrialization, from 1931 to 1939, Haifa's population more than doubled, from 50,000 to 106,000. Of that increase, 50,000 were Jews.[23]

Moreover, by creating the distribution system as a countrywide "loop," it would be possible to create additional powerhouses throughout the country that would "later permit the transmission of [their] power to the great circuit." Most immediately, Rutenberg envisaged the construction of several more powerhouses along the Jordan River, which justified the high-tension line running up along the Jordan Valley from Jerusalem to the Sea of Galilee. He also planned to build an installation in Beer Sheba, another point he envisioned as a future center of Jewish immigration, justifying the inclusion of the town in the "great circuit."[24]

The economic benefits of the scheme were significant, according to the report. On its own, it stood to generate a 10 percent profit on an initial outlay of £2 million. And if the Palestine Government agreed to electrify the railway, the scheme, according to Rutenberg, would be "exceedingly lucrative" from the start. In his proposal, Rutenberg again held out the promise of Jewish capitalists poised to invest in Palestine. But first, the scheme had to be approved. Only with "an actual beginning of important and productive work by Jews" would sympathizers in Europe and America be moved to "contribute the enormous sums required for work in Palestine."[25]

In February 1921, two months after Rutenberg had presented his plans, the British submitted his proposal for external review by the engineer John Snell. Snell reported that he was "greatly impressed with the general conception." He concluded: "If the scheme is sound and commercially possible it ought to have an enormously beneficial effect upon the restoration of Palestine as an important agricultural country, and in the development of industries within the country."[26] Another important issue was the strong relationship between the availability of accurate numbers and the issue of trust. Snell affirmed that the key prerequisite of realizing the scheme was accurate numbers, and he commended Rutenberg for having begun the important work of obtaining them. This, however, did not resolve the question of how to produce accurate assessments of Rutenberg's work if he was the one supplying all the data. Snell and other engineers made occasional calls for independent data collection, but nothing ever came of it.[27] Assessments of Rutenberg's scheme remained entirely dependent on the numbers Rutenberg himself provided. To compensate for this obvious problem, Rutenberg's character became of signal importance. Snell, for instance, concluded his report by calling both the scheme and its author "statesmanlike."[28] In short, trust in Rutenberg's person offered the solution to the problem of not having access to firsthand knowledge of the land.[29]

BRITISH REACTIONS

Rutenberg's proposal elicited considerable excitement among British officials. In his dispatches, High Commissioner Samuel outlined the benefits that he expected the project to bring to agricultural and industrial development, and stressed the need to "render every assistance to Mr. Rutenberg."[30] London echoed the excitement. The head of the newly created Middle East Department of the Colonial Office, John Shuckburgh, wrote to Colonial Secretary Winston Churchill: "This scheme is the main economic pivot of the whole Zionist programme.... The Department regard the early initiation of the Rutenberg scheme as a potent factor, if not the most potent factor toward the successful development of Palestine."[31] Major Hubert Young, also of the Middle East Department, concurred: "The successful inauguration of Mr. Rutenberg's schemes will do more than anything else to pacify Palestine, facilitate immigration, and develop the country."[32]

The enthusiasm over the power system went beyond British government circles. In the popular press at the time, the excitement was palpable. The *Daily Chronicle* predicted that Rutenberg's scheme would give Palestine a valuable position as a commercial and cultural bridge between East and West, providing a staging post for Mediterranean and Mesopotamian trade. "Work of the sort that is to be carried out in the Jordan valley," the article concluded, "is a good example of the great benefits that the Zionists can bring to the entire country, including the Arabs."[33] Zionist activities in Palestine were construed as part and parcel of the imperative to develop, which in turn would facilitate the corollary obligation of providing for the well-being of the entire population. In the effort to electrify Palestine, the dual mandate fused into a single project.

At a parliamentary debate in 1922, Churchill defended the concession by claiming that Rutenberg's plan was "the main and principal means" by which they would honor their commitment both to the Jewish national home and to the well-being of the Arabs of Palestine. The "development of the waterways and the water power of Palestine" was the sine qua non that would allow the Jews to create "a new world entirely, a new means of existence" in Palestine, thus claiming none of the existing resources, and thus taking nothing away from the Palestinians, a claim that would be repeated often by the 1930s. The concession, explained Churchill, was a means "to interest Zionists in the creation of this new Palestine world which, without injustice to a single individual, without taking away one scrap of what was there before, would endow the whole country with the assurance of a greater prosperity and the means of a higher economic and social life."[34]

The economism of the framing also shielded the British government from accusations of partisanship. Presenting their Palestine policy as guided solely by scientific logic allowed the British government to negotiate what was fast becoming a political minefield, by largely sidestepping it. Yet one of the principal ways it did so presents one of the most curious paradoxes of the story. In his report, Rutenberg did not hide his commitment to Zionism. On the contrary, he flaunted it. "National enthusiasm and self-sacrifice have an important part in the productivity of Jewish labour in Palestine," he declared. Yet such ideological pronouncements offered the British a technical and economic argument for supporting Zionism: the ideological underpinnings of Zionists' desire to develop the land would cause them to work harder, and demand less in return, than people motivated by financial profit alone. In effect, the source of Rutenberg's appeal and the appeal of his project were connected both to the way the project embodied scientific reason and to the way it flouted a purely rational economic and scientific calculus, since his ideological commitment to Zionism would cause him and his people to work harder than "objective" reasoning would seem to justify.

The argument was effective with the British both because of the anticipated windfall of an especially motivated developer and because this same reasoning could then be used, paradoxically, to deflect accusations of political bias. Hitching their wagon to an accomplished engineer who was also on the record as a committed Zionist was presented by the British government merely as good business sense, not an indication of pro-Zionist sympathies. In a parliamentary debate, Churchill described Rutenberg in the following way: "He is a man of exceptional ability and personal force. He is a Zionist." Rutenberg would not have been able to raise sufficient capital if it had not been for "sentimental and quasi-religious emotions."[35] In a handwritten memo attached to a file discussing electrification in Palestine, Middle East Department head John Shuckburgh explained the logic of his department's policy with respect to electrification:

> The grant of the Rutenberg concession was part and parcel of our Zionist policy. The assumption underlying it was that Jewish enthusiasm for Zionism was such as to exclude all difficulty in financing Zionist projects, even when (as in the case of Rutenberg) undertaking by the concessionaire is on disadvantageous terms. So long as we can take this ground we have a good answer to critics. Palestine, we can say, profits by Zionist enthusiasm, because Zionists,

thanks to their enthusiasm, will submit to more onerous terms than other people in carrying out works of public utility. It is a very strong argument."[36]

In other words, the Zionists' ideological commitment was transformed into a technical rationale for British support. In 1925, Hebert Samuel defended the decision to grant the concession to Rutenberg without considering any other bids by saying, "It was not anticipated that any tender would be forthcoming for an enterprise the financial attractions of which were not very great."[37] At the same time, the Zionists' technical capabilities were also essential to lending support to their activities. Rutenberg embodied this curious admixture. He possessed both technological expertise and ideological fervor. Both postures, though seemingly contradictory, constituted essential assets in his dealings with the British. And both functioned as tools of depoliticization.

Rutenberg's proposal also received attention outside of Britain. The US Department of Commerce published a booklet describing Rutenberg's venture as "by far the most important project which exists for the economic rehabilitation of Palestine."[38] In October 1922, a consulting engineer at AEG, the German electrical manufacturer, deemed the project "the most important condition for the development of Palestine" and "the basis of the future economical life" that would serve as "a stimulus for many other projects and as a signal for the realisation of other enterprises of importance."[39] In 1929, the founder and chairman of the General Electric Company (GEC) in Britain predicted that thanks to the electrification work being undertaken in Palestine, in which the GEC had recently become a substantial investor, "within a generation Palestine will become the gate to the East, the principal centre of communication with India through Iraq."[40]

RUTENBERG'S THREE WISHES

In late March 1921, three months after personally delivering his glossy report to the high commissioner in Jerusalem, Rutenberg wrote to Samuel with a few requests. He was preparing for a fund-raising trip to America, and in order to maximize the fund-raising potential, he needed three things: first, consent to electrify the Palestinian railway; second, an in-principle grant of the countrywide Jordan concession; third, a signed concession for a local electrification scheme in Jaffa.[41] Within a year, he would have all three, an extraordinary accomplishment given the timescales on which colonial administrators usually operated. Let us look at each in order.

Electrifying the Railway: A Solution in Search of a Problem

On the face of it, the decision of whether to electrifying the railway was strictly a financial matter. Snell's report from March 1921, like most engineering reports to follow, contained a significant qualification in the phrase "if the scheme is commercially possible." This part, however, British officials were all too happy to ignore. But in fact there was good reason to doubt the commercial viability of the scheme. In Rutenberg's original proposal from December 1920, he had presented his plans as a boon to the railway. According to his estimates, electrifying the Jaffa-to-Jerusalem portion of the line alone would save the British government in Palestine £E50,000, by eliminating the need for coal, while doubling traffic capacity.[42] But in a series of subtle shifts manifest in the correspondence between Rutenberg and the British through the spring of 1921, he began conveying a different message, one that ultimately turned the dependency on its head.

In May 1921, less than six months after submitting his initial proposal, Rutenberg submitted detailed specifications of the initial stages of the plan: the first hydroelectric station on the Jordan and an assessment of the merits of railway electrification. From this report it was clear that the railway would constitute, for the foreseeable future and by some margin, the main consumer of electricity. Rutenberg's assessment concluded that "electrification of the railways, i.e. the erection of a large consumer of energy is an *essential condition* of the realization of my Jordan project." Even if the Palestine Government would agree to electrify only the Jerusalem-Jaffa branch line, the demand would reach 4 million kilowatt-hours per year, which was more than three times greater than the second-largest single consumer, which was for private industry in Haifa, at 1.6 million kilowatt-hours. And that demand, we recall, was not real, but aspirational.

Although Rutenberg still claimed some benefits for the railway—such as a 20 percent increase in carrying capacity, the need for fewer locomotives, shorter stops at stations, and a reduction of annual expenditure—the central issue was no longer of hydroelectrification's benefiting the railway, but of the railway making hydroelectrification possible. Without a large consumer like the railway, Rutenberg now said, the scheme would not be financially viable. Nevertheless, Rutenberg wrote, "the realization of the projected Hydroelectric station at the earliest possible time is of the utmost importance for the development of Palestine." He concluded by requesting a commitment from the government to electrify the Jaffa-Jerusalem line and guarantee an electricity purchase of at least £E40,000 annually.[43]

Arguably, the conclusion closest to hand was that there was a flaw in the original plan. The British, however, seemed fully onboard with Rutenberg's conclusion: the fault lay not with an ill-suited solution (large-scale hydroelectrification) but with the lack of an adequate problem (sufficient demand). In June 1921, a month after Rutenberg published his revised assessment, the high commissioner wrote to the colonial secretary regarding the question of whether to electrify the railway: "The demand for energy from other consumers is, and will probably long remain, only a small fraction of the total supply which will have been made available. This aspect of the question must be borne in mind when considering the narrower issue of the advantages and disadvantages from a purely railway point of view."[44] To Samuel, and indeed the entire Palestine Government, this was an argument in favor of electrifying the railway.[45] Another official concurred that Rutenberg's plan "cannot be judged purely on its technical merits. It is the back-bone of the Zionist policy of offering employment to Jewish immigrants."[46] Many more colonial officials and consulting engineers agreed: Rutenberg's plan was too important to be hindered by narrowly technical or economic considerations.[47]

There were a minority of dissenting voices, however, expressing the belief that Rutenberg's works were not worth an expenditure of £250,000 to convert the fifty-six miles of the Jaffa-Jerusalem branch line to run on electrical traction.[48] The opposition notwithstanding, in November 1921 the Palestine Government approved the electrification of the Jaffa-Jerusalem line "in principle." All this, with no hard evidence that Rutenberg's scheme stood to bring any economic advantage at all to Palestine, at least as compared to a more modest fuel-powered scheme.[49] And so, Rutenberg's first wish had been granted.

The Auja Scheme: Marshaling Technological Exigency

Despite the great enthusiasm with which Rutenberg's proposal was received by British officials, there was a lingering apprehension among them to approve a scheme of such heroic proportions even before the mandate had been officially confirmed by the League of Nations. Sensing the British reticence, Rutenberg developed a smaller auxiliary scheme, involving a single hydroelectric powerhouse on the Auja River for the electrification of the

Jaffa District. In April 1921, he sent a draft of the plan to the high commissioner, and after a meeting a few weeks later with several senior government officials at Government House in Jerusalem, he attached it to the larger scheme that he submitted in May 1921. Its threefold purpose was to provide street and private lighting to Jaffa, Tel Aviv, and Petah Tikvah; pump water to Jaffa and Tel Aviv; and facilitate large-scale irrigation of Petah Tikvah's orange groves.[50]

Because the source of the motive power was water, requiring the construction of an intricate system of dams and canals, Rutenberg included a scheme for the irrigation of the whole Auja basin in the proposal, on the argument that electrification and irrigation naturally went hand in hand. The Auja plan called for utilizing the fall of the river by placing a power station by the dilapidated Hadrah Dam, located about 5 miles inland from the Mediterranean. From there, a canal, 1.6 miles long, would issue and loop back onto the Auja a ways downstream, just below the Jerishe Mill. The estimated level difference of 20.5 feet and water flow of 300 cubic feet per second would be capable of generating 530 horsepower. A 6,000-volt high-tension main would lead the current the 3 miles southwest to Tel Aviv and a transformer station on Allenby Road. From there an underground cable would carry the current to Jaffa. Rutenberg explicitly pitched the project as a stopgap that would power "the more thickly populated centres in Palestine . . . until the main project [i.e., the Jordan scheme] is carried into effect."[51]

Electrification itself, however, may well have been the least important of the scheme's objectives. To anyone who cared to look, it ought to have been evident that the Auja River was a poor source of power. The modest 530 horsepower that the plan stood to produce at a maximum would be generated at considerable expense, and as sure as demand was likely to grow, there was no hope of increasing output. This much was plain to the British government's consulting engineers, Messrs. Preece, Cardew, and Rider. In their otherwise laudatory report from 1921, they noted in passing that, "At present, demand will be easily satisfied by the 350 kw [i.e., 530 horsepower] generated by the plant, but if demand grows, it won't be possible to extract more power from the Auja, and it will be necessary to supplement the power by Diesel or other engines."[52] Rutenberg's own experts reached the same conclusion. So, too, did many external observers, including the US Department of Commerce.[53] Indeed, at a meeting of the Tel Aviv municipal council in the spring of 1921, a local engineer noted the diminutive character of the project: "Electric supply by means of the Yarkon (Auja River) will definitely not be sufficient in order to also produce power for industrial purposes," he noted.[54] For Rutenberg, however, there were many other benefits to going hydroelectric, outweighing the plan's technological deficiencies.

A key element necessary for understanding the process was Rutenberg's ability, whatever the challenge of a given moment, to marshal the exigencies of the technology to his commercial and political advantage. In the first instance, the British were the target of his technopolitical strategy. The technological properties of hydropower offered a powerful argument for awarding a single, exclusive concession for the entire country: the heavy capital expenditure on the front end and the vast engineering work required made sense only if carried out on a large scale. This had been a major consideration in designing the original Jordan plan. A thermal power plant run on oil, by contrast, was far less costly and labor-intensive and required only a small patch of land to house the fuel sets. The technology itself did not call for extensive monopolies. Furthermore, since the Auja scheme was deliberately designed to ease the British into the idea of awarding a countrywide concession based on the hydroelectric Jordan scheme, it was in Rutenberg's interest to make the two technological kin. Finally, a scheme based on waterpower necessitated control of vast swaths of land in the Auja basin, which in turn enabled Rutenberg to ask for (and subsequently receive) exclusive rights to irrigation, in addition to the electricity monopoly. In other words, the Auja scheme was a cat's paw, primarily designed not to bring electricity to Jaffa but to furnish Rutenberg with a claim to a countrywide energy monopoly on the strength of technological exigency.

To the British, too, the significance of the Auja project was not primarily lighting the streets of Jaffa and Tel Aviv. Through the winter, spring, and summer of 1921, negotiations were ongoing for both the countrywide Jordan scheme and the Auja works. In March, Rutenberg managed, with Samuel's backing, to convince the Foreign Office to rush through the Auja concession before the ratification of the mandate. The winning argument, as far as the British were concerned, was that tentative approval of the smaller Auja scheme would enable Rutenberg to raise large sums for both schemes on his upcoming trip to America.[55]

As mentioned earlier, it was British policy that no commercial concessions would be granted in the British-controlled Middle East before the peace treaty with the Ottoman Empire had been concluded. It was with reference to this policy that the British had rejected all other applications for commercial licenses in Palestine, including the many electrification schemes that the

British had received since war's end. The real reason for the policy was British fears that opening up the territories for commercial exploitation at this early stage would leave them vulnerable to domination by non-British business interests, primarily American. Such was the level of excitement over Rutenberg's proposals, however, that Colonial Secretary Churchill went to the Board of Trade and the Foreign Office to advocate a change of policy. Churchill pointed out that British support for a Jewish national home in Palestine had been expected to yield a considerable "influx of capital into Palestine and further economic development which would increase the wealth and prosperity of all classes and sections of the population." The sources of this capital, it was always assumed, would be European and American Jews. Unfortunately, the letter continued, the slow progress of the peace negotiations had prevented the British government in Palestine from granting commercial concessions, which in turn had stalled this process to the point of risking its undoing. In order to clear a path for Jewish investment, Churchill proposed to grant "gradually and within certain limits a few other concessions of a nature to create employment." Churchill got his way, and so did Rutenberg: his was the only concession explicitly mentioned in the letter.[56]

The strong ulterior motives on both sides likely explain why neither Rutenberg nor the British nor any of the experts they hired to consult on the scheme questioned the wisdom of spending £E100,000 on a hydroelectric scheme whose maximum output was capped at 530 horsepower, the equivalent of a modern sports car. The fact that both parties agreed on these numbers calls into question Ronen Shamir's claim that the concession was awarded because of a discrepancy between local knowledge and imperial ignorance. In Shamir's telling, the British did not object to the scheme because they lacked local knowledge. The Tel Aviv–based engineer, mentioned earlier, who had identified the limited potential of the Auja scheme, had had local knowledge, whereas the London-based consulting engineers had not. This, according to Shamir, is why the British engineers supported the scheme, while the local engineer was critical of it.[57] We have already seen that the problem of local knowledge was resolved by building trust in the person of Rutenberg, and Shamir arguably misreads the primary motivations behind the British position. Neither they nor Rutenberg was concerned exclusively with the technological virtues of the initial scheme. Rutenberg was interested in getting control of land, and for the Auja scheme to serve as a cat's paw for the Jordan scheme. The British wanted to approve a commercial concession—any concession—to attract capital from Europe and the

United States. Tel Aviv, naturally, regarded the Auja scheme through a narrower lens, related exclusively to the electrification of the district. Moreover, the figures regarding the river's flow, the amount of energy available, and so forth, that Rutenberg presented to the consulting engineers and others were never in dispute. In other words, there was no controversy over the figures and no claim of deeper knowledge based on being physically present on the land. The consulting engineers in London based their assessment on the same figures the local engineer cited. The claim was never made that the British experts lacked any critical piece of data, which the local possessed by virtue of being present on the land. Indeed, as we saw, the consulting engineers did note that the Auja works would have to be supplemented by diesel engines. The divergent conclusions were the result not of conflicting readings of the technical data but of divergent scales and priorities.

The Auja scheme was submitted for review by the government's legal consultants, Messrs. Burchells & Co. in June 1921.[58] In early July, Rutenberg received provisional approval for the Auja scheme.[59] On September 12, 1921, the concession for the electrification of the Jaffa District by means of a hydroelectric power station on the Auja River was signed. It not only gave Rutenberg exclusive rights to generate and distribute electricity in the District of Jaffa but also granted him sole proprietorship of the river's water for irrigation purposes for the whole of the Auja basin's 355 square miles, almost all of which were Palestinian owned.[60] Rutenberg's second wish had been granted.

Negotiating the Jordan Agreement: Technological Utopianism

The third wish, we recall, was an "in-principle" agreement for the larger Jordan scheme. The technical specifications that Rutenberg presented to the British in May 1921 contained slight modifications to the original hydroelectric scheme that Rutenberg had submitted in December 1920. The first phase of the scheme, the power station at the confluence of the Jordan and Yarmuk Rivers, still stood to generate energy vastly in excess of demand: according to Rutenberg's own figures, the largest area of consumption in Palestine, private industry, could muster a demand of roughly 3.2 million kilowatt-hours annually, and water supply another 2.4 million. Yet the powerhouse on the Jordan stood to generate 90 million kilowatt-hours a year.[61]

The terms of the concession were first negotiated in Palestine, beginning in mid-May 1921. In June, it was forwarded to London after receiving the approval of all the department heads and the high commissioner. In addition to the technical and legal consultations, Whitehall was also concerned with the compatibility of the scheme with the border negotiations between the British and the French in northern Palestine, the legal standing of the concessions in the context of the still-pending ratification of the Treaty of Sèvres, and, to a lesser degree, opposition from the Arab Palestinian community.[62]

In early July, Rutenberg received provisional approval of his scheme, following a conference at the Colonial Office and a few amendments to the concession text.[63] In July and August, the Jordan plan was studied by legal and engineering consultants. In early September, the Colonial Office forwarded the finished versions of the Auja concession and Jordan scheme to Churchill for approval. In the cover letter, Shuckburg, the Middle East Department head, again emphasized that Rutenberg was working closely with the Zionist Organization, which would ensure his ability to raise the necessary funds for the venture.[64] On September 21, 1921, nine days after the signing of the Auja concession, the Government of Palestine and Rutenberg concluded the preliminary Jordan agreement, thereby fulfilling Rutenberg's third wish.[65]

Under the concession, a limited liability company with an authorized capital of £1 million was to be formed within two years. Rutenberg was then to transfer his rights and obligations under the concession to the company. The concession also specified the maximum allowable rates for different sectors. Half of any profit after expenditure above 10 percent and all amounts above 15 percent had to be put toward lowering electricity rates. The concession term was for sixty years from the day the first powerhouse started generating electricity. This agreement was closest to the German model, welcoming private enterprise while reserving a significant portion of the benefits of the venture to the municipality as a whole.[66] Among the notable features of the concessions were the extensive powers of expropriation that came with them. "I do not quite like the look of this Clause," wrote Gerard Clauson at the Middle East Department, "It appears to be in Mr. Rutenberg's power to point to any building or piece of ground and say that it is essential for him to have it, and the Government has then got to buy out the owner compulsorily if he will not sell it willingly. This is putting an extremely powerful weapon in Mr. Rutenberg's hands, and also into the hands of the Company which is succeeding Mr. Rutenberg." Clauson acknowledged that the expropriation clause could not be watered down to a fair-weather entitlement, contingent on universal consent. Nevertheless, he suggested that the clause should be rephrased such that any expropriation had to be effected through the

Palestine Government, which was obliged to carry out the expropriations only when they were deemed "reasonable."[67] A version of Clauson's suggestion made it into the finished concession, but Rutenberg's powers in this regard remained considerable. With the expropriation and exclusivity clauses, the British had bestowed significant powers on Rutenberg, and they knew it. In a letter to Colonial Secretary Churchill a few weeks before the concession was granted, R. V. Vernon, a senior Colonial Office official, wrote, "Considering the small area of Palestine, its sparse population, its backward economic development, and the peculiar conditions of its water supply, I feel no doubt that a policy of granting a monopoly is not only justifiable but necessary."[68]

With the signing of the Auja concession on September 12 and the Jordan agreement on September 21, and the approval in principle of railway electrification in November, Rutenberg's three wishes had been granted. He could now depart for America with several tangible achievements at his back. In late November 1921, Samuel drafted a letter in support of Rutenberg's upcoming trip to further boost Rutenberg's standing. The wording of the letter is telling with respect to both the British view on electrification and the significant role they saw for American Jewish capital in the scheme:

> The mission on which you will be engaged in America is one of the greatest importance to Palestine. The plan for the utilisation of the principal sources of water power in the country for the production of electric energy, combined with the irrigation and colonization of certain of the lands now vacant or sparsely cultivated, is one of the most valuable measures that could be devised for the economic regeneration of the country. His Majesty's government, as well as the administration of Palestine, are greatly interested in the scheme, and desirous of extending to it every encouragement. Its successful execution, by increasing the prosperity of the country, will benefit all sections of the community. It will be a striking example of the advantage to be received from the introduction of Jewish enterprise and capital, and the establishment of additional colonies of Jewish settlers. I would express the very earnest hope that you may secure in America the measure of financial support that is essential to enable you to carry into execution your most valuable project.[69]

THE BORDERS AND THE POWER SYSTEM

In what is surely a unique case in history, Rutenberg's electrification proposal specified his vision not only for the electric grid but also for the state fit to contain it. In his proposal from 1920, he wrote: "The foregoing scheme is defining the boundaries of Palestine. . . . This work cannot be accomplished on a territory that is not under Palestinian control."[70] The northern border was determined in two stages. It was initially decided in an Anglo-French Agreement in December 1920, but the final demarcation process was not completed until March 1923, at which point, as Gideon Biger has observed, "the geographic principle 'water for Palestine, access roads to Syria' had replaced the historic biblical formula 'from Dan to Beersheba.'"[71] Rutenberg was a constant presence during the Anglo-French negotiations, and he influenced the process in both stages. He served as an expert on the Anglo-French Water Commission, and the record makes it clear that his opinion weighed heavily throughout the proceedings.[72] For a moment, the whole Zionist movement was mobilized for the border question, not least in America, where the Zionists even managed to sway Woodrow Wilson in their favor. In a telegram to the British government, Wilson called for the Litani River and Jordan tributaries to be included in Palestine. "The Zionist project depends on a rational border in the north and east in order for the country to develop economically and be able to support itself," he wrote.[73] Meanwhile, all other experts, according to instructions from Churchill, "should of course be conversant with the Rutenberg scheme."[74] The British constantly and explicitly weighed their own political and military interests against Rutenberg's scheme, often finding, as in one example from Churchill, that advantages "from an imperial point of view" were "more than counter-balanced by the disadvantages from the Zionist [i.e., Rutenberg's] point of view." The border as it was ultimately set down did not correspond to any strategic military considerations, something that was noted repeatedly at the time.[75] In 1937, in the context of another boundary discussion, Rutenberg stated to the British authorities that, in his capacity as water expert in the early 1920s, he had "succeeded in convincing the French Authorities of the necessity to modify the Northern boundaries so as to conform with the economic and topographic requirements of Palestine."[76]

This is not to say that Rutenberg was able to dictate the border definition or even single-handedly shape the British negotiating position vis-à-vis the French. As the increasingly frustrated British representative on the boundary commission wrote in a private note to his superiors at the Colonial Office, "I find that each individual has his own interests."[77] Rutenberg was ultimately forced to abandon all but three of the fourteen powerhouses of his grand original vision that were supposed to run from the Litani River (in present-

MAP 1. Borders and power lines at the start of the mandate.

day Lebanon) southward through the entire Jordan Valley. Unsurprisingly, the three power stations that remained, and to which the boundary concessions were made, were ones for which Rutenberg could supply detailed technical plans, backed up by topographical, hydrological, and meteorological data.

The starting point of negotiations between the French and the British was the 1916 Sykes-Picot Agreement, according to which the border between the French and British spheres of influence would run from a point just north of Acre eastward to the Sea of Galilee, placing the entire Upper Galilee, including the town of Safed and Lake Huleh, on the French side of the border. Such a border, Weizmann cabled a confidant in early 1920, "would render impossible the development of a modern economic state."[78] In the negotiations from 1920 to 1923, the British successfully pushed the border significantly northward, especially in the eastern section. This was motivated by the desire to include as much as possible of the riparian system that fed the Jordan River from its inception below Lake Huleh. The entire Yarmuk Valley up to al-Hamma was included in Palestine, to enable the construction of a canal between the Yarmuk River and the Sea of Galilee, and "because the foothills of the Litani River," as the *Jewish Telegraphic Agency* explained at the time, "are claimed by Pinhas Rutenberg as essential to his Hydraulic project."[79]

Based on those plans, the British also successfully moved the border to secure the Dan springs for Palestine. Lake Huleh was included, in the words of one government report, to enable "the eventual construction of a dam at the point where the River Jordan issues from the Lake."[80] Without it, as Weizmann told a confidant, Rutenberg's project would be "rendered impossible."[81] This was done over the objections of the chief British negotiator. His growing frustration with Rutenberg's influence was manifest in a letter to the Colonial Office in which he likened the inclusion of Lake Huleh in Palestine to "an obnoxious intrusion, or an irritant boil."[82] The Sea of Galilee, bisected along a north-south line under the Sykes-Picot Agreement, was included in its entirety in Palestine, along with a strip of land around its eastern shore, widening as it traveled south, in order to include the point where the Yarmuk River debouches onto the Jordan River.[83]

Rutenberg's influence on Palestine's eastern border was no less significant. At the Cairo Conference in spring 1921, the British decided to split the Palestine mandate in two, designating the western territory as the fulfillment of Britain's pledge to the Jews in the Balfour Declaration and the eastern territory as the fulfillment of Britain's pledge to the Arabs in the Husayn-McMahon Correspondence. The border between the territories was roughly imagined to run along the Jordan River but was not defined until the following summer, shortly before being submitted to the League of Nations for approval in September 1922.[84] As in the case of the northern border, Rutenberg was actively involved in the process of delineation.[85] As Major Hubert Young of the Colonial Office explained, when the border was defined in the summer of 1922, it contained one major irregularity in that "the triangle of land between the Yarmuk and the Jordan (which is, of course, really across the Jordan and should therefore, by any ordinary acceptance of the term, lie within Transjordan) was retained for Palestine," because "it was necessary for the purposes of the Rutenberg Concession."[86] The border was submitted to the League of Nations as the Transjordan Memorandum and passed by the League Council in September 1922.[87] Emir Abdullah, the ruler of Transjordan, was not included in the proceedings until after the border had already been submitted to the League. In October, the high commissioner met with Abdullah to discuss "informally" the question of borders. It was not much of a discussion, of course, since the matter had already been settled. Abdullah nevertheless expressed his consternation regarding the exclusion from Transjordan of Samakh and other areas to the east of the Jordan River. Abdullah claimed that according to older definitions of Palestine, Samakh should not be included, though in reality his chief concern was that the area's fertile soils would have generated much-needed revenue for the fledgling Transjordanian monarchy.[88]

PALESTINE MAPPED AND IMAGINED

As we have already seen, the forward-looking vision for Palestine that Rutenberg's project embodied had a receptive audience in Britain. This was true not least of the popular science publications. For example, a lengthy article on Rutenberg's scheme in the *Engineering News-Record* from 1922 stated: "The enterprise is of vast extent, covering practically the whole length of Palestine from north to south and the major portion of its width, from the Mediterranean to the eastern border of the Jordan valley."[89] These are interesting words, given that they assume a precise length and width at a time when Palestine in fact had no precisely defined boundaries. At the time the article appeared in print, the demarcation of the Palestinian-Transjordanian border was still a couple of months away, and the Anglo-French border demarcation would not be completed for another eight months. In fact, it was in connection with Rutenberg's scheme that references to a precisely defined Palestine first started to emerge. Though never expressly acknowledged, the power station and grid shaped out the contours of something that had not existed before: a clearly defined Palestinian territory. Furthermore, the projected grid did not simply coextend with the pattern of Jewish settlement at the time, the famous N-shape.[90] The works delineated something more than that: not an ethnonational community but a country and state, and, equally important, an economy.

In other words, the power system's role in the making of modern Palestine was not simply or most importantly a matter of borders. The electrical works also contained within them a vision for the character of the territory whose most telling expression is found in the projected power station's scale. As mentioned, the first stage of the scheme was expected to generate fully ten times the present demand. "The fundamental principle underlying the plans of the electric part of the work," Rutenberg explained in his initial proposal from December 1920, "has been, that. . . . a relatively considerable quantity of energy should be supplied to centres chosen as points of intense immigration and colonisation, even before there is a demand for such energy. . . . Owing to this consideration, the distribution has been calculated on a large scale—even at the risk of increasing the expense of the first establishment—in order to make possible, in the future, a development worthy of the enterprise in question."[91] The article in the *Engineering News-Record* from June 1922 went on to discuss the grand scale in relation to Palestine's present and future: "The conditions in the country at present make it impracticable to execute the whole project at once, but it will have to be constructed in successive stages, in a measure keeping pace with the upbuilding of the country."[92]

The power system conjured a unified territorial scale where one had not existed before. It also provided it with a social and economic character that bore little relation to the present situation in Palestine. The projected power system was a scheme tailored not for the present but for a particular future: it was a countrywide, not to say national, technological system, whose ultimate realization depended on the ultimate realization of an industrialized Palestine, capable of mounting energy demand at civilized levels for the national economy and the private household. Its technical blueprint staked out an end goal that was essential in calling modern Palestine into being, both as an imaginary and a material reality.

The power system's function in this process was twofold. On the one hand, its output was effectively constituted as a scientific metric that indexed

Palestine's degree of civilization.[93] In the introduction, we saw several general examples of the notion of a direct link between the level of civilization and the level of electricity consumption. These notions were manifest in the context of Palestine too. In an address to the Anglo-Palestine Club in London in 1921, Sir Philip Dawson, a Conservative MP and a leading expert on electrical engineering, claimed that Palestine could sustain five times its current population, provided that "modern" farming methods were introduced. As far as electricity consumption was concerned, Dawson maintained, the biggest obstacles were the "primitive methods of agriculture" and "inefficient labour," which translated into a "small field for the use of electricity."[94] The *Engineering News-Record* was even more explicit about the link between the level of civilizational progress and the growth of the grid: "Immigration and modernization of methods and living conditions are considered by the promoters of the enterprise to be prime essentials of the successful realization of the ultimate project."[95] In an interview with the *Sidney Morning Herald* in 1929, Sir Hugo Hirst, the founder of the General Electric Company in Britain, said: "The financial success of the Palestine Electrical Corporation depends ... upon the development of the country as a whole.... Industries set up by new settlers, are also developing. This can be ascertained from figures relating to the consumption of electrical power."[96] On the other hand, the power system's function was not just indexical; it was also heralded as the motor of Palestine's development. The system would provide an impetus for extensive housing construction, which would require cement mills, brick works, transportation, and so on. The British had lent their support to the scheme, the article explained, "because its successful execution will increase the general prosperity of the country."[97] This was entirely in line with Rutenberg's own thinking. In his original proposal, he stated: "The very fact of the existence in these regions of a great quantity of cheap energy will be the best and the only true stimulant for the development of agriculture and industry, and consequently for the development of Jewish colonisation on a large scale." And he estimated that ten to fifteen thousand people would find employment as a direct or indirect result of his undertaking.[98] In the statistical abstract put out by the Zionist organization Keren Hayesod in 1929, the matter was put even more directly: "The possibility of obtaining cheap current will certainly give an all-round strong impetus to industrial development."[99]

In 1930, the *Financial Times* noted that there was still reason to doubt whether Rutenberg's megaproject was appropriately scaled for the as-yet-underdeveloped Palestine: "For at the present moment, Palestine is barely ready even for the half capacity output of the first stage." But the article immediately went on to list what the author called Palestine's "potential possibilities." Exploitation of the Dead Sea minerals was imminent; there were calls for a Jordan Valley railway link to Haifa; and there were plans to divert the oil from Mesopotamia to that city from to the Persian Gulf. The construction of a deep-water port in Haifa would itself be a significant consumer of electricity. "Nature has contrived this steep fall of the waters of the Jordan and the Yarmuk, on a site which would be remarkably convenient as a distributing centre of electrical supply for all the developments which I have mentioned," the writer concluded, underscoring the centrality of the power system to the developmental vision for Palestine, but also for the wider Middle East and British Empire.[100] Rutenberg himself often referred to his project as "the future nerve-centre" of the region.[101]

CONCLUSION

By following the evolution of Rutenberg's plan for the electrification of Palestine—the surveys, the negotiations, and the expert assessments—we begin to sense the outlines of a new object, the technocapitalist entity I have been referring to as electrical Palestine. It was forged like the region of overlap in a Venn diagram: it arose where a technological zone, a national market, and a nationalist ideology met. The territory that emerged captured the imagination of statesmen, engineers, and the wider public in much of Europe and America, not least Palestine's British rulers.

When the consulting engineer John Snell delivered his verdict on the Rutenberg plan, he praised it as essential to the "restoration of Palestine." Almost every single activity in Palestine undertaken by the Zionists was fitted with that curious yet ubiquitous prefix. Phrases such as "the work of reconstruction," "reclamation of land," and so on were commonplace. This, of course, fitted well with the notion that the Zionists were engaged in a "re-" activity of their own, namely, "the return to Palestine," "the return to the land," and "the rebuilding of the Jewish National Home." The ideas were not unique to Zionism. Both the British and the French conceived of their development work as applying modern technologies toward the restoration of the region to a supposed prelapsarian bountifulness.[102] It illustrates, then, how well Zionism linked up with other ideologies of its day—not just nationalism but also modernism and the technocapitalism of the civilizing mission. It was precisely because Zionism and British colonialism were creatures of the same modernizing paradigm that they worked so well together.[103]

This goes some way toward explaining why nowhere in the voluminous correspondence, reports, meeting minutes, and other related documents anyone ever criticized the technological aspects of Rutenberg's plan. This despite the fact that it was set to generate electrical energy vastly in excess of current or foreseeable demand, at a cost that was not in fact viable. Instead, when there was a mismatch between the projected output capacity of the installations and energy demand in Palestine, it was the latter that was deemed deficient. It was the essence of what Thomas Hughes has described as "supply in search of demand"—a solution in search of a problem.[104]

As a result, the record leaves a superficial appearance of a technocratic process centered on how best to effect economic development in Palestine. Indeed, it was in this vein that both the British and the Zionists responded to their critics, as I will explore more fully in the next chapter. But the putatively technocratic nature of British rule in Palestine was in fact governed by a host of extraneous considerations and assumptions that it shared with Zionism, whose program happened also to be the main beneficiary of the dominant schema. The overreach had a profound impact on the history of the area. Conceiving of and building the powerhouse helped create Palestine as a bounded entity—a precisely defined, national territory—with a population, topography, natural resources, and various other quantifiable and calculable objects.

Consequently, and contrary to the way it was and still tends to be described, the power system was not a neutral object. Instead, like all technological objects, it embodied a range of agendas. This vision was born of a particular definition and grouping of needs. The current state of hydraulic technology, the data of the riparian system of the Jordan Valley, of rainfall and topography, together with the plans for industrialization in Haifa and Tel Aviv, the capital-intensive, high-yield citriculture along the coastal plain, and other potentially cultivable areas, requiring the use of electrical water pumps for irrigation each constituted a variable of technopolitics. Together they made up the technocapitalist complex of electrical Palestine, which at once defined the critical problems and their solutions. That is to say, while in retrospect the definition of goals, the identification of obstacles, and the solutions they implied appear as products of disinterested technological analysis, in fact each element was composed of a heterogeneous assemblage of mutually influential factors ranging far and wide beyond a homogeneous domain of technics. None of the moves recounted here was foregone, and none pure.

And so, we have added another element to the complex this book refers to as electrical Palestine. Through electrification, the British and the Zionists established a common language and shared practices, and they worked out an agreed-upon vision for Palestine that provided the basis for relations between them, and their relations, in turn, with the Palestinian Arabs, whose conspicuous absence so far is a reflection of their absence in the considerations of the powers that be. They make their first appearance in the next chapter.

THREE

The Politics of Thin Circuitries

> We are giving to a Jewish organisation a grip over the whole economic life of Palestine.
>
> —*Middle East Department memo (September 1921)*[1]

> Minor obstacles of political nature exist, but these have no place in a commercial report.
>
> —ADDISON E. SOUTHARD, *American Consul at Jerusalem (October 1922)*[2]

> Rutenberg's lampposts are the gallows of our nation!
>
> —PALESTINIAN PROTEST CHANT, JAFFA *(1923)*

IN SUMMER 1923, PINHAS RUTENBERG threw the switch on mandatory Palestine's first electrical distribution system, lighting up a portion of Allenby Street in Tel Aviv.[3] The British mandatory, as noted, had high hopes for the scheme. The excitement was echoed among the Jewish residents of Tel Aviv. To them, the roadside pylons could not multiply fast enough.[4] To the Palestinians in Jaffa, however, the grid's expansion was a mixed blessing. The high-tension cable wound its way into town with promises of modernity and the creature comforts of civilized life, but it also signaled the encroachment of Jewish nationalism on Arab Palestine.

The mayor of Jaffa and most of the municipal council were cautiously supportive of Rutenberg's undertaking, in large part because so, too, was most of Jaffa's commercial class.[5] But a significant contingent within the Palestinian Arab community was staunchly opposed to Rutenberg's undertaking.[6] They remained unconvinced in the face of British and Zionist assurances that Rutenberg's electrification scheme was an apolitical project for everyone's benefit. To them, it was part and parcel of the Zionist project, and proof positive of Britain's Zionist bias. In a letter to the British government in 1922, the Palestinian Arab Executive protested that "the Zionists, through Mr. Rutenberg, are aiming at getting a stranglehold on the economics of Palestine, and once that is in their hands they will become virtual masters of the country."[7]

Rutenberg's activities and the significance that Palestinian political leaders attributed to them at the time have been almost entirely overlooked in the sholarship. But, in fact, the work to electrify Jaffa involved intense negotiations between Rutenberg, backed by Zionist organizations, and the Palestinian Arab community, and it left an enduring mark on the two national movements and their relations with each other and the British. Those engaged in the struggle against Rutenberg invariably presented the issue as a national concern, a threat to the land and people of Arab Palestine. At one anti-Rutenberg rally a few weeks before the lights were turned on along Allenby Street, the crowd chanted, "Rutenberg's lampposts are the gallows of our nation!"[8] As their words indicate, the grid took on a metonymic function, related to both Zionism and modernity. For the Jewish community in Palestine, these were mutually complementary concepts, while for the Palestinians they were fundamentally at odds.

In their work on social norms and knowledge production, Steven Shapin and Simon Schaffer famously conclude that "solutions to the problem of knowledge are solutions to the problem of social order."[9] The problem of Palestinian electrification was precisely such a problem of knowledge, and settling the controversy surrounding it was primarily a question of *boundary-work*.[10] That is, the controversy was not fought and settled on a ready-made arena. Instead, the struggle was over the arena itself. At the heart of the dispute lay the question: What sort of activity is electrification?

Here I will follow the Palestinian opposition to Rutenberg's electrification project and the strategies Rutenberg in turn devised to overcome it. Both the nature of the opposition and the character of Rutenberg's response flowed from the interaction of political circumstances with the technological properties of the power system. The Palestinian Arabs grounded their opposition to the system—and the Zionism they saw it as representing—in their ability to disrupt it, that is, in (threats of) sabotage. In response, Rutenberg and the British drove a hard wedge between politics and technology, placing Palestinian opposition with the former and the grid with the latter. Through their success, the grid moved out of reach of politics, as Palestinian opposition was dismissed as a category error.

This sort of boundary-work was a natural outcome of the elevation of science as the unfailing guide of colonial development. The corollary of this "science for development" doctrine was the denigration of the things that were deemed its opposite: subjectivity, ideology, passion, politics. The mandatory record abounds with this sentiment. For instance, after the conclusion of the agreement at San Remo in 1920, Louis Bols, chief administrator of Palestine during the period of military rule, called a meeting with local Arab dignitaries and explained to them that the Balfour Declaration was entirely compatible with the desires of the Arab population. If everyone cooperated, the economy would grow and everyone would flourish. "Politics," he warned by contrast, "are the enemy of industry."[11]

INITIAL CONTACT WITH THE PALESTINIANS

The British and Rutenberg had entered into negotiations over an electricity concession in the fall of 1920, as we saw in the previous chapter. In December 1920, Rutenberg presented his final proposal for a hydrostation on the Jordan River. To it, he attached a subsidiary plan for three auxiliary diesel-powered stations in Jaffa, Haifa, and Jerusalem. In the following May, he elaborated on his initial proposal by submitting a detailed technical schematic for a hydroelectric power station on the Auja River (the Yarkon in Hebrew). According to the plan, a dam, canal, and power station would be built a few miles from the coast, and the power would be conveyed to Tel Aviv by means of a high-tension line, and from there to Jaffa. The project also contained an irrigation plan for the 355 square miles of the Auja basin.

Initially, the Palestinian Arabs hardly figured at all in the deliberations over electrification, which is why they have been absent from the story so far. The outbreak of violence in May 1921, however, had a transformative impact on the proceedings. The violent outbreak on May 1 began with fisticuffs between two Jewish May Day parades, one organized by the far-left (Jewish) Communist Party, the other by Ahdut HaAvoda, the mainstream organization of labor Zionism. The brawl soon evolved into an Arab-Jewish conflagration and spread from Jaffa to other parts of the country. By the time the fighting wound down on May 7, scores had been killed, and High Commissioner Herbert Samuel's thinking with respect to electrification had undergone a major transformation.[12]

Before the "disturbances," as they became known to British officialdom, Samuel had rejected Colonial Secretary Churchill's suggestion to run Rutenberg's scheme by the Advisory Council in Palestine, the consultative body staffed by eleven British officials and ten local representatives.[13] Instead, he recommended that the concession be negotiated in secrecy. Only upon its signing, he advised, should the terms be publicized "together with [a] full statement of advantages of the scheme to Palestine."[14] After the May riots, Samuel's commitment to Rutenberg's project remained undiminished, but his fear of further violence resulting from Arab disaffection, a threat he now considered acute, caused him to part ways with the developmentalists at the Colonial Office.[15] For all its merit, Samuel now believed, electrification could not be carried out in the teeth of Arab opposition in hopes of securing their acquiescence later on.[16]

Like the British, Rutenberg had initially been blind to the challenge that swaying the Palestinian Arab community posed. By the time of the violence in May 1921, however, he was already wise to the fact that the Palestinian Arabs, as he put it, considered his undertaking "as the most dangerous factor in the realization of Jewish claims in Palestine.... Arab Meetings and Conferences," he noted, had "resolved to boycott the Scheme and to fight against it with all might."[17] As the realization sank in, Rutenberg became increasingly cognizant of the practical difficulties presented by Arab hostility to his undertaking.

Like Samuel, Rutenberg realized that his project could not be carried out over the objections of the Palestinians. But Rutenberg's concerns were different from those that animated the high commissioner. To begin, Rutenberg feared that strenuous and sustained Arab resistance would undermine the fundamental assumption undergirding British support for his scheme, namely, that economic development would bring peace and prosperity to the country. But Arab opposition presented another, more immediate problem for Rutenberg: sabotage. In an intelligence report sent back to Rutenberg in January 1922, Ibrahim Shammas, a member of the Arab Executive delegation to London, was quoted as saying, "Even if Rutenberg succeeds in getting the money, his work will not progress. We will not sell him land, we will disturb his work, we will destroy his machinery."[18]

Of course, electrification was not the only Zionist venture to have to contend with sabotage, but it was uniquely vulnerable to it. The grid's sprawling character made it almost impossible to safeguard by force of arms, and local disruptions risked causing system-wide breakdown, especially in a young grid that had not yet acquired the ability to reroute its current through auxiliary pathways. Timothy Mitchell has argued that, in the twentieth century, people's ability to make effective political claims is largely determined by their position within the prevailing energy system. Specifically, according to Mitchell, the early twentieth-century shift from coal to oil as the principal source of energy was a condition of possibility for modern politics. Whereas the movement of coal was such that it generated choke points, subject to exploitation by workers whose physical control over those points gave purchase to their political claims, the more diffuse movement of oil offered no such avenue. Mitchell likens the movement of oil to the pattern of an electric grid, writing that the grid's multiple pathways eliminate choke points, and thus the leverage they afford oppositional politics.[19]

This book and its author owe a large intellectual debt to Mitchell. But the grid analogy works only for mature power systems. To understand how electrification interacts with politics—and indeed how any technology interacts with politics—we need to be more sensitive to system evolution and contingency. The story of Palestine's electrification during the mandate period is largely a story about the gradual thickening of the grid, and the forms of politics, oppositional or otherwise, that were opened up and foreclosed as the grid grew. Paying close attention to the evolution of the grid alongside the evolution of the conflict between Arabs and Jews suggests that Mitchell's macro-level claim needs to be nuanced, by paying greater attention to the internal evolution of a given energy technology. In the following, I will explore the first part of this history, the grid at its thinnest.

When we think of physical exposure in the context of Zionism in Palestine, the first thing that comes to mind is Jewish agricultural settlements, a subject that has received considerable scholarly attention.[20] Yet, in contrast to the electric grid, a few armed guards were usually sufficient to safeguard such settlements from sabotage. And even a successful attack did not threaten to disrupt the settlement enterprise as a whole. By contrast, given the techno-spatial properties of electric grids, if the Palestinians were directly hostile to the venture on political grounds, it would be unworkable on technical grounds. Only by gaining at least the passive consent of all of Palestine's inhabitants could the grid be viably secured. This profoundly impacted the nature of Rutenberg's efforts, the strategies he adopted, and the progress of the endeavor. By the same token, the Palestinian campaign against Rutenberg was also critically shaped by an arena that electricity participated in creating, as Palestinians' ability to significantly disrupt the progress of Rutenberg's work became the collateral with which they were able to press their political demands and resist Zionism.

Rutenberg established contact with leaders of the Palestinian community early on. Whereas the Palestinian Arab Executive repeatedly rebuffed him through 1921, he had better luck with local bodies, such as the municipality in Jaffa.[21] He first made contact with the municipality in the fall of 1920 and received a favorable reply from the mayor, 'Issam al-Sa'id, in early 1921. He continued to correspond with the mayor through the spring.[22] From the start, the municipality found itself in a difficult position. While elements within the political class exerted strong, continuous pressure on the municipality not to cooperate with Rutenberg, a substantial proportion of the town's commercial class was eager to receive electricity. Their desire for this prerequisite of modern living apparently trumped whatever misgivings they may have had about the source of the current.[23]

Nevertheless, there were several reasons for the Palestinian Arabs to be wary of the project. In the summer of 1921, the Jaffa Municipal Council engaged an engineer based in Alexandria, Egypt, to assess Rutenberg's scheme. In his report, the engineer criticized the plan on several counts, citing among other things political resistance and inflated rates.[24] It was unwelcome news to Rutenberg, but it worried Samuel all the more. Although he dismissed the report as "superficial," he noted with concern that it had made the Jaffa municipality leery of the scheme.[25] Identifying the primary source of Arab resistance, perhaps wishfully, to be "technical considerations" rather than political ones, Samuel resolved to fight fire with fire: "a favourable report from experts in England would probably modify" the critical assessment, he reckoned, and allay the Palestinians' concerns.[26]

As noted earlier, in contrast to the Jordan scheme, the British did not submit the plans for the smaller Jaffa scheme for technical review before it was provisionally approved in July 1921.[27] But in September, only days before the concession was to be signed, Samuel changed his mind and insisted on sending the Jaffa scheme to the government's own engineering consultants, Preece, Cardew and Rider.[28] This, as we saw, had not been deemed necessary earlier and was only undertaken at this point in an attempt to throw cold water on the Arab opposition. The consulting engineers issued their report three days later, an exceptionally quick turnaround. They concluded: "We are of the opinion that the proposed scheme is a practical one and that it has the possibilities of being a sound commercial success."[29] In their assessment, "The system proposed is thoroughly modern in all respects."[30]

Unsurprisingly, perhaps, the report did not assuage the Arab opposition, and Samuel continued to insist on formal Arab approval before officially awarding the concession to Rutenberg. Rutenberg, for his part, emphasized that overcoming Palestinian opposition was vital, since failure to do so would scare off investors and make the scheme impossible to carry out. But because of the practical significance of Palestinian consent, Rutenberg further argued, there was no need to make the agreement conditional on Arab approval. In fact, soliciting formal approval from the Palestinians would lessen the chances of obtaining it, as "demagogic agitators," hostile to the undertaking for political reasons, would capitalize on such a stipulation. Rather, Rutenberg argued, Palestinian consent needed to be built up over time. As the benefits of electricity became tangible, he assured the British, the legitimacy of the politically motivated rabble-rousers would start to erode.[31]

Whitehall was highly motivated to believe Rutenberg, not least, as we saw, because of the keen desire to promote commercial activity in Palestine and attract Jewish capital. Consequently, the British ended up adopting Rutenberg's view, seemingly without recognizing that, in fact, that was what they had done. Major Young at the Colonial Office, for instance, stated, "My own view is that we should proceed boldly without obtaining 'Arab' consent. Without the agreement, Mr. Rutenberg will not be in a position financially to take the necessary steps to secure Arab co-operation."[32]

This put the Colonial Office at odds with the high commissioner's revised position following the riots in May 1921. In a note to Churchill, Major Young of the Middle East Department cautiously suggested that it might be necessary "to overrule Sir H. Samuel ... I think we should ... press Sir Herbert to consent in [the concession's] signature."[33] Most senior Colonial Office officials, as well as the colonial secretary, favored Rutenberg's proposed course of action, and in the end Samuel was forced to relent.[34] Rather than being party to the agreement, then, the Palestinian Arabs were in effect given a status more like that of the other elements that had to be tamed in modern man's conquest of nature. The concession for the Auja undertaking was signed on September 12, 1921.

JAFFA'S DEMANDS

Soon after the concession was signed, Major James Campbell, assistant district governor of Jaffa, convened a conference at his offices with Arab representatives of Jaffa and Shaykh Muwannis (alt. Sheikh Munes), the village holding title to most of the land in the Auja basin required for Rutenberg's hydroelectric installation.[35] The Arab representatives raised a number of objections, but Rutenberg's conciliatory approach apparently assuaged their concerns, and by the end of the meeting the parties had produced a document, dated October 15, 1921, that was signed by all present, in which the Palestinian representatives affirmed that "the Scheme of Mr. Rutenberg is of benefit to the inhabitants of Jaffa and Sheikh Munes." There is no evidence that bribes were involved, though they may have been, as they were in other cases (notably Tiberias, discussed later). Most likely, however, a combination of heavy British pressure and the promise of economic growth resulting from access to cheap power convinced the attendees to attach their names to the agreement.[36]

It was a first important step on the path to securing Palestinian Arab acceptance of the scheme, and Rutenberg immediately proceeded to the next one. In his original proposal for the Auja scheme, he had emphasized the importance of giving it "the character of a municipalised undertaking," by which he meant that the communities taking current "should participate in the financing of the project."[37] Calculated on the cost of erecting a low-tension distribution system in each town, Rutenberg suggested municipal investments in the amounts of £E6,000 and £E25,000 for Tel Aviv and Jaffa, respectively.

There were many benefits to such an arrangement, the most obvious—financing—being the least important.[38] To be sure, it was not inconsequential that it spared Rutenberg from having to finance construction of the local distribution systems himself. But the real significance lay in relieving him and Jaffa from having to do business directly with one another, an unwelcome hassle for Rutenberg and a political liability for the Jaffa municipal council. More important still, it introduced a vital safeguard against sabotage. "I could without great difficulty find the entire sum required," Rutenberg wrote to Samuel, "but as Your Excellency has been first to realise, the financial participation of the Municipality of Jaffa is important."[39] Though outwardly framed as a mutually beneficial business partnership, for Rutenberg the plan for municipal coinvestment was designed primarily to neutralize the recalcitrant elements within the Arab community, who might otherwise "disturb his work" and "destroy his machinery," as Shammas of the Arab Executive had pledged.[40] "As I explained to you," Rutenberg wrote in a letter to the head of the Middle East Department, "the importance of the financial participation of the Palestine Municipalities is not so much their pecuniary assistance as the fact that through their financial participation the Arab Municipalities will be induced to provide for the safety and development of the project."⁴¹ If Jaffa invested its own money in the scheme, it would have a vested interest in safeguarding it from sabotage. "Mr. Rutenberg," one government official wrote, "is very anxious that the scheme by which the Municipalities should become the eventual owners of the system in their respective towns should be carried through. The reasons for this are obvious in that municipalities would have a direct interest in maintaining the security of the installations."⁴²

This was not how Rutenberg presented the plan to the Palestinians. To the mayor of Jaffa, Rutenberg wrote in the spring: "I wish to make it quite clear that my proposal to the Jaffa Municipality and other local bodies to contribute in the above described proportion has no other motive than my belief that it is in the interests of the country that they be given the possibility to become the owners of their own distribution systems."⁴³ But there was an important obstacle in the way of municipal coinvestment in the poor financial state that most towns found themselves in during the aftermath of World War I. None had sums of the required magnitude at their disposal, and no right-thinking bank would grant a loan under the circumstances. Anticipating this obstacle, Rutenberg prevailed on the British to back the municipal loans in a manifest breach of British imperial practice.⁴⁴ Major Vernon of the Middle East Department captured the widespread British view in a dispatch when he wrote, "The future prosperity of Palestine is so much bound up with the success of the Rutenberg scheme that we ought in the best interests of the country to do all we properly can to bring the scheme into operation."⁴⁵

As for the policy's primary target, the Jaffa municipality, the terms of the proposed loan guarantee offered a way to negotiate the conflicting demands of the oppositional faction, on the one hand, and the widespread clamor for electrical power among its constituency, on the other. On November 16, 1921, the Jaffa municipal council formally announced its willingness to participate in Rutenberg's scheme, contingent on a number of conditions. Two stood out. First, the council demanded to increase its portion of the investment in the company from £E25,000 to £E60,000, and thus for an owner's share of the company of 60 percent. According to the council's second demand, there should be no company activity within Jaffa's boundaries. The electricity company would lead current up to a transformer station located outside the town boundaries and leave it to Jaffa to erect its own distribution network and oversee provision.⁴⁶

In reality, both demands were political theater. Even if the cash-poor municipality had been able to raise the £E60,000 (an exceedingly unlikely prospect), it was no doubt aware that Rutenberg would not agree to surrender control of his company. Indeed, according to the assistant governor of Jaffa, the first demand was "inserted for political reasons without any real hope that any quantity of shares will be taken up by the local inhabitants."[47] The second demand was no more realistic, since controlling the electricity infrastructure within a town's boundaries effectively constitutes no control at all. Electricity cannot be stored economically in anything but very small quantities; without continuous electrical current, an electric grid is only so many pylons and miles of copper wiring. Inevitably, the hand resting on the switch back at the powerhouse has veto power over the entire system. Considering the municipality's demands in light of the electric grid's technological properties suggests that we read them not as blunt anti-Zionism but as a subtler tactic of negotiating multiple interests and pressures. As the assistant governor of Jaffa commented regarding the declaration: "The Mayor has had considerable difficulty, I know, in bringing his council with him to the point now reached."[48]

The way that the British and Rutenberg chose to handle Jaffa's demands was equally telling of their concerns, and it signaled a decisive shift in their modus operandi. The stakes were high, as the outcome of the negotiations with Jaffa would not be limited to that district alone. Shortly after Jaffa presented its demands, the Haifa municipality followed suit, demanding the right to buy a majority share of its local distribution system. Rutenberg and the British reckoned that "serious obstruction [was] expected" should they be denied it.[49] Aware of the need to respond to these demands in a manner that avoided provoking acts of sabotage, the British and Rutenberg shuttled drafts of their replies to the Jaffa municipality back and forth between them. The British even asked for Rutenberg's approval of their official reply before finally sending it to Jaffa.[50]

The resulting tactic was one in which the Palestinian demands were not confronted head-on but instead were neutralized by other means. In his response, Rutenberg stated, "The inhabitants of Jaffa will have the same unlimited right to apply for shares of the company as any other persons, on the terms of the issue." And as far as the placement of the transformer station and distribution system, Rutenberg responded that it "is purely a technical question."[51]

In other words, both demands were met with reference to putatively objective laws of nature: the free market and technological exigency. Rutenberg made these promises knowing full well that Jaffa would have no means of coming up with £E60,000 without some sort of preferential deal. And in the event, technical considerations would indeed require the presence of the power company inside the town. Consequently, by deferring to the apparently apolitical demands of the technical and the market, Rutenberg effectively neutralized the political claims of the municipality. Of course, the free-market rationale was not consistent. Jaffa *would* get preferential treatment, and the rules of the free market *would* be suspended, but only up to a value of £E25,000, and a quarter ownership in the venture. Once Jaffa's participation as a substantial coinvestor was secured—that is to say, when the town had a sufficient incentive to prevent sabotage—the rules of the free market would reactivate, ensuring that the municipality would remain effectively powerless.[52]

THE POLITICS OF NON-POLITICS

Presenting one's own position as mandated by science created a highly useful corollary: it followed by definition that anyone who opposed Rutenberg was working against the objective good. Opposing claims were *political*—the enemy of industry. As we have seen, this technocratic framing had not always been Rutenberg's preferred modus operandi. In his original concession application, Rutenberg explicitly linked electrification to Jewish colonization.[53] The British initially brought up Rutenberg's Zionism as a point in his favor, on the rationale that it would compel him to work harder under worse conditions than someone motivated by financial profit alone. Furthermore, as we have seen, the Zionist commitment to Palestine would not merely spur hard work on the ground. It would also help raise the capital necessary for the project, because wealthy Jews around the world would be particularly amenable to investing in a Jewish-run enterprise in Palestine. Those investors, it was believed, would be willing to take a slightly larger risk, or accept a lower return on their investment, because of the project's role in the creation of a Jewish refuge in Palestine. As late as August 1921, less than a month before the concession was signed, R. V. Vernon of the Colonial Office referred to Rutenberg as someone "not out to make money so much as to make the Zionist policy a success."[54]

The negotiations over municipal investment in late 1921 were the first expression of a shift on Rutenberg's part away from any overtly political word or deed to a neutral language of economic and technological exigency. At the end of 1923, Rutenberg reported to his board that mounting opposition to his scheme in the fall of 1921 had brought him to the conclusion that "the only sane policy was to conduct the work as a purely business proposition, outside the Zionist Organization as such and outside politics of any sort, welcoming Arab collaboration on purely economic grounds."[55] The shift is well documented in the sources from 1922 on, mainly because of Rutenberg's efforts to police the language of his associates when talking about his project. In June 1922, he scolded the Zionist Organization of America (ZOA) in a letter for "connecting the 'Rutenberg scheme' with the 'Jewish Agency' and the future control of the Z. O. [Zionist Organization] over it." As a result, Rutenberg asserted, the ZOA was "compromising myself and the whole scheme."[56] In a particularly vitriolic exchange with Frederick Kisch, a member of the Zionist Executive, Rutenberg wrote, "On several occasions I have explained to you that under the present extremely difficult conditions I am bound to conduct my work on strict business lines and decidedly separate myself from any connection with the Zionist Organization."[57]

Like many who protest too much, however, Rutenberg was not telling the whole story. In fact, the language of scientific reason was strategically deployed as the situation called for it, and in different circumstances he slipped into other registers. In the end-of-year report for 1923, for instance, Rutenberg also asserted that his undertaking was "the solution of the Jewish Palestine problem."[58] In effect, then, Rutenberg positioned himself so that he could pursue his interests by way of objective exigency. It was not the absence of politics; it was the pursuit of political ends by supposedly nonpolitical means, a politics of non-politics.

It would be a mistake, however, to understand Rutenberg's maneuvering as simply a function of Zionist scheming. His actions, often carefully workshopped together with the British, did not simply amount to politics by another name; they represented a displacement of political power onto technical things, a typical move for someone with Rutenberg's training and experience as a systems entrepreneur. In this sense, Rutenberg's actions were no different from those of the storied systems builders of the twentieth century, the likes of Samuel Insull in Chicago, Oskar von Miller in Germany, and Charles Merz in Britain, all of whom were primarily confronted with the challenge of stitching together technical and nontechnical elements, and of figuring out how to advance their projects by shaping them and their environments, so that they appeared to hold the solution to a critical problem.[59] The primary challenge confronting Rutenberg was how to make his venture an obligatory passage point on the path to an economically prosperous and therefore civilized Palestine. It was certainly no accident that the technical and economic logics that guided Rutenberg aligned so neatly with the effort to create a Jewish state in Palestine, but Zionism alone does not fully account for the strategies he pursued or the forces and logics he was marshaling in pursuit of his project. Indeed, Zionism ended up as much a product of Rutenberg's technocapitalist logic as ever Zionism shaped the land and its people.

As Rutenberg excised politics from his vocabulary, at least in public, so did the British, who also benefited from the technical framing. It allowed them to retain a neutral posture in Palestine, providing them with a measure of protection from criticism at home and at the Permanent Mandates Commission of the League of Nations. The framing grew out of a dovetailing of needs between Rutenberg and the Zionists, on the one hand, and the British government, on the other.[60] And it was the product of active collaboration; the notion that Rutenberg's project was apolitical appeared at the same time in internal British and Zionist records.[61] By excising politics from the process, while making sure that their political interests were being satisfied, the Zionists and British were able to triumph over the Palestinians without ever having to acknowledge trying. For the British government, the value of the business framing became increasingly important as its Palestine policy became the target of growing criticism from the spring of 1922.[62]

BAIT AND SWITCH

Sticking his opponents with the label "political," and contrasting it to his own actions of scientific rationality, was a device Rutenberg employed often and with great success. The most remarkable success of his politics of nonpolitics involved a process that unfolded between the summers of 1921 and 1922, when Rutenberg managed to completely overturn the conditions of the concession, while retaining all the benefits under the original agreement. When Rutenberg received the provisional go-ahead for the Auja scheme from the British in July 1921, he immediately moved to acquire the 150 dunams he needed in the Auja basin. Most of the land belonged to inhabitants of the Palestinian village Shaykh Muwannis. When Rutenberg approached the Palestinian landowners, however, they demanded prices significantly above market value.[63] The summer of 1921 had seen an extensive campaign against

Rutenberg in Jaffa and around the country, and Auja landowners were urged not to sell their land to him.[64] Most of them, however, were not so categorical about the matter. They were willing to sell their land, but at a price that reflected not just the value of it but also the social cost to the seller. Baruch Kimmerling, in his landmark study of Zionism and territory, refers to it as the added "political value" of land.[65] Rutenberg was not just any buyer but a fiercely contentious character at the head of a deeply controversial enterprise. The landowners may have explicitly told Rutenberg's agents that the price included not just the land but also their pledge not to interfere with the construction and running of the powerhouse. The power company's internal records contain a few oblique comments that could be interpreted in that way. References to requests for payouts are legion in the records from the period, which should not surprise us given that there was a long-standing Ottoman precedent of paying tribes to keep them from sabotaging railways, roads, and telegraph lines. In the fall of 1922, Harry Sacher wrote to Rutenberg. He told him that the work in Jaffa was going well and added: "Of course, we run up against people who try to get in our way in the hope of blackmailing us, but we must and do expect this."[66] Rutenberg was generally loath to make any payouts. Referring to his troubles in the Auja basin, he wrote in a letter to a confidant that "influential leaders of the Arab people are insistently asking for tangible compensations, and I do not want to incur new expenses."[67]

No doubt, it was within Rutenberg's means simply to pay the inflated prices, if indeed it is accurate to describe them as inflated. It was not hard to see that the land was on the cusp of a significant value increase, or that it was particularly valuable to Rutenberg.[68] Despite what Rutenberg would later tell the British, it was not primarily the high cost that made him reluctant to go ahead with the land transactions. It was out of concern for the long-term security of his works that he hesitated. Given the widespread and growing opposition to his scheme among the Arabs (more on this later), building a sprawling hydroelectric installation complete with a complex and expansive system of feeding canals and dams in the Arab-dominated Auja basin had come to look like a significant liability. Even if Rutenberg paid off the landowners at the stage of acquiring the land, the Palestinians would continue to hold an effective veto over the proper functioning of the physically exposed installation.[69]

It was already evident to Rutenberg that the output of the hydrostation would have to be supplemented by a fuel station to meet the expected rise in demand. Encountering resistance in the basin, Rutenberg thus decided to switch the timetable around. In the early fall of 1921, soon after signing the concession, he put the plans for the hydrostation on hold and instead began making preparations for a fuel station.[70] He negotiated with the Tel Aviv municipal council to buy a plot of Jewish-owned land on the outskirts of the township. In contrast to the territorially expansive water-powered generating plant, a fuel-powered station merely required a structure to house the fuel sets, and predominantly Jewish Tel Aviv offered a far more hospitable environment than the predominantly Arab-populated basin of the Auja River. This lessened the political leverage that could be generated from threats of sabotage. Rutenberg made arrangements with the township for the purchase of the land, and at the end of the year he traveled to Germany, where he bought two 500 horsepower diesel generators from AEG.[71]

There was a significant risk to the path Rutenberg had chosen. The Auja concession explicitly called for the construction of a hydroelectric plant on the Auja River. Under its terms, construction had to begin within nine months of signing the concession and be completed in another two on pain of forfeiting the licence. Starting construction on the fuel station and putting the hydroelectric station on hold would make it impossible to meet the deadline. For the switch to work, Rutenberg had to persuade the British to let the fuel station take the hydrostation's place in the timetable. Switching technologies, there was also the risk of undermining the Auja scheme's function as a cat's paw for the larger Jordan scheme. His closest adviser, the lawyer Harry Sacher, expressed deep concern in a letter to Rutenberg: "I am quite sure that the effect upon the mind of the government of Palestine would be exceedingly unfortunate."[72] Sacher told Rutenberg that he had consulted with David Eder of the Zionist Commission, who agreed with Sacher that the switch was inadvisable. "We urge you to reconsider the whole matter," Sacher wrote, "and we are confident that you will ultimately come to the same decision as ourselves."[73]

Not so. Rutenberg saw other benefits to the switch that, in his assessment, trumped his advisers' concerns. Most important, the British were no longer the most significant obstacle; it was the Palestinian Arab opposition and its threat of sabotage that constituted the reverse salient.[74] "The Arabs on hearing of the fuel station will no doubt become more reasonable," he wrote to the American Zionists in defense of his strategy.[75] A small fuel station in the midst of Jewish Tel Aviv rather than a sprawling hydroelectric installation in the predominantly Palestinian-populated Auja basin, Rutenberg reasoned, would drastically reduce the Palestinians' ability to sabotage the electrical equipment, and thus reduce the efficacy of any political claim grounded in that ability. "Later on," he wrote, "the Arabs will be begging us to negotiate."[76]

Between the summers of 1921 and 1922, Rutenberg sent several reports to the British regarding his attempts to buy land in the Auja basin. In October 1921, he informed the Palestine Government that "certain Arab landowners have raised their prices by three or four hundred percent in order to stultify the Auja scheme."[77] It was the work, Rutenberg told the British, of a small number of "demagogic agitators" who, despite Rutenberg's "clear business exposition of the matter" had "incited Arab landowners to oppose the scheme" by "demand[ing] exorbitant prices for the land, making the transaction commercially prohibitive."[78] Blaming the price hike on a small number of extremists was a familiar Zionist strategy to minimize the significance of anti-Zionist activities among Palestinian Arabs.[79] Rutenberg told the British that buying the land at the inflated prices was inadvisable for two reasons. First, it would enrich precisely the contingent of Palestinians that opposed his work (and by extension, so the logic went, stood in the way of progress). Second, and more important, the exorbitant cost of the land would have to be passed on to the consumers, raising electricity rates to the point of impeding economic development.[80] The British accepted Rutenberg's version of events, another instance in which they unreflectively reproduced his language in their own internal reports, which in turn laid the foundation for future policy.

In the spring of 1922, as the German fuel sets arrived off the coast of Jaffa, Rutenberg presented his alternative solution to the British, characteristically preferring to present an accomplished fact over a work in progress. Given the Palestinians' politically motivated demand for sums "out of all proportion to the market prices" and the inadvisability of exercising his right of expropriation, Rutenberg explained, he had decided to put the Auja scheme on hold and instead construct a diesel-powered generating station on the outskirts of Tel Aviv. Since the diesel station would produce at least as much electricity as the hydroelectric station, Rutenberg asked that the postponement of the latter be approved. His request was granted almost immediately.[81] For a number of years, the company was forced to apply for extensions, usually granted in six-month increments.[82] It was not always a smooth process. There was a great deal of frustration directed toward Rutenberg, particularly among local officials, for upending the terms of the agreement.[83] At one point, the attorney general suggested that the agreed-upon sums for the municipal loans should be reduced, given the lower costs of a fuel system over waterpower.[84] Nothing came of it, however, and finally, in 1926, the power company was released from the obligation altogether, in exchange for allowing the government to draw water from the Auja toward Jerusalem's water supply.[85] Rutenberg had managed to effect the change of locations and technologies while retaining his concession in full, including the exclusive irrigation rights over the 355 square miles of land in the Auja basin, as well as the municipal investment scheme and electricity rates, both calculated on the technological exigencies and far greater cost of the original hydroelectric scheme. It had been a high-risk move with an equally high reward, allowing Rutenberg to have his cake and eat it too.

Even after Rutenberg received approval from the British to transfer the concession timetable onto the fuel station, he continued his efforts to acquire land in the basin. Rutenberg figured he might yet want to build his hydroplant in the future, but more important, the irrigation rights that he still held stood to bring in a nice profit.[86] Indeed, by the decade's end, he had formed a subsidiary of the power company devoted to irrigating the basin.[87] Yet only 2 of the 355 square miles of the basin were ever irrigated, namely, the area of the high-lying Jewish settlements.[88] Criticizing this state of affairs in 1932, the Palestinian Arab newspaper *Filastin* employed rhetoric not unlike that of colonial developmentalism, concluding: "We must ask how the Government, and the Jews even, can recognize this sterilization of a source of a country's development."[89] Despite such calls, and despite several committees of inquiry faulting the government for not undertaking more irrigation work in the country, the Arab residents of the Auja basin remained without proper irrigation for the remainder of the mandate.[90]

OPPOSITION IN LONDON AND FAILURE IN AMERICA

The Rutenberg scheme did not have detractors just in Palestine. Opposition had also begun to build among Conservative MPs and newspapers back in the metropole. This caused the British government to launch its own domestic politics of non-politics. Pressure started to mount in spring 1922, when Samuel traveled back to London to brief Parliament on Palestinian affairs.[91] The Opposition MP, William Joynson-Hicks, brought a motion before the House of Commons proposing to make the Rutenberg concession subject to parliamentary approval. In the motion, Joynson-Hicks charged that Rutenberg's plan was a Zionist scheme supported by British officials who sought to bolster the Zionist movement's political power by granting it economic power. Joynson-Hicks read out a long list of applications for concessions covering parts of the contract that had been granted to Rutenberg, disproving the government's claim that Rutenberg's had been the only application it received. He concluded that "the Rutenberg contract contains the most astonishing concessions I have ever seen or read of in my life. This contract gives over the development of the whole country to Mr. Rutenberg." He furthermore asserted that "Great Britain has no right to hand out such powers, and such vast possibilities of control over the whole development of Palestine" to a man of such suspicious aspect and motivations.[92] Another Opposition MP, J. Butcher, agreed: "These are very large powers, and powers which in their comparatively short reign in Palestine the Jews, who invaded that country and treated the inhabitants in a somewhat abrupt manner, have obtained."[93] Just like the Arab critics, these Opposition MPs insisted on the fundamentally political nature of the matter, and the extraordinary political power that came with the concession. Although Joynson-Hicks's motion was ultimately defeated, 292 to 35, the mounting criticism at home made it all the more important for the British government, in order to bolster the legitimacy of the stance it had taken, to see tangible developments on the ground.[94] It exerted growing pressure on Rutenberg and his associates to expedite construction.[95]

Rutenberg, meanwhile, had left for a fund-raising campaign in the United States in March. The trip was expected to last only a few weeks and bring in considerable funds. But in the event, Rutenberg was obstructed at every turn by the internal fissures of the American Zionist movement and the powerful personalities and vanities on all sides. In a moment of utter exasperation, Rutenberg resorted to outright threats. To Justice Brandeis he wrote: "If I do not now succeed in raising the money needed for my work, i.e. if I do not succeed in creating the new basis for work and life in Palestine, Zionism will remain what it is—a tremendous bluff. . . . Zionism would appear to be the greatest crime in history; and it would be my duty to state it openly."[96] Notwithstanding these desperate measures, Rutenberg was not able to raise the funds he had been hoping for in America. He persisted through May and June 1922, despite pleas from his associates in Palestine and London to return and attend to the Auja scheme. He was coming dangerously close to the deadline for beginning construction in Jaffa. Meanwhile, British correspondence with Rutenberg and his associates assumed an increasingly harsh tone. Finally, he left America in late June.[97]

On September 18, 1922, work began on the Jaffa scheme, now refashioned as a thermoelectric plant on the outskirts of Tel Aviv, in a former orange grove adjoining the Tel Aviv railway station. Labor was primarily supplied by the Jewish Labor Cooperative Association, though as a British government reported added, "Arrangements ... have been made whereby the work of transporting all materials, etc., from Jaffa Port and the Railway Station will be carried out entirely by Arab labour. Arab labour is also employed on the supply of road metal obtained from Beit Nabala quarry which is being used for foundation purposes."[98] Both the British and the power company were at pains to stress the presence of Arabs in the workforce, which testifies to the importance they attributed to it, and to the need for such facts to bolster the idea that Zionist industry would benefit all of Palestine's inhabitants.

MUNICIPAL CONTRACTS

As Arab opposition grew, Rutenberg conducted parallel negotiations over electrification and municipal investments with the municipalities of Jaffa, Haifa, and Tel Aviv.[99] By the spring of 1923, a detailed agreement had been worked out, the upshot of which was that the Palestine Electric Corporation would effectively assume all financial risk.[100] The municipalities would fund their own local grids by means of a loan from the Zionist-owned Anglo-Palestine Company at 8 percent interest.[101] The company would then lease the grid from the municipality for a twenty-five-year period, after which it would be subject to renewal at five-year intervals. The rate of the lease would be set as a percentage of various revenue streams (private lighting, industry, etc.). Furthermore, if the revenue from the lease commission was not sufficient to cover the interest and amortization of the loan taken in order to fund the electricity infrastructure, the company would cover the difference, up to a value of 8 percent of the municipalities' initial investments. Any commission surplus after the loan payments would fall to the municipalities, though the company reserved the right to make deductions from future surpluses as compensation for sums contributed in previous years under the loan guarantee. According to the electricity company's estimates, the initial revenue the municipalities were expected to derive from the commission was about 4 percent, which meant that the power company would have to cover the remaining 4 percent. But by the second half of the 1920s, according to the company's estimates, the revenue from the commission would cover the loan payments in full.[102]

In plainer English, the contract allowed the municipalities to borrow for the construction of the local transmission grids from the company against future profits. Consequently, the scheme involved no capital outlay for the municipalities. In addition, they stood to receive a modest profit from the lease agreement, and by the end of the twenty-five years, ownership of the local grids would transfer into their hands. In addition to getting their electric grids for free, and the prospect of some modest profits on the lease commission, the municipalities also stood to benefit from a reduction in consumption of oil and other fossil fuels.[103]

The agreement provided the template for all contracts between the power company and municipalities in Palestine, whether Jewish or Arab.[104] However, in another arrangement that would prove a model in the company's dealings with Jewish communities, to the agreement with Tel Aviv was added a separate, confidential letter, in which Rutenberg pledged that if the municipality was unable to secure a loan, Rutenberg would provide the money, either as a loan directly from the electricity company or by some other means.[105] The Tel Aviv municipal council passed a resolution on March 16, 1923, to accept Rutenberg's scheme and participate in it with £E10,000.[106]

THIN CIRCUITRIES

As construction of the power station and grid began in September 1922, opposition to electrification hardened in the Palestinian community. One notable case in point was 'Umar al-Baytar, the chairman of the Jaffa chapter of the Muslim-Christian Association (MCA).[107] In the fall of 1921, his had been one of the signatures (the only one in Arabic) attached to the statement declaring that Rutenberg's scheme "is of benefit to the inhabitants of Jaffa and Sheikh Munes."[108] By the following summer, Baytar had reversed his position and become a leading figure in the campaign against Rutenberg. He composed open letters to the mayors of Jerusalem and Jaffa, in which the municipalities were held to account for their dealings with Rutenberg. Baytar demanded to know what arrangements, if any, the mayors had come to with Rutenberg, "being certain," the letters, subsequently serialized in *Filastin*, added, "that a patriot like you would not approve of a scheme which is bound to damage the interests of the country."[109]

Ragheb Nashashibi, the mayor of Jerusalem, falsely claimed not to have entered into any negotiations with Rutenberg,[110] while Mayor 'Issam Sa'id of Jaffa, whose dealings with Rutenberg had left undeniable traces in the public record, had little choice but to respond more frankly.[111] "The municipality," his published response read, "was ... confronted with a fait accompli. Its silence would have been taken for non-conditional acquiescence. Therefore it found itself forced ... to decide in favor of participation in the scheme, in order to protect its interests, as well as those of the Arab citizens." Mayor Sa'id then enumerated the demands discussed earlier, including the increase in the municipality's share of the investment. He neglected to mention, however, that those demands, which Rutenberg had skirted so nimbly with reference to technological exigency and the free market, were dead letters.[112] The mayor may not have been aware that the substance of Rutenberg's acquiescence to Jaffa's demands in effect amounted to rejection. Alternatively, he, like Rutenberg and the British, found it useful to maintain the fiction that Rutenberg's concessions were practically meaningful.

By the summer of 1922, the Jaffa MCA had emerged as the center of anti-Rutenberg politics, mainly as a result of Baytar's advocacy. On May 15, 1923, he drafted a petition, which according to *Filastin* received thousands of signatures. The petition amounted to an indictment not just of electrification (as part of Zionism) but also of British rule, and any Arab cooperation with it: "We make clear to the Municipality that the town does not care for any sum which the Municipality will make on this scheme, and that the town will never be bound by any decision that is carried out by the Municipal council if it is against the will of the people."[113] Throughout April, May, and June 1923, a series of meetings were held between members of the Jaffa municipality and the MCA. In May, a group of Palestinian notables, mostly MCA members, wrote to the mayor of Jaffa, stating that "the native inhabitants are all determined to refuse the scheme from its foundation as it is prejudicial to their national and political rights."[114] Several rallies took place in Jaffa in the period.[115] The great pressure that was brought on the Jaffa municipal council ultimately forced it to retract its previous approval of the electrification plans. On April 1, 1923, it appointed a commission, which included Baytar, to study the matter, and in May the commission published a report that noted "the increase of discontent among the natives against this [Rutenberg's] scheme" and concluded by advising the municipality to reject it: "We consider such a scheme as a trespass on municipal rights, as clearly defined by laws, and we believe that the government respects the law."[116] In a meeting with Rutenberg and District Commissioner Campbell, representatives of the municipality insisted on their right to undertake their own electrification, in accordance with "the laws of civilised countries." Rutenberg's concession was "against the law, against public policy and against nationalism."[117] The legal argument was important and often invoked. The letter from the municipal committee cited earlier pointed to the "existing laws of the country," referring to the Ottoman legal code, and a subsequent letter from the same committee specified that Rutenberg's concession violated the Vilayet Administrative Laws, as well as Article 31 of the electrification concession itself, which contained the stipulation regarding waterpower. Rutenberg's abandonment of the hydroscheme was particularly problematic from the Arab point of view, given that the Auja basin was predominantly populated by Arab farmers who had a considerable interest in seeing the area irrigated. Under the electrification concession, the only person legally entitled to undertake any irrigation was Rutenberg. By abandoning the electrification and irrigation scheme in the Auja basin, while retaining the exclusive right to those activities, Rutenberg effectively blocked an important avenue of economic growth for the basin's mostly Arab inhabitants. As the municipal committee's letter stated, "The irrigation project if carried out properly and on conditions suitable to the country, will no doubt be of use."[118] Yet a full irrigation plan was never carried out, providing a constant source of friction between the company and the inhabitants of the basin.[119]

From that moment in the summer of 1923, the Jaffa municipality was officially opposed to the scheme. Crowds poured into the streets chanting anti-Rutenberg slogans, extolling the municipality's steadfast rejection.[120] Opposition to Rutenberg in Jaffa was invariably expressed in national terms, and Jaffans were not the only ones engaged in the cause. News articles, pamphlets, and letters from around the country made it clear that electrification had become a matter of national significance all over Arab Palestine. By 1923, it was widely known that the British had also given Rutenberg preliminary approval for a countrywide concession. Consequently, the solidarity of Palestine's Arabs with Jaffa was a natural outgrowth of the concession's reach. In Tulkarm, Nazareth, Tiberias, Jerusalem, and elsewhere, Palestinians were engaged in their own local struggles against what they saw as the threat of Zionist current. Like the grid itself, it was a struggle in which the local and the national were inseparably linked.

As a result of the intensified opposition, negotiations with both Jaffa and Haifa stalled in the spring of 1923. In Haifa, Rutenberg's proposal made a "not unfavourable impression on the more progressive elements" of the local population, according to G. S. Symes, the district governor of the north. The municipal council's ultimate refusal to go forward with the plans, he explained, were based on "motives ... [that] were (admittedly) political."[121] Symes's assessment, then, reflected growing antagonism toward the proposal but also the established trend of branding any opposition to the scheme as political, while locating the scheme itself outside of politics.

Palestinian agitation against Rutenberg revolved around two themes: first, the "national" scope of the threat, which thus also served to perform the nation itself at a formative stage; second, the insistence on the political nature of electrification, what might be called oppositional boundary-work. In May, the Jerusalem-based First Arabic Economic Council published a pamphlet addressed, "To you, the victims of Rutenberg's greed." The pamphlet thanked the inhabitants of Jaffa for their "gallant resistance" in the face of Rutenberg's attempt to "enslave" them, and implored them to remain steadfast.[122] In July, *Filastin* triumphantly reported that "Mr. Rutenberg travelled from one city to the another in Palestine and tried, with the help of the Government, to obtain loans from the Municipalities in order to be able to execute his project; but he was rejected everywhere."[123] From Haifa, the lawyer and Arab Executive member Wadi' al-Bustani wrote an open letter to Mayor Sa'id, urging him to intensify resistance in Jaffa: "Is it not the imperative duty of the Jaffa municipality which possess such rights by law to preserve them by taking up a lawsuit?" Bustani demanded. Addressing Sa'id directly, he continued, "And you, strong and public spirited man whose justice is sharper than a double-edged sword, why should you not cut with this sword these poles each one of which is like a nail in one's heart and a sore on the eye rather than light."[124]

In late May 1923, a crowd of more than one thousand people attended a meeting in Tulkarm to discuss Rutenberg's scheme.[125] The number is staggering, given that the town had fewer than four thousand inhabitants at the time. Many more meetings followed. In early June, at a meeting in Jaffa attended by representatives from all parts of Palestine, Kamal al-Dajani asserted that this was not an economic issue at all but a political one: "No profit can be expected from the Rutenberg scheme, but it lays down the foundation of the Jewish National Home."[126] In mid-June, the Sixth Arab Palestinian Congress adopted a resolution that condemned "the erection of poles and extension of wire" and called for a countrywide boycott of the works.[127] In November, Samuel received a telegram from the MCA in Nazareth, which stated, "The efforts made by the Government to execute Rutenberg's project at Jaffa after it has been rejected by the native inhabitants are considered as an insult to the Arab population as a whole. We strongly protest against this action and demand that the national feelings be safeguarded in accordance with British justice."[128]

As one would expect, the threat of sabotage was ever-present. In May, Rutenberg expressed his concern to the British that the network being constructed in Jaffa would become the target of sabotage, and he asked the chief secretary to issue a public warning against such acts. They promptly obliged and enlisted the attorney general's help in drafting the notice and then instructed the assistant district governor of Jaffa to distribute the note widely through local channels. The note read: "The public are hereby warned that it is an offence under the Ottoman Penal Coder for a person to destroy or damage property of which he is not the owner. If, therefore, any person destroys or damages the Power House, engines, poles, wires or any other part of the system for the generation and supply of electricity in Jaffa which is now being constructed by the Jaffa Electric Company, Limited, he is committing an offence under the Ottoman Penal Code for which he can be prosecuted."[129] In the fall, the British authorities arrested two "Arab agitators" alleged to have incited a crowd at the Mosque of Omar in Jerusalem against Rutenberg's works, and exhorted it to sabotage the electric power station in Jaffa.[130]

LIGHT IN TEL AVIV, DARKNESS IN JAFFA

On June 10, 1923, the first powerhouse in Palestine was inaugurated; seventeen days later, it was put into commission around the clock. Located on the Jaffa-Sarona Road, just north of the border of Tel Aviv Township, the station consisted of two vertical four-cylinder diesel engines of 500 horsepower each. From it issued two high-tension cables, each transmitting 6,000 volts. One ran along Allenby Road for a stretch of 1 mile, linking up to two transformer stations. The second high-tension cable ran for 1.2 miles, reaching a first transformer station on the Jaffa–Tel Aviv Road and a second on King George Avenue in Jaffa. From these load centers, low-tension distribution fanned out over Tel Aviv and Jaffa. But while Tel Aviv already had a street lighting

FIGURE 5. Jewish electricity workers in Jaffa, ca. 1922. Courtesy of the British National Archives, CO 1069/731.

scheme in place, Jaffa had only two lights of 2,000 candlepower each, one located at the Governate Building and the other at the prison.[131]

With the growing unrest, the politics of non-politics became increasingly strained. On June 24, Jamal al-Husayni, secretary of the Executive Committee of the Sixth Palestine Arab Congress, made a statement that linked the growing unrest (and surge in violent crime) to the loss of legitimacy that British rule was suffering. "People hate and despise the Government which they call a Jewish [i.e., Zionist] government, and it is natural that one who despises a person will not fear him."[132] The following month, Mr. Becke in a parliamentary session asked the colonial secretary, William Ormsby-Gore, whether "his attention has been called to the fact that the Rutenberg Company in Palestine is erecting electric standards in Jaffa against the wishes of the municipality; is this being done with the Government's consent; and will he take steps to inquire into this owing to the unrest that is being caused by this action." Ormsby-Gore, however, denied knowledge of any opposition to the Rutenberg scheme, or of any unrest caused by such opposition. The scheme, he added, "has the full consent of the Palestine Government."[133]

To be sure, opposition to Rutenberg's scheme was not universal among Palestinian Arabs. In July 1923, the municipality in Jaffa received a letter from a group of Arab merchants who had businesses located on the main thoroughfares of the town, such as King George, Mustakim, and Suk al-Dayr. The letter protested the municipality's "offending resolution" to reject Rutenberg's scheme. Highlighting the symbolic power of electric light and echoing the power company's own business framing, they protested "the reign of darkness in our streets which are in the main business section of Jaffa": "We oppose the fact that you have adjoined politics to matters of economic importance. We need the illumination as light is needed for our eyes. It is one of the most important things that a man of culture cannot do without. It is a well known fact that electric light revives business.... A modern man tries to make his work easier, using all the modern means of technology. Therefore, we cannot consent to remaining in darkness and that our businesses, which are already suffering, will also suffer from the absence of light."

The letter went on to state that the merchants had long awaited the arrival of electricity, which would enable them to "make use of the latest technical improvements which are the only stimulants for the development of a country." They concluded the letter: "We demand that you shall not refuse us the electric light of Rutenberg's plant."[134] Shortly thereafter, the Jaffa Chamber of Commerce wrote to the assistant district governor, declaring that the chamber had voted "to give the merchants their full support in their righteous demands for electric light in streets of the main business center of the town." The merchants, they wrote, paid both direct and indirect taxes and had the right to demand services, such as electricity. Also: "They have righteously mentioned in their petition that politics should not interfere with business." The letter concluded by asking the assistant governor to forward it, with his endorsement of its contents, to the Jaffa municipality.[135] Some quarters of the Palestinian community thus began to adopt the technocapitalist logic that initially had been the domain of the British and the Zionists. The way they framed their demand for power was channeled through the boundaries established, opposition being political, while support was neutral and scientific.

Confronted in 1923 with a Palestinian opposition growing increasingly active and vocal, Rutenberg was forced to find a way to neutralize it. Sensing that there was no progress to be made in the major Palestinian centers, like Haifa and Jaffa, Rutenberg began looking for another entry point, by targeting the weakest component in the Palestinian national circuit.[136] Two places soon emerged as the most attractive, the Arab towns of Nazareth and Tiberias. Opinion within the electricity company was split on where to proceed, and initially Rutenberg resolved to go ahead in both towns.[137] Rutenberg's right-hand man, Yakutiel Baharaw, favored Nazareth, "the lighting of which," he wrote, "is of great political significance to us."[138] So did the director of the Department of Public Works.[139] Ultimately, however, the high commissioner urged Rutenberg to focus on Tiberias. To sweeten the deal he got the department to commit to funding the installation of a water pumping station there, which would consume large amounts of electricity.[140] Given that, and the "agitation raised by the Moslem Christian Association" in Nazareth, Rutenberg consented.[141]

Rutenberg had made provisions for such a course of action. For the past year, he had been cultivating relations with the long-serving mufti of Tiberias, Abd al-Salam al-Tabari, the mayor, Zaki al-Hudayf, and other prominent Muslim leaders through a local agent, referred to in the records of the power company as "Mr. X." Those efforts paid dividends in mid-October 1923, at the height of antielectrification agitation, when the Tiberias municipality voted unanimously to accept Rutenberg's scheme for electrification.[142] In this instance, it was clear that money had changed hands. It was a tenuous relationship, Rutenberg knew, between his outfit and "people who can be bought by our enemies as easily as they are bought now."[143] But it was all it took. The first defection from the ranks of Palestinian resistance proved the unraveling of the opposition as a whole. "Tiberias was an important strategic move," Rutenberg later explained to his board, "its unanimous favourable decision supported by influential members of the Moslem Christian Association being of great help."[144] Indeed, the rationale behind building a powerhouse in Tiberias was exclusively political, as Rutenberg himself admitted. In a report from 1927, he wrote that the Tiberias powerhouse "was erected at the commencement of our operations in Palestine for reasons of policy more than for business reasons, the aim being to overcome the opposition of the Arab population to our Works."[145] At the time, Rutenberg mentioned being approached by Arabs demanding "compensation," and continued, "the breaking of the Arab front is very important and I am doing my best to obtain this first important practical result."[146]

The agreement in Tiberias proved catastrophic to the structural integrity of the opposition, and the national resistance crumbled in short order. Rutenberg, who proceeded directly from Tiberias to Jaffa, managed through another series of intense negotiations to get an oral commitment from the mayor to agree to the scheme on November 12, 1923. The published statement of the Jaffa municipal council read: "In compliance with the wishes and demand of the Government to light the streets of the town with electricity, and in view of the fact that this is suitable also from the economical point of view it has been decided to light the streets of Jaffa with electricity." In a brief reference to its earlier demand, the municipality reserved the right to "buy the network later on."[147] On November 13, before the agreement had been signed, Rutenberg turned on the switch in Jaffa, having done all the preparatory work for just such an occasion. Rutenberg explained: "The Arab political organisations were thus prevented from using their pressure for the decision of the Municipality to be reversed."[148] He promptly expanded the grid by another six miles in Jaffa, for a total length of twenty miles. Also, a fifth transformer station was erected in Ajami, an entirely Arab neighborhood, and the local residents were reported to be "very satisfied." By the end of November, 852 applications had been received from Arab residents requesting to be connected to the grid.[149] In January 1924, Ali Mustakim, one of the richest notables in Jaffa, came to an agreement with the power company to build new shops in Jaffa and Tel Aviv, lit by the company's electricity.[150]

In instructions to the Jaffa office of the power company, Rutenberg took great care to underscore the importance of giving the first Arab consumers special treatment and attention. All work teams in Jaffa should have as many Arab workers as possible, and they were to engage a "reliable good Moslem foreman." The Jewish workers in Jaffa should as far as possible be Arabic speakers; there should be at least one Jewish Arabic speaker on each work team. In general, "Persons coming in touch with Arabs on house connections, meter reading, collection of money, etc. to speak Arabic and treat

Arab consumers and population in general, in the most correct and friendly manner." He instructed that the application forms be amended so that the Arabic appeared first, Hebrew second. Finally, they should do "everything possible" to get advertisements published in the Jaffa newspapers, especially *Filastin*; "if necessary demand assistance from the Municipality," he instructed.[151]

So keen was Rutenberg to establish facts on the ground that he immediately wrote to the British, securing a loan guarantee for Jaffa's share of the investment, promising secretly to guarantee, in his turn, the British loan guarantee.[152] Expanding the grid was, of course, in the nature of the enterprise, but Rutenberg's alacrity at this critical phase should also be understood in the context of Palestinian oppositional politics. By thickening the grid— creating auxiliary power generators and alternate pathways for the current— Rutenberg was effectively reducing its vulnerability to sabotage, thereby reducing the force of Palestinian oppositional politics. Rutenberg's eagerness in this regard was noted by the Palestine Government at the time. "The reasons for this are obvious," one official noted, echoing the technopolitical logic explored earlier, "in that Municipalities would have a direct interest in maintaining the security of the installations."[153]

The sudden collapse of Palestinian resistance to Rutenberg was the result of effective pressure being brought to bear on a thin circuitry of another kind—that of a nascent national movement. Just as the thin circuitry of the grid opened up the possibility for an oppositional politics grounded in the threat of disrupting transmission, Palestinian nationalist politics had not yet acquired the thickness of firm institutional grounding. By Rutenberg's own reckoning, "the decision of the most influential Jaffa Municipality to collaborate with the Company" had been key.[154] The pressure on Jaffa had come from all directions. By his successful framing of the issue, Rutenberg had managed to equate opposition to his scheme with backwardness. It severely weakened the Palestinians' standing with respect to the British, who exerted strong pressure on Jaffa and Haifa to take Rutenberg's current. But pressure also emanated from within the Arab community. As we have seen, some politicians, like the mayor of Jaffa, never opposed the scheme, in large part because important sectors of his constituency wanted electricity and were less particular than the political elite about the source of the power.

The same was true in Haifa. Although the municipal council there maintained a united front against Rutenberg for a while, it was under constant pressure from within the Arab community to provide electricity. The district governor of the Northern District reported to the high commissioner in late 1922 that the oil lamps currently in use for street lighting in Haifa were "inadequate and unsatisfactory from every point of view (except that of burglars and foot-pads)." He went on to report that "the townspeople feel that whilst in many respects Haifa is gradually assuming the appearance of a civilized town no steps are being taken to provide one of the first, and least expensive, requirements of such a town, viz., an electrical installation." The current state of affairs had given rise to "a good deal of real and not unreasonable bitterness," according to him.[155] Along similar lines, the Haifa-based journal *al-Nafīr* pointedly asked in an editorial from April 1923: "Is darkness preferable to light?" It went on: "We are not acquainted with the cause that makes them [the municipal council] oppose the [Rutenberg] scheme. Is darkness better for us than light? Until when shall we be sunk in the obscurity of generations and dark hateful intolerance?"[156] The greatest sign of Rutenberg's success was not the number of consumers added or kilowatt-hours produced and sold but the consolidation of his boundary-work. With his framing so widely accepted, the actual growth of the grid was a foregone conclusion.

In the event, no formal agreement was ever signed between the power company and Jaffa. The negotiations continued through 1924, but ultimately Rutenberg backed out, citing the need to await the promulgation of the government's electricity ordinance. The ordinance was held up for more than a year at the Colonial Office, and then in various negotiations, to the consternation of everyone involved, and of no one more so than Mayor Sa'id.[157] Opposition never disappeared completely. After mounting public pressure, the Jaffa municipality voted in late 1929 to end its relationship with the power company and instead install a system of lux lamps. Nothing ever came of it, however, a fact that was not passed over in silence by the Palestinian press. "The nation is waiting," *al-Jami'a al-Islamiya* reminded the municipality a few months later.[158] The Yishuv was not slow to fit the episode into the civilizational narrative we have become so familiar with. "What is positive electricity?" went a joke popular in the Jewish community at the time. "Well, in Jaffa, it is really an old fashioned 'lux' lamp."[159]

THE GRID THICKENS

In a triumphant letter to a confidant at the end of 1923, Rutenberg wrote about the "good results" obtained in Palestine: "The hostile Arab front has been broken. Arabs are now applying for energy. The Jaffa business is already proving to be a highly lucrative undertaking.... Everything points to the Haifa Power House proving to be at least as good as Jaffa."[160] The report summing up the year noted that "the opposition of the Arab Population which was encountered by the Corporation as a result of political agitation no longer exists."[161] Four years after arriving in Palestine, and two years after winning the concession, after "most complicated, difficult and dangerous work," Rutenberg had obtained the consent of virtually the entire country for his scheme. He had acquired land in the Jordan Valley, Tiberias, and Haifa. Major Arab towns, such as Ramleh and Acre, had put in unofficial applications for electricity shortly after the lights went on in Jaffa.[162] Making inroads in Arab towns was "very important politically" to Rutenberg because "it would make influential Arabs in these ... towns interested in our concern and would break the entire Haifa congress front."[163] After concluding the agreement with Jaffa, Rutenberg was approached by the governor of Haifa, who informed him that the Haifa municipal council was now willing to accept the deal Rutenberg had offered the previous year.[164]

Furthermore, Rutenberg had concluded an agreement with the British to supply their planned railway workshops in Haifa, the energy needs of which alone would cover the operating expenses of the Haifa station. In a twist that kept repeating itself in the history of Palestinian electrification, the British seemed only dimly aware of the fact that in agreeing to let Rutenberg build a power station in Haifa, they were preempting the outcome of the negotiations for the countrywide concession, which had not yet begun when the Haifa agreement was made. Too late, as we will see in chapter 4, the British would realize that they had in effect green-lighted an extralegal power station and grid, with the effect that they found themselves locked into an arrangement with Rutenberg before having had a chance to negotiate its terms.[165] The benefit of this was not lost on Rutenberg, however, who made the offer already in early 1922, fully prepared to sacrifice one of the generators designated for Jaffa for a toehold in Haifa.

In doing so, Rutenberg was motivated by his systems thinking. In accordance with his training at the St. Petersburg Polytechnical Institute and reigning technological fashions, and like all the great systems builders in twentieth-century Western society, Rutenberg built big.[166] And he did so even when it meant jeopardizing short-term goals, as when he extended his fund-raising trip in America through the summer of 1922 to secure the funds for his Jordan project, coming dangerously close to missing the deadlines under the Jaffa agreement and thus forfeiting the concession.

By late 1923, with an informal agreement in hand and construction of a Haifa powerhouse soon to begin, Rutenberg judged the conditions in Haifa "exceedingly favourable."[167] The Haifa powerhouse was inaugurated on June 24, 1925, in a ceremony presided over by the high commissioner. Construction had cost the company almost £E85,000. The plant did not begin regular operations until September 1, 1925. Lingering resistance within the municipality to the power company prolonged negotiations until December, when the municipality and the power company worked out a short-term agreement. By that time, the predominantly Jewish Suburb Council of Hadar Hacarmel had already concluded a separate agreement with the company. Although the agreement was meant to be in force for only a few months, due to the prolonged negotiations over the Jordan concession, it remained in force for more than a year.[168] The Haifa plant was immediately profitable.[169]

Meanwhile, the power system in Jaffa and Tel Aviv continued to grow. New generators were soon added to the powerhouse. In anticipation of being connected to the grid, whole neighborhoods sprung up in the rapidly expanding White City with houses prefitted with electrical wiring. Already by the end of 1923, Rutenberg advised his board of directors to invest another £E5,000 in extending the grid by a total of five miles in these neighborhoods. As of December 1, 1923, 40 percent of the population in Tel Aviv were consuming Rutenberg's electricity—"only," the ambitious Rutenberg griped.[170] In addition, Rutenberg concluded an agreement with the Air Ministry to extend the grid to the British army base at Sarafand for an estimated extra annual revenue of £E5,000. The eighteen-mile 15,000-volt high-tension line, built in the fall of 1924, extended two-thirds of the way to Rishon Letzion and passed close by Mikveh Israel, Rehovot, Sarona, Ramleh, and Lydda, covering the entire southern portion of the Jaffa District and putting thousands of potential new consumers within reach of the wires.[171] By early 1925, the company branched off from the Sarafand mainline to supply Mikveh Israel and Rishon Letzion.[172] The two lines were completed by the summer, and another high-tension line was extended the three miles to Petah Tikvah. The maximum capacity at the Jaffa powerhouse had been gradually raised to 2,750 horsepower. Six more substations had been added to the original four, and another 1.9 miles of underground cables were added to the original 2.5. The length of the overhead lines had more than doubled, from 11 miles to 24, and another 3 were under construction.[173]

Although the original estimates held that it would take five years before the company saw profit, in his first annual report at the end of 1923, Rutenberg expected the venture to pay dividends on its shares already by the end of its first financial year, in October 1924.[174] In the year and a half after Rutenberg concluded the agreement with Jaffa, electricity consumption increased by 456 percent.[175] Although private consumption was still a minor proportion of the total, the number of private consumers increased several times over. Street lighting represented only about one-tenth of total consumption. The largest growth sector by far was industry, constituting about two-thirds of total kilowatt-hours sold.[176] By the end of 1925, Rutenberg's current was lighting and powering Tel Aviv, Jaffa, Haifa, and Tiberias, all of which either broke even or ran at a profit with the expectation—correct, as it turned out—that all powerhouses would turn a profit within the year.[177] Business was good, the grid was fast thickening.

For a while, it looked as if the dual mandate were indeed going to be compatible. A report of the Permanent Mandates Commission's meeting in Geneva on December 2, 1925, stated: "Palestine and Transjordan have occupied a great deal of the attention of the commission. It is obvious that the task of developing the country with due regard to the dual principles embodied in the Mandate, does not, at least for the moment, permit of rapid development in the political field, but it is satisfactory to note that peace and order reign in the territory, that political agitation has diminished."[178] To the PMC, as to most contemporaries, it made perfect sense to distinguish between political development, of which there had been none, and economic development, of which there had been plenty.

ELECTRIFICATION AND DEPRESSION

From 1926 to 1928, the Yishuv suffered a severe economic crisis. It was brought on by an economic crisis in Poland that stanched the flow of capital to Palestine. By 1928, twice as many Jews left Palestine as arrived. Economic recovery began in 1929, returning to precrisis levels by 1931. In many quarters of the Zionist movement, the economic downturn brought on a crisis of confidence.[179] Electricity consumption, however, continued to grow, as did the grid. Profits grew by about 10 percent per year through the crisis, and the power company paid out dividends on its shares.[180] Indeed, the electricity works proved essential in the depression years, especially, as we will see in the next chapter, the large-scale construction on the Jordan River, which employed close to one thousand workers at the height of the depression.[181]

According to the government-issued *Blue Book on Palestine* for the years 1927 and 1928, "The most important stimulant to local industry has been the establishment of electrical power stations at Tel-Aviv, Haifa and Tiberias."[182] In fact, profits continued to rise throughout the 1920s.[183] By the end of the decade, Palestine had some fifteen thousand electricity consumers, and the powerhouses had increased their output capacity by a factor of five to more than 5,000 horsepower, or close to 4,000 kilowatts.[184] The revenue from the powerhouse in Haifa also rose steadily through the crisis, and the company proceeded with plans to extend a high-tension line southward along the coast to supply all the coastal colonies, from which Rutenberg expected a demand of 1.5 million kilowatt-hours a year. His first priority was to completely wire the coastline from Acre to Tel Aviv.

The expansion is another clear indication of Rutenberg's systems approach: the extension would cost the company £E30,000 at the moment of the Yishuv's gravest economic crisis. The primary motivation behind the bold initiative was the desire to stoke demand, especially in agriculture, in anticipation of the huge output of the Jordan hydroelectric plant that was expected to be completed within a few years. Rutenberg even had plans to found additional companies for the easy importation of electrical machinery and appliances to facilitate the growth of demand.[185] By the fall of 1929, a high-tension line between Haifa and Hadera had been completed, and many orange grove wells had been connected to the grid, with the expectation of soon connecting the Haifa harbor quarries and the Shemen factories, which would substantially increase profits for 1930.[186]

By the summer of 1928, the low- and high-tension network around Rishon Letzion and Rehovot were completed.[187] And by the early 1930s, before the hydroelectric station on the Jordan had come online, a slew of local grids were built and connected to the high-tension mains, including a medium-tension line from the Sea of Galilee, via Tiberias to Migdal, powered by the excess energy produced by the diesel generators in operation at the Tel Or work site. Other places connected included Ra'anana and Netanya on the Sharon coastal plain, through an extension of the network in Tel Aviv.[188] By the end of 1930, the output of the powerhouses was 4,125 horsepower in Tel Aviv, 1,200 horsepower in Haifa, and 180 horsepower in Tiberias. The Jordan powerhouse was expected to produce 34,000 horsepower. In 1930, the Jaffa section alone earned a net profit of £P40,000, representing a growth over the previous year of more than 25 percent.[189] By this time, electricity powered half of all motors in Palestine.[190]

Industry, unsurprisingly, dropped as a proportion of total energy consumption during the crisis. Major consumers, such as the Silicate brick factory and the Tel Aviv municipality, reduced their electricity consumption. But other sectors, notably private consumers and agriculture, more than made up for the shortfall. Consequently, during the years of the crisis, another 1,125 horsepower were added to the Jaffa station, and high-tension lines were constructed from Rishon Letzion to Nes Ziona, and by early 1927 low-tension networks were constructed in Petah Tikva and Rishon Letzion. A factory for the construction of electricity pylons was also built. In another instance where Rutenberg gave preferential treatment to a Jewish consumer, he wrote to Siegfried Hoofien at the Anglo-Palestine Company, after it had just turned down a loan request for £E2,500 from the municipal council of Rishon Letzion to build its low-tension distribution system. Rutenberg informed Hoofien that his company would "give" Hoofien the money for the loan. "But please remember," Rutenberg admonished him, "that this arrangement must remain absolutely confidential. Nobody even at the bank must know about it because it may compromise my entire policy with regard to similar works." He instructed Hoofien to say that he had received the funds from London.[191]

Rutenberg was hardly flummoxed by the crisis. In fact, he welcomed it. To his mind, the crisis had been brought on by the philanthropism that permeated the economic thinking of the Zionist movement. Such irrationality caused insufficient attention to be paid to the relationship between cost and benefit. In his report to the board in 1927, he stated: "I consider this crisis to be very useful inasmuch [as] people are now forced to adapt themselves to actual conditions of life and sound economic forms of work."[192] In this, he echoed a larger debate within the Zionist movement, adopting the position most closely associated with Vladimir Jabotinsky, the leader of Revisionist Zionism.[193]

CONCLUSION

The story of Palestine's first electric supply system shows how the technological properties of the grid interacted with political and economic factors to shape relations between Zionists, Palestinians, and the British. This was largely a matter of boundary-work, that is, the effort to draw the borders around the activity so as to define it in a way favorable to one's interests. Understood in this sense, boundary-work is a form of politics, although one whose goal is often to deny any social or political influence on the activity being promoted.[194] To say that Rutenberg engaged in boundary-work is not simply to say that he hid his political agenda behind technical laws. His boundary-work constituted a claim for a fundamental conceptual division of Palestine. The nature of this division flowed from the technosocial properties of the system he was seeking to promote, and had consequences far beyond it.

First of all, Rutenberg's boundary-work propelled both him and his strategy to vaunted positions. It catapulted Rutenberg into the role of elder statesman. In 1929, Rutenberg was appointed chairman of the Va'ad Leumi, the executive branch of the Jewish state-in-the-making, and through the 1930s, as Arab-Jewish relations deteriorated, Rutenberg was at the forefront of the efforts to bring about rapprochement. He negotiated with different Palestinian Arab factions and with leaders of surrounding countries, including Nuri al-Said of Iraq. Rutenberg became especially close with Emir Abdullah of Transjordan, a relationship that began over a shared interest in the Jordan River.[195] Rutenberg was also called on to mediate between the labor and revisionist factions of the Zionist movement when, following the assassination of the prominent labor Zionist Haim Arlosoroff, they found themselves on the brink of armed conflict.[196] Indeed, for the last decade of Rutenberg's life, until his death in 1942, his political work largely took him out of the energy business. In short, Rutenberg's boundary-work carved out a political space for him within electrical Palestine, from which he exerted a powerful influence on mandatory politics.

His strategy of politics of non-politics was similarly buoyed. The strategy was not all of his own making, of course. As with so many of the entities in this story, a key condition of its success was the ability to evolve in response to events on the ground in Palestine. It built on late British colonialism's developmentalist rationale, and it was an elaboration on a familiar Zionist discourse that had already emerged in conversation with British imperialism—that the Jews' superior ability to develop Palestine would improve the lives of all the land's inhabitants. As a result of Rutenberg's successful application of it, however, the politics of non-politics was cemented as a primary Zionist strategy throughout the mandate period and into statehood. It also became a central avenue of Zionist claims-making through the mandate period and into statehood. Boundary-work continues to be a critical front in the Arab-Israeli conflict to this day.

As for the Palestinians, electrification constituted one of the first explicitly nationalist struggles. Rutenberg's countrywide electrification scheme called out for resistance of the same geographic scale. Electrification stopped at Palestine's borders, while drawing together local arenas within that territory by virtue of designating them all targets of the grid's growth. All across the country, Palestinians were talking about electrification, urging each other to remain steadfast in the face of Rutenberg's attempts to entice them. In this sense, the grid can be thought of as another technology, besides print capitalism, producing the effect of coevalness at the heart of Benedict Anderson's classical formula for the emergence of national imagined communities.[197] The grid participated in producing a national scale in Palestine, ordering its politics.

The Palestinians' eventual failure to maintain a united front against electrification facilitated the thickening of the grid. That, in turn, reduced its vulnerability to sabotage and changed the conditions for oppositional politics. The failure steered the fractured Palestinian national movement in a new direction. For the remainder of the decade, the Palestinians sought to fight the system from within, as they also fought each other, through the Supreme Muslim Council, the office of the Grand Mufti, and the mayoralties of major cities.[198]

By losing the competition over the definition of electrification's boundaries, the Palestinians lost in other ways too, as they saw their claims moved out to the subjective realm of politics, where they were stripped of epistemic authority. For the duration of the mandate, the Palestinian press would often carry pieces of oppositional boundary-work, insisting on the fundamentally political nature of electricity. One piece from the newspaper *al-Jami'a al-Islamiya* in 1934, for example, excoriated those who took "Rutenberg's Zionist current" as "traitors to their city and homeland."[199]

There is also the question of why Rutenberg opted to promote his venture by means of boundary-work in the first place, and why he was successful. In both cases, the answer lies in the precise properties of the technological project that he was promoting, which both forced and enabled him to make such an argument. Forced, because the vulnerability of the electrical system, due to its techno-spatial properties, compelled Rutenberg to seek to realize his scheme by co-optive rather than coercive means. This is why I believe Shamir is wrong to claim that "Arab agitation ... did not determine the location of the powerhouse and the technology used for generating electrical energy."[200] In fact, the opposite is true. As we have seen, both the change of technologies and the locations of the power station were motivated by ongoing negotiations with the Palestinians and the fear of sabotage. But the grid also enabled Rutenberg's boundary-work, and then critically shaped it throughout. The claims that he was able to make—for irrigation rights, for instance—were tightly linked to the technological circumstances of his project. This observation also speaks to recent concerns regarding the link between energy infrastructure and politics, not least in Timothy Mitchell's work.[201] Rather than locating structuring effects only in the shift from one energy system to another (from coal to oil, in Mitchell's case), here I have sought to bring a greater sense of fluidity to the process, by underscoring how the continual evolution of infrastructure is forever reshaping political possibilities.

The idea of a politics of non-politics, as used here, relates to a host of studies of the disembodied exercise of power, pioneered by Max Weber and Michel Foucault and elaborated upon by many, notably James Scott in *Seeing Like a State* and James Ferguson in *The Anti-politics Machine*. But the dynamics at work here differ in some significant ways from Scott's and Ferguson's accounts. They argue, following Foucault, that ostensibly apolitical policies inevitably end up underwriting certain political agendas and power relations. On this view, however, the reproduction of power relations is an unintentional and essentially uncontrollable effect. In the present case, by contrast, Rutenberg deliberately chose to pursue political goals, or goals that had significant political consequences, by way of a depoliticized language that drew the boundaries around his venture so as to be impervious to politically grounded objections. Rather than turning the crank on an antipolitics machine that produces politics in spite of itself, Rutenberg operated an instrument of his own careful design: he whispered his politics into one end, and out the other came digestible bits of rationally mandated exigency.[202] Or, as he put it himself in a confidential meeting with the Zionist Executive: "The principle of 'money talks' eliminates difficulties with Arabs and workers."[203]

FOUR

The Radiance of the Jewish National Home

war with nature, a war with the circumstances
—YOSEF BARATZ[1]

THE JORDAN RIVER IS A 156-mile-long body of water that runs through a structural depression formed through a sudden Oligocene movement of the earth's crust millions of years ago. In English, the depression is known as the Jordan Valley; locally it is known as the Ghor. The river, as much as any active participant in this story, was always subject to history's twists and turns. As Richard White points out, rivers are contingent, evolving entities; they "constantly adjust; they compensate for events that affect them" and are in this sense historical products of their own and others' making.[2] The Jordan River had not always looked like it did when Rutenberg set his mind to harnessing its motive force in the early 1920s, and the Jordan of that time bears only some resemblance to the Jordan of today.

It rises on the slopes of Mount Hermon on the southern tip of the Anti-Lebanon range. Before Israel drained most of the Huleh marshes in the 1950s, they would collect the ice water making its way down from the Hermon. The water traveled for about 6 miles through the marshes until reaching Lake Huleh, the catchment area for the three principal Jordanian tributaries: the Hasbani, Baniyas, and Dan. At the efflux of the lake, the tributaries merge to form the Jordan. From there the river drops more than 650 feet in the course of its 16-mile run down into the Sea of Galilee, located some 670 feet below sea level, and with a surface area that, at the time, fluctuated around sixty square miles depending on the season. The highest levels occurred during April and May, the lowest in November and December. According to estimates produced by Rutenberg's survey teams, the river's depth differed some 5 feet between the summer high and winter low.

The final leg of the river's journey begins at the bottom of the Sea of Galilee, where it exits through an outlet whose width also fluctuates with the seasons (back then, from 60 to 100 feet). From there it travels 186 miles, 62 as the crow flies, to its terminus at the Dead Sea. During this last leg, the river undertakes a further level drop of 600 feet, descending to the lowest land elevation on earth. For the first 12 miles the width of the valley is 3 to 4 miles, after which it widens further and merges with the plain of Beisan. Within the valley itself, the river bores down farther, cutting a trench within a trench with steep walls of alluvial soil, braced by Cenomanian and Senonian limestone. A few miles below the Sea of Galilee, the Jordan intersects with the Yarmuk River, which barrels down through many steep falls from the Druze highlands of the Hawran, nearly doubling the water discharge.[3]

During the initial section of the Jordan after the confluence, the river runs 131 feet wide. In 1920, it was lined by oleanders in the upper reaches, poplars in the middle reaches, and tamarisks and poplars in the lower reaches. After the initial stretch of many falls and rapids, the river slowed down and meandered the rest of the way to its terminus at the Dead Sea.

The weight and flow of water is equal to its capacity for doing work, that is, to its energy potential; in its total journey from the mountains of south Lebanon to the Dead Sea, the Jordan falls 2,372 feet through a steep and fairly even gradient. The Jordan Valley has been a site of agriculture for thousands of years.[4]

As we saw in the last chapter, the importance of the electrification of Jaffa lay in the way Rutenberg was able, through his boundary-work, to set up the physical and conceptual framework that would facilitate the continued growth of the system. After that initial success, he turned to the centerpiece of his electrification scheme. As soon as the concession was signed, Rutenberg began preparing the system of hydroelectric power stations on the Jordan River. In the following, we will peer into the organic machinery of the Jordan River at the time that it came to be harnessed for electrical power. We will see how this endeavor staked out Palestine as a Jewish national space.

Earlier generations of scholars of the nation-state have focused on the emergence of discursive and institutional structures that make it possible to imagine the nation. But as the environmental historian Brett Walker has shown, material structures also provide prospective citizens with a means of knowing the state, as it creates national ecosystems.[5] We have seen how the process of forging this "hybrid system" also forged a unified political-geographical unit, in the form of modern Palestine. In this chapter and the next, we will see how the hydroelectric power station and countrywide transmission system had an effect similar to the things more traditionally associated with the emergence of modern nation-states in colonial settings, such as Benedict Anderson's well-known trinity of the census, map, and museum.[6] Both symbolically and materially, the power system bounded and structured the modern "national political field" in Palestine.[7]

On March 29, 1923, the Palestine Electric Corporation was incorporated under Palestinian law with a nominal capital of £1 million, divided into one million one-pound shares, in accordance with the stipulation of the concession signed a year and a half earlier. The board included Rutenberg, Dr. George Halpern, M. O. Nassatisin, Harry Sacher, and Judah Magnes. In April, the high commissioner approved the company's memorandum and articles of association.[8] In August, the Crown Agents declared that they were satisfied with the evidence given of sufficient capital, and the high commissioner was authorized to proceed with negotiations with the company to finalize the concession. Rutenberg transferred his concessionary rights to the company and submitted a detailed technical plan for the first two power stations, as a first step in the final concession negotiations.[9] The reasons for choosing a point just below the Jordan-Yarmuk confluence for the first power station have already been discussed. Although the technical specifics of the new plan were significantly different in parts, the considerations regarding the location remained the same. The foremost of those considerations, we recall, was the advantage of using the Sea of Galilee as a water reservoir, in order to even out the winter and summer difference in rainfall and thus ensure steady power generation.

The first power station, named Naharayim ("two rivers" in Hebrew), would exploit the 170-foot level drop from the Sea of Galilee to Jisr al-Mujami'. Rutenberg proposed to build an earth dam near Delhamieh, close to the Jordan River's efflux from the Sea of Galilee, to retain water for a depth of 20 feet. Another dam was to be built at the Yarmuk waterfall, retaining water for a depth of 26 feet and creating an intermediary Yarmuk reservoir. The bed of the Jordan River would be deepened and straightened, in order to increase the quantity and rapidity of the river's flow. A concrete-lined canal, some 1.2 miles long, would convey the Jordan waters into the Yarmuk reservoir, and a 437-foot-long canal would bring the water down to the headworks by Jisr al-Mujami', which included a pressure reservoir, penstocks, powerhouse, and tailrace. The average yearly quantity of water of the Jordan and Yarmuk together was estimated at 12 gallon per second. The capacity of the turbine shafts would be 24,000 horsepower, and the yearly quantity of energy estimated to be available, allowing for a 25 percent loss in transmission on lines and in transformers, was about sixty megawatts. According to the plan, two turbines of 15,000 horsepower each would be installed initially. Additional turbines would then be added as needed.

The electricity was to be transmitted from the powerhouse by high-tension lines of 66,000 volts to the main transformer stations in Haifa and Jaffa.[10] The regional transmission lines would be 15,000 volts, the municipal areas would be fed by underground cables of 6,300 volts, and the local low-tension distribution networks would have a voltage of 380 between wires and 220 to neutral, as per the European standard set by the turn-of-the-century German precedent.[11] The diesel houses in Jaffa, Haifa, and Tiberias would serve as standby units, producing energy at 6,300 volts, capable of being taken up directly by the municipal cables. The second hydroelectric station would be built at Abadieh, on the Jordan River, below the Sea of Galilee and above the first station. Total cost of construction was estimated at £E750,000, and annual expenditure at £E91,500. Annual income for the first year was estimated at £E147,000.[12] These figures held their own against any electrification scheme in the Western world at the time.[13]

The British government's consulting engineers issued their assessment of Rutenberg's plan in May 1923. It was uniformly favorable, concluding that "the designs are in accordance with accepted standards." The report also commended Rutenberg for the rigor and ingenuity employed in obtaining survey information on which the scheme was based, characterizing his survey works as "thoroughly and exhaustively conducted."[14] The matter of correct figures was crucial, especially with regard to water levels in the various bodies of water. Levels exceeding the estimated maximums would threaten to destroy land used for farming and grazing. Standing water would also expose locals to the risk of malaria by providing hospitable environs for mosquitoes. Moreover, submerging railway and telegraph lines, roads, and other important structures could prove highly disruptive and exceedingly costly to correct after the fact.[15] In the fall of 1924, the director of the Department of Public Works in Palestine, H. B. Lees, prepared a report for the high commissioner after studying Rutenberg's plans and carrying out an inspection of the area designated for the hydroelectric station. Lees also found the plans to meet the highest engineering standards, though he noted, as did most assessors of Rutenberg's plans, that there was no real means of examining the raw data that Rutenberg had provided.[16]

Comparing this plan with the one from December 1920 gives some credence to the critics of the concession, such as Lord Raglan who claimed in Parliament that the original plans "were not really genuine plans at all," but rather hastily put together on the basis of faulty and incomplete information. It is true that the plan Rutenberg submitted in 1923 differed significantly from the one for which he had obtained the preliminary contract in September 1921.[17] The British government, however, was unmoved by such criticisms. Lord Arnold, undersecretary of state for the colonies, dismissed Raglan by stating simply that "enterprises in Palestine connected with the name of Mr. Rutenberg represent the greatest practical effort that has yet been made to develop some of the natural resources of the country."[18] To the British government, Rutenberg was unimpeachable. By this point, he had ascended to a position of such authority that he seemed impervious to criticism, no matter how factual.

Rutenberg's proposal was the target of a different line of criticism from another quarter, namely, the Zionist Organization. Both American and British Zionists had doubts about the scheme. In April 1922, the American Zionist organization the Palestine Development Council submitted the latest version of Rutenberg's scheme for review to M. L. C. Lowenstein, the chairman of the General Electric Company of New York. And around the same time, the London-based Economic Board for Palestine sent Rutenberg's proposal to the British consulting engineers Charles H. Merz and William McLellan. Both reports were favorable overall, and positively laudatory on the technical aspects of the scheme. But both also expressed concern regarding its economic viability. Specifically, its vast scope raised concerns about overproduction, at least initially.[19] Lowenstein compared Palestine to India, where consumption per capita was about half of the demand that Rutenberg had estimated for Palestine. Lowenstein suggested that Rutenberg's estimates "should be divided by half."[20] Similarly, Merz and McLellan also considered its vast scope to be unwarranted and possibly harmful.[21]

It bears noting that these consulting engineers belonged to the same exclusive club of internationally renowned large-systems builders in which Rutenberg was claiming membership. In other words, conceptually they were of one mind with Rutenberg about the virtues of large-scale systems, for all the reasons that have already been discussed, including the benefits of centralization, scale economies, and the technical conditions for electricity generation and transmission. Charles H. Merz was the founder and chairman of the Newcastle upon Tyne Electric Supply Company in Britain. In the chaotic and small-scale British market, the Newcastle system stood out as the only electric supply system before 1920 to be organized as a coherent regional transmission network.[22] Starting in 1904, Merz worked for more than two decades and in the face of strong opposition from the technologically conservative political class to organize London's and Britain's electric supply systems on the large-scale model of places like Berlin and Chicago, as well as his own Newcastle system.[23] During World War I, Merz and McLellan were brought on as consultants for the government, and when the technological and political winds shifted following the war, they were instrumental in planning Britain's National Grid, which was finally enacted in Parliament in 1926 and completed some ten years later.[24]

Partly as a result of the negative economic assessments of the reports, the two Zionist organizations had declined to invest in the scheme back in 1922 when Rutenberg was trying to raise the £1 million in required capital.[25] In 1924, when Rutenberg renewed his attempts to raise funds among the American Zionists, they turned to Loewenstein for an assessment of the revised scheme.[26] Loewenstein's report again extolled the technical virtues of the plan and again raised concerns about the huge scale of the project, given the modest present demand in Palestine. He suggested that demand could be satisfied by two turbines of 4,000 horsepower each, rather than the 15,000-horsepower turbines that Rutenberg was proposing. That way, Loewenstein explained, it would be possible to keep initial costs down and add machinery to the powerhouse later on as demand increased.[27] This would have put it more in line with other contemporary hydroelectric stations in areas of moderate electricity demand.[28]

Rutenberg, exhibiting the defining character of a systems entrepreneur, consistently put his grand vision for a countrywide system before any single intermediary step that would get him there. As we saw, in the summer of 1922 he had gambled with the Jaffa concession against the pleading of his closest associates, by extending his American fund-raising campaign for the Jordan project, thereby coming dangerously close to missing the deadline for commencement of work on the Jaffa powerhouse. In 1922, putting the big picture before the small had nearly cost him the concession. He was no more likely now, in 1924, to downscale his hydroelectric installation for short-term benefits. The American Zionists forwarded Loewenstein's report to Rutenberg, who defended the use of the 15,000-horsepower units based on the argument that the same standard should be kept across this powerhouse and the one that he intended to build at Abadieh. His second reason for the 15,000-horsepower generators he stated simply as: "The amount of water available."[29] When these arguments failed to persuade the Americans, Rutenberg broke off contact with them.[30]

DOG IN THE MANGER

From the moment that Rutenberg had acquired the countrywide monopoly through the preliminary agreement for electrification on September 21, 1921, he guarded it jealously. The grand scale of the system served him in numerous ways. It also presented two significant difficulties, both of which were clearly on display in Haifa. The northern district governor, as well as the inhabitants of the town, regularly complained about what the former, in a letter to the chief secretary, called the "miserably inadequate system of street lighting," referring to a prewar system of lux lamps.[31] The problem confronting Rutenberg in Haifa and elsewhere stemmed from the fact that the only electrification concession currently in force in Palestine was the one for the Jaffa District. Haifa, along with the rest of the country, fell under the Jordan River concession, for which only a provisional agreement had been signed in September 1921. According to the terms of the concession, we recall, Rutenberg enjoyed exclusive rights to develop electrification schemes in all parts of the country, but he was not permitted to do so until he had fulfilled the stipulations of the preliminary agreement, that is, before he had founded a company with £1 million in capital and negotiated a formal concession with the Colonial Office. Consequently, Rutenberg often referred to the dangers of becoming "the dog in the manger," that is, although he did have a statutory right of exclusivity with respect to electrification in Palestine, he feared a situation in which he would repeatedly have to beat back competing bids with reference to his exclusive rights under the agreement with the British, while being unable to provide power himself. It was liable to result in a rapid and radical shift in fortunes, as his project went from being part of the solution to being part of the problem.[32] This might, as Rutenberg feared, impel the British to ignore their commitments and, in the interest of keeping the peace, authorize the construction of temporary plants throughout Palestine.

This, as Rutenberg well knew, would "undermine the monopoly I have obtained."[33]

The district governor of the Northern District, which included Haifa, was the most vocal of those expressing frustration at the slow pace of electrification. The governor had been in touch with Rutenberg as early as December 1921, and he was under considerable pressure from the residents of the city, having received "several semi-official inquiries" and "tentative proposals" for electrification schemes. Were it not for Rutenberg's concession, the governor claimed in late 1922, "I have very little doubt that the town would by now have been properly lighted."[34] Yet it was not only the legally binding agreement with Rutenberg that prevented the governor from supporting any of the local schemes regularly crossing his desk; he conceded that he also believed that Rutenberg's scheme was the superior one, promising abundant energy at rates far below the other offers.[35] In the event, as we saw in chapter 3, Rutenberg solved the problem by playing on the British desire for cheap electricity, building powerhouses in Haifa and Tiberias long before there was a legal framework for them. That solution came with problems of its own.

The initial assumption was that finalizing and signing the concession would be a relatively easy matter. After having submitted his updated proposal for the first and second hydroelectric stations on the Jordan River, Rutenberg traveled to London in November 1923 to present his plan to the government, with the expectation that the concession would be signed during his stay in London.[36] His hopes were soon dashed, however. Over the course of the next two and a half years, both the British and the power company representatives would raise uncountable objections over matters large and small. They included the maximum and minimum allowable limits of the Sea of Galilee, customs and import duties, the concession's applicability to Transjordan, rights and procedures for compensation to landowners for destruction caused by construction, and water rights for the company and small water users in the area around the Sea of Galilee and the Jordan and Yarmuk Rivers. In the meantime, Rutenberg's legal position with respect to Haifa and Tiberias, as well as the Jaffa network, which soon stretched well beyond the borders of the district, grew increasingly precarious. In Haifa and Tiberias, as we saw in chapter 3, Rutenberg solved the matter by the legally questionable means of unofficial agreements with the municipalities. In the rest of the country, he held his breath and hoped for the best.

THROWING TRANSJORDAN INTO PALESTINE

The experience of reading the thousands of pages of documentation relating to the negotiations between Rutenberg and the British over the Jordan concession, the innumerable meeting minutes detailing painstaking negotiations, and the endless series of engineering reports, surely approximates the pain and frustration experienced by the negotiating parties themselves. The readers of this book will be mostly spared. One issue, however, deserves closer attention: the question of the concession's applicability to Transjordan. When, in September 1921, the British gave preliminary approval for an exclusive countrywide concession, the area was still all one Palestine.[37] A year later, however, the League of Nations acceded to a British request to divide the territory in two and designated the Jordan and Yarmuk Rivers as the boundary line. Beyond that line, the British-backed endeavor to establish a Jewish national home would not go.

The division of the area into Palestine and Transjordan, with the Jordan and Yarmuk Rivers marking the border, effectively transnationalized Rutenberg's works. The powerhouse itself would be located east of the Jordan River and south of the Yarkon, in Transjordanian territory, but the works in their entirety would straddle both sides of the river along an eight-mile stretch. But electric grids rarely respect political boundaries. More often, they make them. In this instance, the powerhouse's location straddling the border stood in tension with the natural inclination of electric grids to sprawl. The border demanded separation and sanctification of national space, while the grid demanded endlessly multiplying connections, the production and expansion of a unified technological zone.[38] This led to a curious situation, described by one observer, "Physically you are in Transjordan; politically, you are in Palestine."[39] Many scholars, including Mary C. Wilson, have noted that "the carving up of geographic Syria [into Syria, Lebanon, Palestine, and Transjordan] ... severely disrupted normal economic, political, and social ties wrought by centuries of habit and usage."[40] But the act of division did not merely bring about the end of such ties; it also created the conditions and need for new ones.

The border between Palestine and Transjordan came about in an attempt to square the three territorial commitments that Great Britain had made during World War I: to Sherif Husayin of Mecca through the Husayn-McMahon Correspondence, to the French under the Sykes-Picot Agreement, and to the Jews in the Balfour Declaration. Palestine and Transjordan were constituted in fulfillment of those pledges, so the British maintained, by supporting the creation of a Jewish national home west of the Jordan River, and an Arab kingdom east of it. The creation of Transjordan also served British imperial aims as a "buffer and bridge," between the French and Saudis, and Iraq and Palestine, respectively.[41] Mostly, however, the making of Transjordan was the product of a series of interim ad-libs, starting with the feckless territorial designation whence Transjordan derived its name.

After a few initial stumbles, the British built a workable administration for Transjordan around Sherif Husayn's second son, Abdullah, who arrived in the country from the Hijaz in late 1920. The British seized on his arrival and his claim to the Syrian throne by making him the figurehead around which to build a centralized Transjordanian authority. Britain formalized the decision to split the territory designated for the Palestinian mandate in two at the Cairo Conference in spring 1921. Palestine would be governed from Jerusalem, under a British mandatory apparatus, while Transjordan, though under the authority of the same high commissioner, was given its own local government at whose head was Abdullah, designated as the emir.[42]

The actual borders were not set until a year later, when the Palestine mandate charter was about to be submitted to the League of Nations for approval. The general idea was clear enough: Transjordan, being *trans*-Jordan, would be composed of all the territory of mandatory Palestine that lay beyond the Jordan River. For a couple of months starting in summer 1922, British officials in London and Palestine carried on a "fevered correspondence" to determine the exact placement of the border.[43] The most significant issue, as we already saw, concerned Rutenberg's scheme, for which the British were at pains to clear a path even before they had formally granted any concessions to him. The single significant departure from the notion that the Jordan River would constitute Transjordan's western border was motivated by Rutenberg's proposal: the inclusion of the Samakh Triangle in Palestine.

Samakh is bounded by the shore of the Sea of Galilee in the north, the Jordan River in the west, and the Yarmuk River in the east. The confluence of the two rivers eight miles south of the Sea of Galilee forms the triangle's southern tip. The proposed site for the powerhouse lay just below the triangle, east of the Jordan and south of the Yarkon (see map 1 in chapter 2). As we saw in an earlier chapter, in the interest of easing Rutenberg's way, the high commissioner had determined that the Samakh Triangle should be included in Palestine. For reasons that remain unclear, however, he stopped short of giving Rutenberg all the land he needed, and indeed the powerhouse itself, whose intended location was immediately southeast of the triangle, ended up on the Transjordanian side of the border.[44]

Only later did the British realize that by giving Rutenberg a concession for both Palestine and Transjordan, and making border adjustments on the basis of his scheme, they had created a situation rife with complications.[45] A few people sounded warnings at the time, including the Opposition MP W. Joynson-Hicks. His letter to *The Times* in summer 1922, calling attention to the wide scope of Rutenberg's concessionary rights, set off protests in Transjordan, especially among Bedouin and landowners on the Jordan River's eastern bank. They sent a formal petition to the authorities, which motivated the chief British representative in Amman, Harry St. John Philby, to bring the matter up with the high commissioner: "There would appear to be a conflict of ideas," he wrote, "between, on the one hand, our intention with regard to maintaining our promise to support Arab independence in Trans-Jordan, and on the other hand, the terms of the Rutenberg Concession." Philby proposed to solve the situation by canceling the concession in Transjordan.[46]

Not until 1924, however, when the British began negotiations with Rutenberg over the final terms of the concession, did they start to grasp the full significance of the matter. Deviating from "the governing fact that what is across the Jordan is Transjordan," and thus "throwing part of Transjordan into Palestine," as Major Hubert Young of the Middle East Department put it in a report that year, was an "extreme step," liable to cause serious difficulties.[47] Young blamed the Palestine Government, and especially the high commissioner, for what soon came to be regarded as a serious blunder. Despite the border adjustments, he noted, a large portion of the Jordan works would be located in Transjordan. In Young's view, the high commissioner either should have recommended no diversion at all from the natural border of the Jordan River or should have suggested a far more extensive adjustment, so that all of Rutenberg's works had ended up in Palestine.[48]

In May, the legal councilor of Palestine, Norman Bentwich, wrote a report that effectively shifted the blame back onto the Colonial Office. It did so by bringing another border issue into relief. The problem, according to Bentwich, was not that the relatively small area of the Samakh Triangle had been included in Palestine but that Rutenberg's concession had been made to apply to Transjordan as well as Palestine. And this, he noted, had been the doing of the Colonial Office.[49] The Rutenberg concession thus became the subject of two boundary disputes. One issue was the inclusion in Palestine of the Samakh Triangle, which, as Young had pointed out, "by any ordinary acceptance of the term," ought to belong to Transjordan.[50] The second issue was that Rutenberg had been granted an exclusive concession not just for the territory of the Jewish national home but also for that part of Palestine that had subsequently been excluded from it.

The Transjordanian issue was one of the most intractable sticking points in the concession negotiations. During that time, a number of solutions were floated. The inhabitants of Transjordan and the British administrators there favored giving Samakh back to the emirate. In return, Transjordan would be more amenable to accepting the electricity concession's application to its territory.[51] Others, including Rutenberg and some at the Colonial Office, preferred a more clear-cut separation between the two territories, while ensuring the viability of the electrification venture. They suggested further expanding Palestine's borders eastward, out to the foothills of the Transjordanian highlands, while canceling the concession in the remaining parts of Transjordan.[52]

In the event, the status quo prevailed. The current borders and scope of the concession were left unaltered. To appease the Transjordanians, a clause was added to the concession whereby the farmers and Bedouin living close to the Jordan and Yarmuk were guaranteed a minimum amount of water annually, which they would either draw from the rivers or receive from the PEC. The concession, for its part, would remain valid in Transjordan, but any electrification activity there would be subject to approval by the Transjordanian government. Finally, the British persuaded Rutenberg to agree that the towns of Amman, Salt, and Karak would be exempt from the exclusivity clause of the concession, and thus free to carry out their own electrification schemes.[53]

BRIDGE OVER TROUBLED WATERS

Parallel to the concession negotiations with the British, the power company was busy seeking to acquire the land necessary for the hydroelectric works. In the event, the company's definition of "necessary" included a settlement enterprise. In the summer of 1921, Rutenberg began lobbying the British to sell him upwards of one hundred thousand dunums around the works. As he stated in his requests, he needed enough land not just to house the works themselves but also to ensure a measure of protection for them. This he envisioned doing by creating a protective buffer of Jewish settlements.[54] As

Rutenberg explained to the British: "The only and safest form of protection for this station, the nerve centre of the economic life of Palestine, can be afforded by the settlement on surrounding lands of selected agricultural workers who shall be self sustaining, and in possession of sufficient arms for self defence against raids."[55]

Rutenberg's request was initially met with sympathy at both Government House in Jerusalem and the Colonial Office in London. But as Chief Secretary Wyndham Deedes's report from August 1921 made clear, there were political considerations militating against granting a government lease of land to Rutenberg. His project was receiving negative attention in the Arab press, and a particular focus of the opposition was on the prospects of Rutenberg's laying claim to land.[56] Instead, Rutenberg and the British found another solution. Citing an Ottoman precedent, the Palestinian government had taken possession of some 250,000 dunums in the area around Jisr al-Mujami', in the Beisan area.[57] The British had indicated that the PEC would be able to lease as much as 50,000 dunums. But the British first worked out an agreement with the inhabitants of Beisan, the Ghor Mudawara Agreement, whereby they were allowed to buy individual parcels of land for a nominal fee.[58] The offer proved so popular and the individual plots so generous that when the government had completed most of its transactions, there was little land left for the power company.[59]

The Beisan land agreement was met with outrage by the Zionists. Chaim Weizmann, president of the Zionist Organization, denounced it in the strongest terms, accusing the high commissioner personally of having "condemned" the land to "stagnation and sterility" by carelessly giving it to "certain Bedouin" who "frequented" the area.[60] By the same token, the political leadership of the Palestinian Arabs regarded the move as an important victory in the fight against Zionism.[61] The view that this was bad for the Jews and good for the Arabs persists in the scholarly record.[62]

But Rutenberg, on learning the news from the chief secretary of the Palestine Government, was not overly concerned. He met with representatives of the main Zionist settlement organizations, the Palestine Jewish Colonization Association (PICA), and the Jewish National Fund (JNF) and convinced them to allocate £E45,000 to land purchases in the area around the works.[63] Representatives of the power company then established contact with the Nazarene Tuma al-Sabbagh, who served as the agent for the land acquisitions.[64] In accordance with the plan, PICA then leased one thousand dunums to Degania, the Jewish settlement group.[65] Rutenberg was intimately involved in this process and demanded the right of prior consultation in all settlement endeavors around his works. In effect, then, far from being a harsh blow to Zionism, the British land policy was a boon to the PEC and Jewish settlement in the area. Rutenberg was pleased, though not with the pace of the proceedings; he repeatedly wrote to the government with calls to expedite the process of deciding claims and registering lands.[66] All in all, of the fifteen thousand dunums he sought to acquire around his works, twelve thousand dunums were secured in this way.[67] The rest he procured through expropriation, which he was able to do with ease after the Palestine Government gave his company the status of a public utility body in January 1927.[68] This also enabled the government to claim, as it did in the report on the "disturbances" of August 1929, that with a few exceptions, such as the Kabbara concession, "no State land has been provided for the purpose of close settlement by the Jews." Yet the report went on to acknowledge that Jews owned "more than 1,000,000 dunoms (nearly 400 square miles) of land."[69]

As Rutenberg was completing the land transactions on the Palestinian side of the border, he began to look eastward across the river. In early 1926, he wrote to the Palestinian government, asking it to press the Transjordanians to follow Palestine's lead by selling off the lands around Jisr al-Mujami' to the inhabitants, in order that his company might buy it in turn.[70] Yet, by the end of the year, Transjordan had made no efforts to begin registering the lands. The PEC again turned to the Palestinian government.[71] Finally, the chief British representative in Amman, Colonel Henry Cox, arranged a meeting between Rutenberg, Emir Abdullah, and the chief minister of Transjordan, Hassan Pasha Abu al-Huda, to discuss the lands Rutenberg required. Rutenberg had in mind some thirteen hundred dunums belonging to Delhamieh village, and he went to the meeting assuming that the same arrangement would be made in Transjordan as had already been done in Palestine. The chief minister, however, surprised Rutenberg by informing him that the Transjordanian government did not recognize the inhabitants' right to the land and considered it state domain. Instead, he offered to sell the land to the power company at £E3.5 per dunum. It was more than twice the price the power company had paid the landowners on the Palestinian side. Rutenberg nevertheless agreed immediately. In subsequent meetings in January 1927, many of which were with the director of the land registry and future prime minister, Tawfiq Abu al-Huda, it transpired that the Transjordanians were willing to sell much more than just the Delhamieh

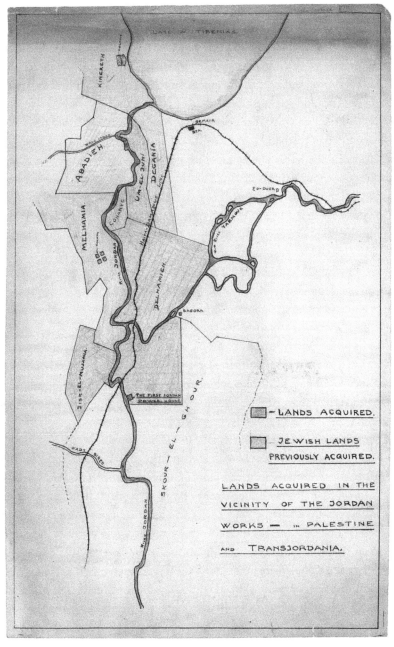

FIGURE 6. Map of land needed around the Naharayim powerhouse. Courtesy of the Israel Electric Corporation Archives.

lands. They were offering up to 6,000 dunums.[72] This was vastly more than the roughly 600 dunums Rutenberg needed for the works themselves. On January 26, the Transjordanian Executive Council passed a decision—which the emir later passed into law—that circumvented the Ottoman-era proscription against land sales to foreigners, specifically for the purpose of being able to sell the land to the power company.[73]

On February 22, 1927, Yakutiel Baharaw signed the deeds of sale, which were recorded in the Irbid Land Registry. The total cost, including taxes and fees, was £E17,145. This was by far the biggest land deal the power company had ever concluded, since the lands on the Palestinian side had mostly been paid for by the Zionist settlement organizations. It was also the only significant land deal that any Zionist group was able to conclude in Transjordan during the mandate period. Attending the transaction was an agreement, signed on March 5, 1927, between the Transjordanian government and the power company. According to it, the Transjordanian government undertook to clear the lands from all cultivators within a period of two months from August 31, 1927, and designate £E3,700 of the sale amount as compensation to be paid out to the inhabitants.[74]

News of the land transaction was picked up by several Arabic newspapers, which were uniformly critical. The Jaffa-based *Alif Ba* ran a screed under the headline "All That Is Hidden Is Revealed," claiming that the Transjordanian government had given Rutenberg the six thousand dunums as a "present" and branding the transaction as a form of "colonialism" that had paved the way for the expansion of Jewish settlement into Transjordan.[75]

Rutenberg was very pleased with the unexpected windfall. It was the first achievement to come out of the power company's efforts to build bridges across the Jordan. It saw the beginning of many beautiful friendships, including an especially close one with the land registry director and later prime minister, Tawfiq Abu al-Huda. Another special relationship developed between Rutenberg and Emir Abdullah, who by all accounts became very close. For the rest of their lives, they would send each other birthday greetings, and the emir often summered with his family in Rutenberg's white villa at Tel Or. When Rutenberg fell ill in his later years, Abdullah sent him several get-well cards.[76]

The agreement also gave Rutenberg the confidence to move the location of the powerhouse slightly farther into Transjordanian territory, for an estimated 20 percent increase in generating capacity. The successful land transaction, and the relationships that the company had formed in pursuit of it, whetted Rutenberg's appetite. In a report to the board of directors later in the year, he held out hope that the PEC would be able to acquire up to three hundred thousand dunums of land along both sides of the Jordan River between the Sea of Galilee and the Dead Sea. He also contemplated a large irrigation scheme that would make the entire Jordan Valley bloom, through cultivation of oranges, sugarcane, and bananas. Finally, the deal imbued Rutenberg with great optimism for the future of Arab-Jewish relations in Palestine and the region. Rutenberg's faith in creating harmonious Jewish-Arab relations through personal relationships motivated him to organize evening courses in Arabic and lectures on Arab customs for his employees. As he explained, "My experience justifies me in stating that the relations between the two sections of the population can be easily improved were the Arabs approached by the Jews in a friendly spirit."[77]

The land deal was the PEC's most notable success in its attempts to bridge the Jordan, but it was not the only one. The next two years would prove to be the high point in the power company's relations with the Hashemite emirate. The Transjordanian government passed several pieces of legislation that were tailor-made to facilitate the power company's operations in the country, including the Foreign Corporation (Immovable Property) Law of 1927. In January 1928, an ordinance was promulgated in the *Official Gazette* that validated the PEC concession, and later in the year the company was registered at the Ministry of Justice, with the diligent assistance of the ministry superintendent, Abdullah Nusayr.[78]

The power company had managed what no other Jewish or Zionist body had: it had bought a substantial plot of land across the Jordan River, in the area that by all accounts was out of bounds of Zionist activity. As soon as the PEC had concluded its purchase of the six thousand dunums, Rutenberg contacted PICA, offering to sell part of the land to it to be used for Jewish settlements. At the time Rutenberg contacted PICA, the power company was in the process of building a workers' camp to house the construction crews for his electrical works. But as soon as the most intense construction phase had been concluded, the workers' camps would be available to Jewish settlers. In Rutenberg's letter to PICA, he stressed the site's advantages: "The colonists who will live in these houses will be, of course, in exceptionally good conditions, having at their disposal electricity and water as much as they like and very cheap." He concluded the letter by saying, "It is desirable to use this opportunity, which I have with such difficulties obtained, to colonise Jews in Transjordan territory."[79] Maintaining good relations, Rutenberg insisted to

PICA, required "that Jews should be in Transjordan." So confident was Rutenberg in those relations, he wrote, that he no longer desired to change the border at all.[80] And why would he, if the border for all intents and purposes could be ignored? Rutenberg was not alone in regarding his achievement as significant. A few years later, at the fourth convention of the Histadrut, David Ben-Gurion, the secretary-general, gave a long opening speech, a significant portion of which was devoted to the Jordan works: "The plant became the first breach [*pirtza*] in the artificial partition that was carried out by the mandatory government of the land [of Israel] into two parts.... At Naharayim the first Hebrew position [*nekudah*]—the first and not the last—was established in Transjordan, situated under the authority of Emir Abdullah."[81]

SIGNED CONCESSION

By 1925, after more than a year of intense negotiations, both the British and the power company were growing desperate to finalize the concession. The British were motivated, most immediately, by their realization that the ever-expanding electric grid in Palestine was increasingly breaking out of the confines of the only concession as yet in force, the one for the Jaffa District. As mentioned earlier, there were three areas not covered by the concession. First, there was the Tiberias powerhouse, which the British had agreed to as a means of breaking the Arab national opposition. Second, there was Haifa, where the British had been taken in by the prospect of cheap electricity for their railway workshops. Finally, there was the expansion of the grid in Jaffa, which by the second half of the 1920s was growing beyond the borders of the district in all directions.[82] In early 1925, before the Haifa and Tiberias powerhouses were operational, Rutenberg had rejected the idea that a special agreement should be worked out between him and the British in order to legally cover those powerhouses. Instead, he wanted to proceed as quickly as possible with the countrywide Jordan concession. This way, he would not risk holding up the larger project, and he might even press the British into expediting approval of it.[83] It was a characteristically high-risk gamble, since it also meant that the power company was expending huge resources on projects with no legal sanction.

Whereas Rutenberg charged into this legal no-man's-land with eyes wide open, the British were characteristically late to the realization of the fine mess they had gotten themselves into. In March 1925, Samuel wrote to Colonial Secretary Leo Amery. He stressed the importance of signing the Jordan concession quickly, since there was "no legal authority sanctioning the actions of the Company in breaking up streets and laying wires etc." In addition, the Palestinians were growing increasingly restive, according to Samuel, and it would be preferable to deal with the Arab opposition on a firmer legal footing. Finally, Samuel concluded, four and a half years had passed since the agreement was signed to grant the concession, and the PEC had been formed two years ago. "I cannot too strongly urge the desirability of definitely confirming the Concession and validating it by the necessary legislation as early as possible," the high commissioner wrote.[84]

The British were also aware that as long as the concession was not finalized, elements within the Palestinian Arab community would seek to animate the masses in opposition to it. This was true as well for the government's domestic critics. Under the heading "The Rutenberg Concession: Court Finds a Right Wrongly Granted," the *Daily Mail* lambasted the government's dealings with Rutenberg: "It is well to mention that a number of genuine Arab endeavours to utilise water-power, such as Nablus Municipality scheme, have been prevented or forbidden in the interest of the Rutenberg scheme."[85] In another article in the *Daily Mail*, J. M. N. Jeffries assailed the concession as "against the will of the people of Palestine and at the expense of the British taxpayer."[86] Finally, the agreement made with Rutenberg in 1921 expressly prohibited any other agent from undertaking any form of electrification during the period of negotiation. For all these reasons, it appeared that the only possibility of seeing Palestine electrified was through an agreement with Rutenberg.[87]

Rutenberg, for his part, was on the verge of a nervous breakdown. The pressure on him was enormous, having already spent £E300,000 on the Jordan works with no firm guarantee that the concession would ultimately fall to him.[88] In late September 1925, as the negotiations dragged on with no end in sight, he complained to Colonel Symes: "My dreams of taking a holiday have, of course, vanished and I have to remain in this gloomy and bleak London atmosphere until things are settled. My only consolation are the Pavlova ballets."[89] In short, there was no going back for either the British or Rutenberg. For better or worse, they were stuck with each other.

By the fall and winter of 1925 and at his wit's end, Rutenberg began to make ever greater concessions. By that point, the pressure on the Middle East Department at the Colonial Office from Rutenberg and his associates, including his London lawyer, was unrelenting. In the fall of 1925, they took to calling up Undersecretary of State John Hathorn Hall at the Middle East Department at two-hour intervals asking for the concession to be signed.[90] In the final two months before the concession was signed, Clauson at the Middle East Department went into hiding to avoid the power company's insistent entreaties.[91]

On March 5, 1926, the "Concession for the utilization of the waters of the rivers Jordan and Yarmuk and their affluents for generating and supplying electrical energy" was finally signed.[92] The concession included an array of technical clauses. It stipulated maximum tariffs to be charged in different sectors. The Transjordan issue was resolved as described earlier in this chapter. It also recognized the Palestine Electric Corporation as a "public utility body under Government control," which entitled the company and its property special protection. In the quiet period of the late 1920s, this may have seemed a minor point. But a decade later and for a decade to come, protecting the grid, first from Arab rebels and then, during World War II, from German aerial bombardment, would prove a significant challenge.

CONSTRUCTION AT TEL OR

In 1882, the settlers at the newly founded Rishon Letzion settlement found themselves in dire financial straits. They decided to sell some land and summoned a surveyor. The surveyor, whom the settlers had engaged once before, found their holdings fifty dunums smaller. "How can you possibly get a smaller figure now than you did the last time?" asked one of the residents. The surveyor's answer, as recorded in the journal of one of the residents, Chaim Chissin, left the inhabitants mystified: "Look, it was summer then, and it is winter now."

> "What difference does the winter make?"
> "Our Eretz Yisrael has a way of expanding during the summer and shrinking during the winter," the surveyor replied, deep in thought.
> "We're lucky it shrank so little," the settlers muttered, dissatisfied with this new characteristic of the Holy Land.[93]

This sort of natural mystery was precisely what the Zionist project, steeped in European modernism, was fixing to dispel. The power system embodied this effort and put it on wide display. The transformation of the area around the Jordan-Yarmuk confluence, which was given its Hebrew name, Tel Or, in the 1920s, was central to that effort. In the Hebrew press, the work and the site's transformation were portrayed as the embodiment of the transformative impact of Zionism. Previously known by its old Arabic name, Jisr al-Mujami', the place was poised to undergo a radical transformation under a new name. The Arabic denoted Oriental inscrutability; the new, Hebrew name heralded a parting of the enchanted veil. According to one account in the Hebrew press, "The waters of the Yarmuk and the Jordan are to obey the word of Rutenberg and flow together tamely and obediently into channels mapped out for them by the Palestine Electric Corporation."[94] Meanwhile, "the Arabs" looked on and whispered among themselves of "black magic."[95] As with the more famous Tel Aviv, the name Tel Or contained a reference to the ancient past in *Tel*, meaning archaeological mound. *Or* (light), like *aviv* (spring), signaled regeneration and a new dawn. Thus, Tel Or, like Tel Aviv, was a reference to the title of Theodor Herzl's novelistic representation of his Zionist vision in *Altneuland*, a work that stands out in the Zionist canon for its unmitigated technological optimism.[96] Work began in the fall of 1927. The prominent Hebrew writer Haim Shorer documented the start of construction at Tel Or in the pages of the labor Zionist newspaper *Hapoel Hatsair*. Under its new name, Shorer wrote, a "large operation mercilessly banishes all shades of wilderness and mystery."[97] At Tel Or, as elsewhere, the Hebrew pioneers were engaged in what another writer, Yosef Baratz, possibly recalling the Scottish essayist Thomas Carlyle, described as a "war with nature, a war with the circumstances."[98]

Rutenberg's hydroscheme was both a product of this technocapitalist vision and an active shaper of it. We saw that the power system, like all human artifacts, contained political and other agendas. The system embodied assumptions about science, technology, and economic development, which it passed on to the territorial entity it participated in producing. We saw that hydropower was at the center of the process that made Palestine legible and thus accessible to the modernizing British colonial state and Zionist settler colonialism.[99] Most obviously, as we saw, the plans for the power system helped set the borders of modern Palestine.

These things together produced a means of grasping Palestine as a national whole, through a process that also disenchanted the land and refashioned it as a free-market landscape.[100] But the power system was also a producer of Jewish national space. In both the Hebrew and British press, Rutenberg's work at the projected site of the powerhouse was the subject of a great deal of interest. Both, as we have already seen, saw in Rutenberg's works the embodiment of the promise of Zionism: Jews would bring their industry and technological know-how to bear on the development of Palestine for the benefit of all its inhabitants, the British Empire, and the world.

The effort to make the land legible was at the heart of the electrification venture. This critically involved transforming a complex Palestinian reality into numbers. We already know about the great efforts Rutenberg and the power company invested in surveying the land, as part of the general effort to project a scientific attitude to its undertaking in general. Immediately upon the signing of the concession, the power company sent survey teams to the area of the future power station. They measured water flows and quantities of the rivers, conducted borings of the riverbed and shores, and carried out topographical surveys around Jisr al-Mujami' and elsewhere. The company produced detailed reports, which included surveys "with complete accuracy as to detail" to be plotted to a scale of 1:2000 for the length of the Jordan and Yarmuk Rivers.

Within weeks of signing the concession in March 1926, the area was swarming with civil engineers, hydraulic engineers, cartographers, and their security details.[101] After some initial troubles with the local population and the Jordanian police force, Emir Abdullah of Transjordan personally sent an officer to accompany the surveyors.[102] The company dispatched sanitation teams to the region tasked with fighting malaria.[103] At the PEC's expense and under Government Health Department supervision, sixty people worked on both banks of the Jordan to destroy mosquito breeding grounds.[104] A year later, however, it was still impossible to work after dark because of the heightened risk of infection, and malaria continued to plague the area for years to come. Nonetheless, the area was soon abuzz not just with mosquitoes and surveyors but also with workers. By May 1927, there were more than four hundred workers at the construction site, with the expectation of soon bringing in another two hundred.[105] These surveyors, public health experts, engineers, and manual laborers constituted the advance force in the Zionists' war against nature.

As construction at Tel Or got under way in 1927, several articles appeared reporting on the abilities of Jewish workers, who were able to handle the technical tools and carry out their tasks in spite of the inhospitable climate. An essential component of assessing the value of the hydroelectric works was the kind of productive changes they made in the land. In 1929, *The Times* reported that while as many as 90 percent of the Arab inhabitants of the area suffered from malaria, the Tel Or workers' camp had recently reduced its proportion of afflicted from 17 to 1.4 percent. "What has always been a plague spot is now a place where men enjoy real health and happiness; science and sensible feeding has done this and unremitting attention to details."[106]

This was a trope that was echoed all over the world throughout the first half of the twentieth century, from Southeast Asia and Africa to the South Pacific and North America. According to David Ekbladh's account of the construction of the Tennessee Valley Authority, Arthur E. Morgan, the first chairman, had a "moral vision for an 'integrated social and economic order' to the valley." Morgan advocated "actual changes in deep-seated habits, social, economic and personal," through "conscious deliberate effort." Morgan thus proposed an ethical code for the employees of the TVA that included proscriptions against "intemperance, lax sexuality, gambling, and . . . habit forming drugs." Morgan sought to "use the program to create new social relationships."[107] In Ekbladh's telling, however, Morgan's efforts represented a "new strand of reform." But precisely the same kinds of ideas circulated around Tel Or and the electrification endeavor in Palestine ten years earlier. Nor was Palestine the first of its kind. Similar ideas had already been circulating widely in the non-Western world, following the movements of Western-trained engineers.[108]

By the fall of 1926, Rutenberg had secured a wealth of funding, including the participation of British General Electric, despite some grumblings in Zionist circles about letting in "'goyish' capital." Sir Hugo Hirst, the GE chairman, took up a position on the PEC's board of directors.[109] The PEC secured another important source of funding through a British government loan of £250,000 through the Trade Facilities Act.[110] The money would come in handy. After less than two years of construction, the company had spent £E90,000, with the expectation of expenses in the near future amounting to an additional £E210,000.[111]

By the end of 1928, the Yarmuk River had been diverted to its new riverbed, and the work crews had begun pouring the concrete dam blocking the old bed. The expectation was that it would be completed early the following year, after which they would begin construction of the head and tail races. The workers were busily concreting the pressure basin and powerhouse foundation, and the excavation work for the Jordan-to-Yarmuk canal was begun. The work site was visited by numerous dignitaries, including the high commissioner, Parliament members, and Emir Abdullah.[112] "At the moment," the *Financial Times* reported in May 1930, "although the Jordan and Yarmuk still flow unchecked in their immemorial beds, everything is approaching a stage when their waters will be harnessed and Palestine will be enjoying—at a far cheaper rate than at present and with a doubled potentiality of supply—the luxury of self-made electricity." But like many before it, the newspaper also wondered: "Is it conceived on lines far too vast for the needs of Palestine? For, at the present moment, Palestine is barely ready even for the half capacity output of the first stage."[113]

In late February 1930, company engineers successfully tested the generators at the Jordan station, which had been installed in the previous year, and the Haifa–Tel Or, sixty-six-kilowatt high-tension line, which had also been completed, was successfully put under tension for the first time in March.[114] A year later, the installation was all but completed. The company expected to begin a full systems test by the beginning of February 1931.[115]

DELUGE

Nature, however, would turn out to be more difficult to know and control than expected, and in the winter of 1931 the power company, having staked its authority on its ability to predict nature's regularities and bend them to its will, suffered a serious setback. On the evening of February 13, 1931, a flood "of extraordinary violence" descended on the works. It did not come as a complete surprise, and the sluice gates had been left fully open in anticipation. Yet the spillways of both the Yarmuk dam and the Jordan-Yarmuk canal overflowed. At 7:45 p.m. the fill of the headrace canal near the pressure basin gave way. The water tore through the crack, widening it to a hundred-foot rent in minutes and advanced to the next wadi, immediately upstream of the transformer station. Unable to contain the water, the wadi's edges overflowed. The water rose onto the basalt platform above the transformer station, whence it washed down forty-six feet onto the station, bringing a barrage of loose objects in its train, including several giant boulders. The mass dove onto the headworks below, smashing the transformer stations and washing the debris away. By morning, the levels had fallen slightly, receding below the Yarmuk dam sluices. The elevated levels persisted through Saturday and Sunday before the water started to draw back. The retreating water revealed three out of six transformers remaining in place, protected by a mass of stone. Three transformers had been washed away, along with the oil switches, cubicles, isolating links, almost the entire steel structure, pipes, cables, and much else besides. Almost nothing was salvageable. "All the piping of the transformers was broken, the oil flowed out even from the three remaining transformers, water and mud penetrating into them instead," read one somber PEC report. In addition to the damage caused to the transformer station, lighter damage was done along the canal and to the Jordan floodwater outlet.[116]

The repair work was expected to take about four months and cost £P40,000, after the arrival of the new equipment, which would take about eight months. But the real setback was the undeniable and disastrous failure of calculation. According to the company's measurements, the maximum discharging capacity of the rivers was 160 gallons per second, and the Yarmuk dam was built to withstand up to 260. But the flood that hit the works in the evening of February 13 had flowed at a rate of 320 gallons per second.[117] The company suffered a loss of prestige, which threatened to hurt it with respect to the mandate authorities but also, and perhaps more important, with investors. One board member called it "a grave factor in the raising of finance."[118] The problem, in other words, was the flip side of the tremendous power that could be derived from manipulating nature. Precisely because the system became an instrument of Jewish state formation, intimately tied to the core justification of the Zionist project, controlling nature became a political problem of statecraft.[119]

The company began the painstaking work of rebuilding. A year later, the station was once again nearing completion. On June 9, 1932, the powerhouse was inaugurated in a grand ceremony, with many prominent guests, including the high commissioner, the chief British resident in Amman, and Emir Abdullah. The high commissioner praised Rutenberg's work and likened him to electricity poles "that are today being distributed the length and width of the country." Like the poles, he said, Rutenberg had cut a straight path through all obstacles.[120]

DIVISION OF LABOR

As in most areas of high policy in Palestine, the Palestinian Arabs shone with their absence in the work on the power station, even when they were physically present. Though often talked about, they were rarely consulted. An article in *Popular Mechanics* from 1930 suggested that "the benefits to be derived from the introduction of machinery should also reach the Arabs, who are expected to develop into good laborers."[121] Such articles both reflected and reinforced the power system's role in equating Zionism with economic development and civilizational progress. This, of course, was an image that the power company itself was actively promoting. In Rutenberg's much-publicized speech at the inauguration of Naharayim in 1932, he described the powerhouse as a "permanent source of light and power, of culture and civilization."[122] By the same token, the division of labor that this implied relegated the Arabs to an afterthought, the mostly reluctant beneficiaries of Zionist productivity. Arabs were represented at nearly all electrical work sites in Palestine, a fact that was carefully documented and publicized by the electricity company itself, as proof of the productive and pacifying effects of the work. Yet there were no Arabs in skilled positions, and the depictions of Jewish and Arab workers, respectively, perfectly capture the tension between, on the one hand, the universal and egalitarian principles in which the venture was couched and, on the other, the division of labor that it engendered, which in turn produced socioeconomic and ethnic differentiation.[123] As an illustration, the *Blue Book for Palestine* from 1932 listed the wages for various professions. It distinguished between "European" and "Asiatic" labor (meaning Jewish and Arab, respectively). According to its figures, there were no "Asiatic" electricians of either the first or second class, which also happened to be two of the highest-paid occupations on the list.[124] "Electricity has two names in the Holy Land," *Popular Mechanics* informed its readers. "One is 'hashmal,' mentioned in Ezekiel . . . and the second name is 'rutenberg.'"[125] Unsurprisingly, *kahraba*, the Arabic word for electricity, was not part of the popular science vocabulary.

The division of labor that these efforts participated in producing and maintaining had far-reaching consequences, and was replicated in other sectors of the Palestinian economy. One notable example was the undertaking to construct a deep-water port in Haifa. It had been on the mind of British planners since 1915 and the publication of the de Bunsen report, which identified the town as a suitable location for a commercial port.[126] During the brief military occupation of Palestine from 1917 to 1920, General Edmund Allenby had been a fervent supporter of the scheme. The British engineering firm Randall, Palmer, and Triton was hired to consult on the plan in 1922, but lack of funds and political difficulties delayed construction until 1927. When construction finally began, the British encountered a delicate problem over the question of differing wage standards for Jewish and Arab labor, given the disparities in costs of living. The solution, proposed by the high commissioner in a letter to the colonial secretary in April 1929, was simple: give highskilled and thus high-paying jobs to Jews, and low-skilled and thus low-paying jobs to Arabs: "I have reached the conclusion that the practical solution of the question of the employment of Arab and Jewish labour is found in the assignment of certain parts of the work to Arab labour and other parts to Jewish labour. For example, the quarrying of stone would be a matter principally for Arab labour while the manufacture of concrete blocks would be a matter principally for Jewish labour. The economic conditions of labour would thus be scaled on an Arab normal fair wage and a Jewish normal fair wage according to the nature of the works."[127]

Managing issues of race through labor division appears to be a universal feature of technocapitalism, and most starkly so in colonial settings. Robert Vitalis and Myrna Santiago have shown, in the context of the Saudi Arabian and Mexican oil industries, respectively, how racial hierarchies were maintained by creating ethnic homogeneity within labor categories. In both cases, white Americans performed the highest-skilled labor, therefore enjoying the highest status and salary, while indigenous groups served in menial capacities. In Palestine, as in Saudi Arabia and Mexico, the practice was justified with reference to the racial suitability of different racial groups for certain kinds of work.[128]

STAGING AND PERFORMING ZIONISM AT TEL OR

It was not only in the European or colonial imagination that Naharayim took on a metonymic function for Palestine as a whole, and where the particular Palestine that was created was bound up with the production of Jewish national space.[129] In the Zionist imagination, too, Naharayim had tremendous significance as the technological system that provided the Jewish national imaginary with a material correlative. The chief significance of the power system stemmed from the fact that it was the first of its kind, a material structure encompassing the entire territory, that is, a national system. Before Naharayim came online in 1932, the three local powerhouses were transmitting electricity by means of local low-tension electric grids that in the main hewed to the municipal boundaries of the three areas, though with some notable spillover. When Naharayim was put into commission in the spring of 1932, it connected to Haifa and from there to Jaffa and Tel Aviv through newly built 66,000-volt high-tension mains. A smaller branchline made its way northward to Tiberias and southward, connecting the agricultural settlements in the Jordan Valley.[130] Rutenberg's original proposal—now twelve years old—had been realized, and for the first time in its history Palestine had a precise national scale that was corroborated by a material reality.

Indeed, the national scale of the project emerged again and again in treatments of Naharayim's opening. In a description of the inaugural ceremony in June 1932, the labor Zionist newspaper *Davar* reported that from the central control room, "power is sent out *to the entire country*."[131] In a speech a few months later, David Ben-Gurion celebrated the accomplishments at Tel Or, which he described as "the first of its kind in Palestine" and deemed to be of "financial and national value of the highest level."[132] As a material reality, the "radiance of Naharayim"[133] projected out to all the large Jewish towns, including Tel Aviv and Haifa, the Jewish settlements along the coast, the Galilee, and the Jordan Valley. Naharayim lighted and powered the burgeoning Zionist industrial complex centered on Haifa and the mineral extraction works at the Dead Sea, run by another Zionist concessionaire, Moshe Novomeysky.[134]

Jewish Palestine was not the first place to attach such significance to a unified electric grid. In Germany just after World War I, for instance, the treasury minister, Wilhelm Mayer, supported an initiative to nationalize the country's electricity supply and unify the system. In addition to a range of technical advantages, Mayer justified the move before the National Assembly as a way of demonstrating to the world the vitality of the new Germany.[135]

As Naharayim produced Zionist space, it also helped produce, organize, and stage a number of Zionism's primary virtues and ideal types. Naharayim came to serve as a microcosm of the Yishuv, or, rather, of what the Yishuv aspired to be. Specifically, in its everyday practices, spatial and otherwise, Naharayim embodied and defined the kind of productive work that lay at the heart of labor Zionism, as well as the kind of Jewish national subject that labor Zionism sought to foster. It may appear counterintuitive to claim that Palestine and the Zionist project were both constituted in technocapitalism's image, given that the dominant strand within the movement was left-wing, if not socialist, labor Zionism. Indeed, by the late 1920s, labor Zionism had become the dominant faction within the Zionist movement, and its model of development became hegemonic through the mandate period and for at least the first three decades of statehood.[136] Consequently, most of the prominent leaders of the Yishuv championed socialism alongside nationalism.

By the time of the Palestine Mandate's establishment, however, labor Zionism had largely shed its socialism the better to pursue its nationalist aims.[137] Although often retained on the rhetorical level, socialist values were effectively replaced by a more hands-on program centered on the concepts of *Hebrew labor* and the *conquest of labor*. The terms originally referred to the labor Zionist aspiration for collective and individual proletarianization, but in the encounter with the labor market in Palestine, especially the abundance of cheap Arab labor, they came to signify the campaign to replace Arab workers with Jewish workers.[138] One scholar defines the "conquest of labor" as "the dispossession of Arab workers in order to take their places" in the interest of establishing "a solid infrastructure for an autonomous Jewish existence."[139] As we will see evidence of later, the campaign was sometimes presented as one in favor of "organized" labor (meaning Jewish labor) against "unorganized" (i.e., Arab) labor.[140] The move away from socialism toward greater ethnonational differentiation indicated in the shifting significance of the "conquest of labor" concept signaled a larger shift within the Yishuv in the 1920s. The Palestinian Zionists moved away from universalistic values based on class struggle to a Palestine-focused effort to create "a self-sufficient and largely labor-controlled Jewish economy in Palestine." According to Zachary Lockman, the effort succeeded in creating a "dynamic motor propelling the development of a self-sufficient Yishuv" in a way that also assured labor Zionism's political dominance.[141]

The hydroelectric station at Naharayim was integral to the effort to construct a new self-sufficient Jewish worker. This had been Rutenberg's stated aim from the outset. In his original proposal from December 1920, he wrote, "Palestine will be Jewish only if the entire work relative to the building up of Jewish life will be carried out by Jewish workers, however difficult or dangerous the work may be, and whatever sacrifices it may require. The rebuilding of Palestine by Arab labour would result in the creation of an Arab and not a Jewish Palestine, irrespective of the amount of Jewish capital drawn in."[142]

In the fall of 1930, the bulk of the construction work was completed, and a large number of the workers were set to leave. To mark the occasion, the power company organized a lavish farewell party and invited all past and present workers along with their families. The list of dignitaries in attendance reads like a who's who of 1930s labor Zionism, and residents from all the nearby colonies flocked to the site. Rutenberg, who could not attend due to his duties as the chairman of the Jewish National Council (Vaad Leumi), sent his greetings in a note full of praise for the workers. He characterized their efforts as "one of the most important conquests of Hebrew labor."[143] Another speaker, a representative of the labor Zionist body Merkaz Haavoda, Berl Repetor, seconded Rutenberg's claim but added that the victory was for *organized* Hebrew labor. The hydroelectric station had concentrated an unprecedented number of Jewish workers in a single place, as many as eight hundred laboring together at one point, presenting them with a unique opportunity to unionize.[144] This was no small thing for the labor Zionist General Organization of Workers in the Land of Israel (the Histadrut for short), for which the centralized allocation of available employment was a key concern in its labor policy and a prerequisite of the political ascendancy that it achieved around this time.[145] These images soon became well-worn tropes in discussions of Naharayim. The prominent labor Zionist leader Ben-Gurion delivered a similar message in his opening speech for the Histadrut's conference a few months later, describing Tel Or as the most sophisticated and efficient work site in Palestine.[146]

Also present at the party was Moshe Shertok of the Histadrut's Executive Committee. Shertok began by elaborating on what the power company's efforts represented: "The works are not just a victory of the Hebrew hand, but also a victory of the Hebrew mind that creates, and the Hebrew will that builds." Naharayim fit squarely into a story of Zionist conquest, he told the audience. It was made possible by the road network built in the early 1920s by what would become the Histadrut's construction arm, Solel Boneh. And Naharayim in turn had advanced the Zionist campaign by providing a setting for labor organizing. Thanks to the work done at Tel Or, Shertok said, there was now an equitable queuing system for work, and formalized procedures for terminating employment. "These works, which provide us with the main arteries of the life of the country, will be a great source of power and light," Shertok concluded.[147] The next person to enter the stage was the famous singer Hana Kipnis. She performed her set and then took requests from the audience, agreeing to sing only "Eretz Israeli songs."[148] In response, according to one report, "the men and women danced wildly."[149]

Naharayim also staged and honed one of Zionism's most cherished ideal types: the technician, and the cult of the engineer, which, as Derek Penslar has shown, served a particularly important role in Zionism for signifying not just "national independence" but also "national existential change."[150] In *Davar*, the official newspaper of the Histadrut, and the most important publication of labor Zionism, a writer profiled the engineers at the head office of

FIGURE 7. Naharayim.

the power company in Haifa. The article described a man who "pores over a map and talks about it with the same enthusiasm as a Haredi Jew would say *ashrei*," referring to a common Jewish prayer. If, the article went on, you were to wake this engineer up in the middle of the night, he would be able to lecture on the ins and outs of the plan immediately.[151] The suggested equivalence between the Orthodox Jew's commitment to prayer and the Zionist's commitment to his technological project scans as a way of articulating new

Jewish-Zionist values. The orthodox Jew of yesterday is the Zionist of today: the engineer as prophet, engineering blueprints substituted for Holy Scripture. Such an attraction did Naharayim hold on inhabitants and visitors to the country that soon after the plant had gone online, the power company took out ads in several Palestinian and British newspapers informing readers that visitors to Tel Or were required to obtain prior admission cards from the head office in Haifa.[152]

CONCLUSION

The signing of the Jordan agreement on September 21, 1921, set in motion a process to grant Rutenberg a countrywide electricity concession, centered on a hydroelectric power station at the confluence of the Jordan and Yarmuk Rivers. Those negotiations took five years, and the concession was finally signed on March 5, 1926. The first "final draft" of the Jordan concession was prepared in April 1924. This was the first of an uncountable number of "final" drafts produced over the course of the following two years. At the midway point between the first "final" draft and the *final* final draft, the government's legal consultants, Messrs. Burchells & Co., commented, referring to the PEC negotiating team, "It really seems extraordinary the number of after-thoughts they have had."[153] Burchells had a point, but the protractedness was a joint achievement, and the British logged no fewer afterthoughts. Sticking points included water levels of the Sea of Galilee, compensation to landowners, the concession's application to Transjordan, water rights, the character of the expropriation clause, taxes, customs duties, and much else besides.[154] What ultimately settled the negotiations in favor of the power company was once again Rutenberg's ability as a systems entrepreneur. Facts on the ground, relationships forged with, among others, British officials, Palestinian businessmen, and the Transjordanian government, and mounds of scientific data all combined to form a politics that allowed for the construction of Naharayim, and then extended its significance to all areas of Palestinian life and politics.

Naharayim was central to the project of making Palestine into a Jewish national home. Borrowing a phrase from Stuart Elden on the closely related subject of territory, Naharayim was "shaped and shaping, active and reactive" to the social, economic, and political world that was being constructed around it.[155] The British approved Rutenberg's scheme, despite some obvious reasons to doubt its soundness and despite the risk of angering the Palestinian population, because of how well it matched the essential components of their vision of colonial development: the "free-market landscape" it presupposed, the legibility it imposed, and the productivization it promised on account of its technological style and robustly national scale. And once the plans were approved and construction began, Naharayim returned the favor by providing a material structure that stabilized the vision to which it owed its existence.

In so doing, it also stabilized a particular set of power relations. The powerhouse was planned and built by people possessing specialized knowledge, notably the engineer, but also the land surveyor and the public health expert. As we may expect, then, while conjured by technocrats and presented as apolitical, it was crucially bound up with ideology and power. In Henri Lefevbre's words, it was "intimately tied to relations of production and to the 'order' those relations impose." Naharayim, the plant itself, was both a producer and an "objective expression" of those relations.[156]

Rutenberg's plant did not single-handedly make electrical Palestine. But Palestine would have been unrecognizable without it. It is a crucial part of the reason why *this* Palestine was the one that ultimately emerged, in competition with alternative visions, some centered on Palestine and others rejecting the entity altogether. Modern Palestine—a precisely defined area nowadays often referred to, ironically, as "historic" Palestine—was constructed quite recently, and on top of other possible Palestines, whose pasts it erased and futures it obviated. Naharayim marked Palestine out as a Jewish national space, and located "the Arabs," in the category of second-order beneficiaries of Zionist development. Through the particular division of labor that Naharayim engendered, the Palestinians appeared to lack entrepreneurial zest and technical skill. Hence, to many at the time the optimistic version of the Palestinians' future, as the *Engineering News-Record* put it, would see them "develop into good laborers," within a Jewish national frame.[157]

FIVE

Industrialization and Revolt

> The Jewish National Home is no longer an experiment. The growth of its population has been accompanied by political, social and economic developments along the lines laid down at the outset. The chief novelty is the urban and industrial development. The contrast between the modern democratic and primarily European character of the National Home and that of the Arab world around it is striking.
>
> — *Palestine Royal Commission Report (1937)*[1]

THE COMBINED EFFECT OF THE developments described up until this point was technological momentum. The heterogeneous sociotechnical arrangement that was the power system had achieved mass movement and direction, making it increasingly resistant to outside influence.[2] The massive physicality of the system militated against change by requiring deconstruction before any new construction; a growing number of employees and consultants had acquired system-specific skills, whether theoretical, tacit, or applied; educational institutions had developed curricula, professional societies had been formed, research institutes and laboratories had been founded to solve the critical problems defined by the existing system; and manufacturers, a growing number of whom were Palestine based, had specialized in producing system-specific equipment, with which homes, factories, and streets all over the country had been fitted for the past decade. In short, the power system constituted an assembly of many different people, objects, ideas, and institutions that combined to give it certain characteristics and a particular directionality. Achieving technological momentum therefore had profound political implications, broadening and deepening the system's imprint on Palestine.

The continued expansion of the system's generating capacity, and especially the continued growth of the transmission system, involved a considerable feat of heterogeneous engineering, integrating a technological system with the physical, social, economic, and political landscape. As with many efforts associated with the system, this was based primarily on the collection of a diverse set of data, enabling the company's growing number of "men on the ground," negotiating land leases and way leaves, and policing the company's concessionary rights, to carry out their work effectively. Consequently, the technological momentum underwrote other kinds of momentum, resulting in rapid economic growth in the Yishuv and a simultaneous contraction of the Palestinian Arab economy. The 1930s thus witnessed the intensification of a process of uneven development that by the latter half of the decade cleaved the Palestinian economy in two and brought about a marked socioeconomic separation between Jews and Arabs. The power supply system was at the center of this process of uneven development, both facilitating and reflecting it. Any change in the direction staked out by the grid would come only from forces powerful enough to disrupt its systemic momentum. By the second half of the decade, that momentum would be put to the test, with the outbreak of the Great Arab Revolt.

The four years following the completion of Naharayim in the summer of 1932 saw the grid grow at breakneck speed. By this point, Rutenberg was largely absent from the day-to-day running of the system. The year 1932 also witnessed the company's largest relative growth, doubling sales from about 10,000 to 20,000 kilowatt-hours (see table 1). Electricity consumption among Jews increased by a factor of ten in the first half of the decade, at which point the Jewish national home was almost completely electrified. For reasons explored later in this chapter, Palestinian Arab society remained largely unelectrified, and its economy contracted, both as a proportion of the countrywide total and in absolute terms.[3]

The economic divergence of the 1930s is often attributed to the Great Arab Revolt of 1936 to 1939. The revolt no doubt played a significant role in severing ties between the two economies, which, together with the unprecedented violence emanating from all sides, impoverished the Palestinian Arab community. But the economic divergence along ethnonational lines was evident already in the years before the revolt, and the Arab rising was arguably more a consequence of existing economic trends than their cause.

In contrast to earlier violent outbreaks, which had targeted the Zionists, the Great Revolt was directed primarily at the British colonial state. Arab rebels set their sights on the symbols and structures of colonial rule, while endeavoring to establish their own symbols and institutions in their place. They attacked police stations, courthouses, railways, and the Mosul-Haifa pipeline, while establishing competing state institutions, such as a rebel-run postal system, complete with their own stamps.[4] The electric grid was also a prominent target of the attacks, and there was a corresponding effort to set up alternative power supply systems locally. The attacks on the grid provoked two reactions from the power company. First, at great cost in both resources and lives, the company endeavored to maintain the integrity of the grid and minimize the disruptions that the sabotage caused. Second, as in Jaffa a decade earlier, the power company designed the grid to make it more resistant to sabotage. As a result, the geography of the grid came to reflect the geography of the revolt, as auxiliary lines, transformer stations, and standby powerhouses were built according to a logic designed to bypass the violence. In contrast to the early 1920s, however, the mature grid of the 1930s was quite capable of absorbing the counteractions of the Palestinian Arabs. In a clear illustration of how the expansion of the grid and the many forms of momentum it had engendered transformed political possibilities, supply continued virtually uninterrupted through the years of the revolt, and growth proceeded apace, though with a marked slowdown in 1937 (table 1). In short, even the Great Arab Revolt was incapable of disrupting the systemic momentum of the electric supply system. The power system continued its work of forging a modern Jewish state, even as the revolt shook colonial rule to its core. The ultimate upshot was that by the decade's end, Zionist state building and Palestinian oppositional politics were materially embedded in the power system, both facilitating and reflecting the increasingly uneven development of the territory.

The changing material circumstances spurred a reconceptualization, in the form of a proliferation of ethnonational objects and domains. As we have seen, by generating electricity and distributing it over an imagined Jewish national space, the imaginary acquired a material dimension. This, in turn, produced a bounded space capable of hosting a number of national entities, such as an economy, industry, agriculture, politics, and culture. The 1930s saw the proliferation of ethnonational categories, such as "Arab" and "Jewish" industry, and by extension the emergence of ethnonational spaces and economies. A critical aspect of this development was the way it was represented by a variety of primarily Zionist and government research institutes in statistical and economic reports. Electrical Palestine was all but an accomplished fact.

JEWISH INDUSTRY

By 1930, labor Zionism had achieved the dominant position in the politics of the Yishuv that it was going to occupy well into Israeli statehood. While the rhetorical focus of the movement was on agriculture, something that most scholars have since echoed, industry was a prominent feature of the labor Zionist state-building project from the start. Commercial concessions, especially for various public works projects, were an early and prominent concern for the Zionist movement. With the growth of the Yishuv, the desire for industrialization grew too, since it was the only viable means of sustaining the large and quickly growing population. Reports produced by various Zionist bodies, such as Keren Hayesod's statistical abstracts, were at pains to stress Palestine's industrial growth and to link it to Jewish immigration. Its report from 1929 lauded the fact that industry was a large and growing item in Palestine's export balance. In an early expression of the growing trend of ethnicizing economic activity, the statistical abstract of 1929 also noted that "industrial labour in the generally accepted sense of the term only exists among the Jewish population." In this report, as in the discussions generally, electrification occupied pride of place in this largely overlooked arena of intense Zionist settlement activity: "The erection, by the Palestine Electric Corporation, of the electric power stations in Jaffa-Tel Aviv, Haifa and Tiberias was chiefly responsible for this extraordinarily speedy development.... The possibility of obtaining cheap current will certainly give an all-round strong impetus to industrial development."[5]

The power company itself, during the first ten years of operation, focused almost exclusively on industry. By its own design, it was the single largest source of revenue for the company through the 1930s. In accordance with the reigning wisdom among large-systems builders, the tariffs were calibrated such that rates for industry were significantly lower than for private consumption.[6] In effect, then, private consumers subsidized industrial development.

The rapid growth of "Jewish" industry was evident already by the late 1920s. By 1928, 2,270 power-driven machines were at work in factories, employing 18,000 people, including about one-tenth of the Jewish workforce.[7] The largest factories were located in ethnically mixed Haifa, Jewish Tel Aviv, and Arab Jaffa. Of these, Jaffa was already a distant third.[8] The vision for Haifa as an industrial hub—an essential prerequisite for which, as demonstrated earlier, was the provision of large amounts of electric power—

was fast becoming reality. Haifa experienced the largest growth of all the towns of mandatory Palestine: its population more than doubled in the course of the 1930s, from 50,000 to 106,000 inhabitants. In a clear indication of the close relationship between electrified space, industrial space, and Jewish space, almost the entire population increase (50,000 of 56,000) was composed of Jews.[9] In the late 1920s, oil companies began to establish facilities in the town, providing a considerable boost to energy consumption.[10]

The growth of the power system played a critical part in furthering industrial expansion. In the years leading up to the completion of the Naharayim power station, the power company gradually reduced electricity rates in order to stoke demand in anticipation of the large quantities of electricity that would become available when the hydroelectric installation came online. By the early 1930s, the power company had more than halved its industrial rate, putting it well below the maximum allowable rate as stipulated in the concession.[11] Further rate reductions were enacted through the 1930s. The power company correctly estimated that the rate reductions would be immediately offset by increased consumption.[12]

Moreover, the company was actively involved in stoking demand, by investing in a variety of high-energy-consuming activities. Rutenberg had recognized the need for this early on. In a 1928 letter to Lord Reading, Rutenberg stated that "under the peculiar conditions of Palestine, the P. E. C. is bound to take an interest in the different branches of Palestinian economic life, in order to increase and create new consumption of electrical energy."[13] Some of these efforts were public and aboveboard, such as the establishment, in 1929, of Jordan Estates Ltd. to buy, "settle, improve and cultivate land" in Palestine and Transjordan. It was staffed entirely by the upper management of the power company, with Rutenberg as the managing director.[14] The power company was also a major investor in several industrial enterprises, such as the Palestine Foundries & Metal Works, the Palestine Construction Company, the Palestine Cold Storage Co. Ltd., and Jordan Investment Ltd.[15]

But Rutenberg also engaged in less public activities. For instance, he worked hard to increase the proportion of Jews in all of the government's public works projects, especially the largest of them all, the construction of an industrial port in Haifa.[16] He maintained close ties with George C. Thompson, the resident engineer at the Haifa Harbor Works Department, facilitating meetings between Thompson and representatives of Jewish labor. As with High Commissioner Samuel in the 1920s, Thompson was grateful for Rutenberg's help. In one letter, Thompson thanked Rutenberg for his "broad outlook and general helpfulness at all times." Thompson told Rutenberg: "I value your opinion on Engineering and administrative matters."[17] This is illustrative of the larger truth that the development of the Haifa port was a joint Anglo-Zionist effort. The British had identified it as a site of economic interest because of its location at the heart of the regional railway network. The Zionists were drawn to it for the same reasons and envisioned making it the center of Jewish industry in Palestine. Economically, demographically, and geographically, the town evolved through an ongoing conversation between British officials and Jewish developers, conversation to which no one thought to invite the Palestinians.[18]

GRID GROWTH AND HETEROGENEOUS ENGINEERING

At the beginning of this period of intense growth, the power company had a generating capacity of 24,000 kilowatts, of which half was supplied by Naharyim.[19] A 66,000-volt high-tension line connected the Jordan powerhouse to Haifa and continued southward as far as Hadera. Between Haifa and Hadera, medium-tension lines branched off eastward from the coastal main line, supplying the communities for the length and width of the coastal plain.[20]

Extending the power lines involved a major operation of heterogeneous engineering. Negotiating the extensions required intimate knowledge not just of the technical requirements of the supply system and the land but also of the social, political, and commercial life of the country at the macro- and micro-level. For starters, the grid had to grow with the seasons. In November, farmers planted the seeds for wheat, barley, and durra, which they harvested in May and June. Starting soon thereafter, the picking of watermelons, cucumbers, kusa, tomatoes, sesame, and other crops began. The picking season and the months before the cereal harvest—in other words, from July to November—were the ideal time to erect power lines, as there were no seeds in the ground to be disturbed, and the work did not interfere with planting, plowing, reaping, or gathering.[21]

The power company employed several full-time staff to literally and figuratively prepare the ground for the power lines, traveling the length and width of the country, negotiating with landowners.[22] This was an extraordinarily complex effort. The company had to keep careful records of the status of the land on which the poles were erected. The power company records

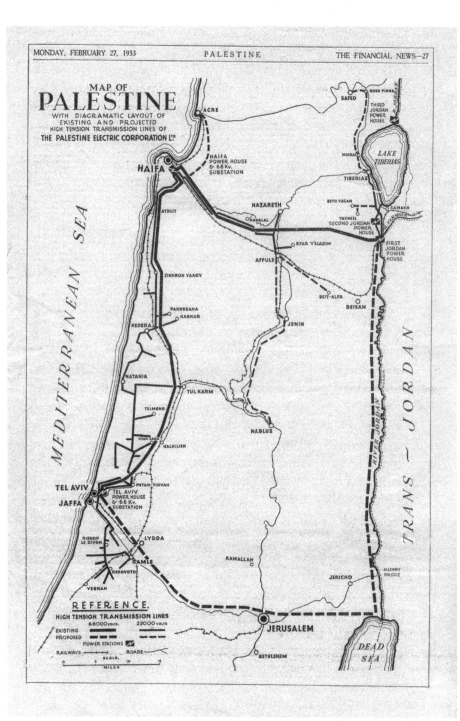

FIGURE 8. The National Grid, 1933. Courtesy of the Israel Electric Corporation Archives.

contain numerous lists of landownership, paired with the number of the pole located on a given plot of land. Given the fragmented nature of landholding, especially among the Palestinian Arabs, virtually every pole required its own negotiating process. One such list produced in the fall of 1930 for poles numbered 650 through 690—40 poles in total—enumerated twenty-four unique landowners.[23] This was no small task; the Haifa-Hadera line alone consisted of 197 poles.[24]

Another reason for keeping detailed records of landownership was to ensure that compensation went to the right people in the right amounts, as individuals occasionally made false claims to landownership to claim compensation from the company. More often disputes arose out of a genuine confusion resulting from the clashing property regimes of Ottoman and British rule.[25] There was also the need to ensure that recognized owners had the proper documentation, in the form of an official land deed (*kushan*). Sometimes, the power company would assist landowners in obtaining such documentation, before concluding a compensation agreement.[26] In some instances, the company was able to negotiate collectively through the village heads (*mukhtars*). But that also required subsequent follow-up with the villagers, lest the village heads keep the financial compensation to themselves, as happened, for example, in one case involving the head of Isfiya village, Maziad Abu Roukon.[27]

The intimate level of knowledge of local conditions that the power company possessed is exemplified in a fairly typical correspondence between the company and a district officer (this one in the Beisan subdistrict) regarding a disputed compensation claim. The company representative stated:

> The monetary value of the damages we estimate as follows: Seeds: 1 kel (sufficient for 6 dunams) produces from 4–10 kel. We accept the maximum of 10, i.e. 6 dunams produce 10 kel of grain. The price of 1 kel of wheat, including harvesting, threshing and transport is 500 mils. The income, therefore, from 6 dunams (10 kel) will be LP. 5.- In order to round out the sum and to allow the Felah [peasant] a fair compensation, our representative estimated the price of one kel at 600 mils (not including the expense of harvesting, threshing and transport to which the damaged crop is not subject, of course). According to said estimation, LP. 1.- covers damages on 1 dunam, and the entire damages caused to 23 dunams and 948 m. will amount to LP.23.948.[28]

By the late 1920s, the company had worked out a standard procedure for dealing with compensation claims. The standard amount it offered was 300 to 500 mils per pole, in return for the landowner's forfeiting any right to bring claims at a later date.²⁹ The standard line of the contract, present in countless power company records, just above signatures by Arabs or Jews, read: "I further declare that I have no claim whatever against" the PEC.³⁰ A common problem in dealing with Palestinians was their reluctance to attach their name to a document formally permitting the power company the use of their land. Arab landowners often requested a modified agreement involving short-term leases or sales mediated by a third party, such as the district offices of the mandatory government. The company's negotiators invariably refused such requests, and most Arab landowners acquiesced in the end.³¹ The British authorities would assist the company in obtaining Arab consent, starting with the district officers and, in cases of greater resistance, working up the chain of command to the director of the public works department, and sometimes all the way to the high commissioner. Ultimately, when a landowner refused to grant a way leave, the power company forced the owner's hand by invoking its right of expropriation under the Electricity Concessions Ordinance of 1927.³²

Policing the concession was another challenge that required intimate knowledge of the land and the people. As the grid grew, the power company continued its efforts to guard its monopoly, even in areas that were not slotted to receive electricity. This, naturally, was a source of considerable friction, which most often played out along ethnic lines, since the power lines, by and large, were coextensive with them. Agents of the power company traveled the land on the lookout for infractions and, when they were found, submitted detailed reports of their observations to the relevant authorities, usually the district governors. As with negotiations over land rights, if the power company failed to get the desired results from the district governors, it continued up the chain of command. Such violations of the concession occurred on a regular basis all over Palestine, in Jewish as well as Arab communities (though, again, as a result of the differentiated geography of the grid, far more Arabs were involved in these disputes than Jews). In one instance, farmers in Samakh on the southeastern shore of the Sea of Galilee dug a small canal to irrigate their land. Near Tapoah, a Jewish farmer installed a pumping plant to irrigate his land.³³ Sometimes cases went to trial, in which case they tended to become flash points of larger Arab-Jewish tensions. In 1932, *Filastin* reported on a court case in which the power company had sued an Arab farmer in the Auja basin for having irrigated his land in spite of the company's exclusive rights under the Jaffa concession. To many members of the Arab community, this appeared as the height of absurdity:

"As part of the evils of concessions in general and in particular to Jews," the article in *Filastin* stated, "we must ask how the Government, and the Jews even, can accept this sterilization of a source of the country's development."[34] The Arab press carried many similar articles criticizing the concession and the power company, which it invariably identified with the figure of Pinhas Rutenberg. The pan-Arab newspaper *al-Jami'a al-Islamiya* wrote around the same time that "when the Palestine Government granted the electricity concession to the celebrated Rutenberg, it deprived the Arab owners of their natural resources, as the first step in realizing the policy of a Jewish National Home." The article went on to ask, "If Uncle 'John Bull' will be indifferent to this question, what faith can we have in the development scheme or any other investment plan advanced by the Palestine Government, and how could the government ask us to agree with a policy for our own destruction."[35]

On the whole, however, as far as acquiring land rights was concerned, the extension of the grid was a rather quiet affair that was quickly routinized.[36] It was the result of a combination of a generous compensation package, which the company rarely diverged from (though in some cases it would quietly pay out much larger sums), and the firm backing of the British authorities, which ensured that stubborn refusal on the part of landowners would ultimately be resolved through forced expropriation.[37]

The growing systemic momentum was reflected in the frequently imperious tone with which power company representatives corresponded with British officials. In the spring of 1930, one of the company directors wrote to the district commissioner in Nazareth following a fracas involving a company work crew in his district, instructing: "Kindly see to it that they should not be disturbed further."[38] The district officers were almost always quick to comply. In this case, only three days later, the head of the village in question (Dabbourieh) received a letter from the district commissioner: "Please notify all concerned that under no circumstances should the work be obstructed and that complaints for damages, if not properly met by the Corporation, should be referred to me for settlement."[39]

This process, too, formed a vital part of the overall systemic momentum to the great benefit of the company. Once the standard was established, it was further reinforced with every landowner who agreed to it. The more entrenched the routine became, the more difficult it became for each subsequent landowner to renegotiate the terms. With time, the process became virtually unstoppable, in large part because the power company was ultimately able to resort to forced expropriation. Moreover, having achieved momentum, the conditions for forced expropriation changed subtly. It was no longer applied when the power company and a landowner failed to reach a negotiated solution, but when the landowner failed to accept compensation according to the power company's preset standard.

The grid's growth relied on the appearance of inevitability, of a routinized operation with predetermined standards for compensation and so on. It needed the appearance, in short, of common sense. As a result, the power company was extremely anxious to avoid the public appearance of a dispute. When the power company reached Qalkilya the landowners demanded compensation of as much as £P5 per pole on the argument that the sum reflected the true value of their land.[40] The case of Qalkilya illustrates the one effective countermeasure that the Palestinian Arabs still possessed. When the landowners stood their ground longer than what was usual in such negotiations, the power company noted with concern how "incitement" and "propaganda" had begun to spread in the neighboring villages, such as Bir-Adam, Kfar Saba, 'Arav, and Abu-Kushk.[41] This was the one thing that would make the company deviate from its standard. In Qalkilya, the PEC paid out as much as £P30 in compensation to a single landowner.[42]

Of course, power lines did not extend just through Arab village lands. There were also Jewish landowners. In an interesting contrast that underscores the uneven nature of the supposedly apolitical grid, the company's negotiations with Jewish communities did not involve the British at all, and when Jewish communities suffered damage as a result of an installation, they turned directly to the company with letters written mostly in Hebrew, but also in Yiddish, German, and other European languages.[43]

BOOM YEARS

Over the course of the 1930s, the power system grew in lockstep with Jewish industry. The first half of the 1930s saw very rapid economic development. The decisive factor was the massive influx of Jewish immigrants in the Fifth Aliya, made up of some 250,000 people, of whom 175,000 arrived in the four years after the Nazis seized power in Germany in 1933.[44] But the economic upturn preceded the bulk of the immigrants; it began in 1932, resulting from rising foreign trade and strong citrus yields in the 1931–32 season. Output increased considerably when Naharayim was put into commission in the summer of 1932. But demand grew even faster. In November 1933, the power company added another turbogenerator to the hydrostation, increasing its total capacity to 25,500 horsepower.[45] But already by January 1934, the company was back to worrying about how to meet the rising demand. It would have to run all stations at maximum capacity to make it through the peak consumption of the summer months. The Haifa powerhouse was already falling short of demand, and the Tel Aviv installation was at maximum capacity. As the year went on, the prospect of shortages grew more alarming.[46]

Relative to demand, everything was in short supply. The power company's work to construct additional high- and medium-tension lines—between Afulah and Beisan, and Haifa and Zichron Ya'akov—as well as a new powerhouse in Haifa were delayed due to a shortage of labor. In Haifa, a thousand households were awaiting connection, as were another fifteen hundred in Jaffa. The company was forced to start turning down new applicants.[47] "We are under exceptional pressure for supply of energy," Rutenberg informed his board in 1934. Several important consumers were about to be added in the fall, and failure to deliver power to them would, according to Rutenberg, compromise the company "both materially and morally."[48]

By 1934, the Jordan powerhouse, containing three turbo generators and generating 25,500 horsepower, transmitted power to Tel Aviv through 66,000-volt mains. From Tel Aviv, a portion of the electricity was carried forward, through 22,000-volt lines, north along the coast to Haifa. In the summer of 1934, the company finished construction on a high-tension 66,000-volt main between Tel Aviv and Haifa, with numerous branch lines of medium tension extending from the north-south main into the country, the chief target of which was the irrigation systems of the citrus orchards in the Jewish colonies. The lines extended through most of the large Jewish settlements in the coastal regions. Just north of Haifa, the Arab town of Acre had also received electricity. Farther inland, most of the Jewish settlements of the Jezreel Valley were electrified, as was the major Arab town of Nazareth, located about midway on a straight east-west line between Haifa and Naharayim.[49]

In the spring of 1934, the company began construction on a new powerhouse in Haifa. It would be Palestine's first steam turbine powerhouse, with a capacity of 18,000 kilowatts, distributed over three turbo generators. When work began, the expectation was that it would be completed by the fall, but labor shortages caused significant delays, and it was not completed until the spring of 1935 (though it began generating current in a limited capacity already in 1934).[50]

At times, the speed of the grid's growth overwhelmed the government departments, which effectively enabled the power company to build with no government oversight.[51] In 1934, the power company added 13,463 new consumers, a number considerably larger than the total number of consumers that the company had supplied only three years earlier. The same year, the company increased its revenue by 61 percent over the previous year.[52] By 1935, new consumers were being connected at a rate of 1,800 a month, a number that continued to grow. The power company concluded an agreement with the Nesher cement factory outside Haifa, guaranteeing a minimum annual income of £P8,000, though estimates for how much revenue it would generate ran closer to £P15,000.[53] Already by March 1935, when the power company's total generating capacity stood at 41,275 kilowatts, it was clear that the system required massive and immediate expansion. Not only did the company carry out thousands of new connections every month, but per capita consumption was increasing rapidly as well. In agriculture, machinery was imported at unprecedented levels, increasing exponentially from year to year. The value of machinery imported in 1935 was nearly double that of the machinery imported in 1933 and 1934 combined.[54] The new 66,000-volt mains from Tel Aviv to Haifa would be insufficient by 1936, and there was an urgent need for another high-tension line from Naharayim to Tel Aviv.[55] The peak load in 1934 had been 10,000 kilowatts, and already by May the following year—that is, before the summer peak—the maximum load had exceeded the previous year's maximum by 60 percent. In Rutenberg's report to the board, he predicted that the peak load would double by 1936. The company rushed the installation of another 12,000-kilowatt turbine in the Haifa powerhouse, bringing the maximum capacity of the station up to 30,000 kilowatts.[56]

THE UNEVEN GROWTH OF THE GRID

In his recent work on electrification in Jaffa and Tel Aviv in the 1920s, Ronen Shamir writes that the expansion of the grid was determined "primarily by commercial considerations and the technological imperatives of machines."[57] Consequently, according to Shamir, economic and technological factors caused ethnonational separatism in Palestine. This is a surprising statement from a scholar so clearly steeped in actor-network theory (ANT). Shamir's presumption of distinct "commercial" and "technological" logics would seem to violate a fundamental tenet of ANT, which does not privilege the agency of machines, but rather promotes a conception whereby nonhuman agency is commensurable with and coproduced by human agency.[58]

Economic interest was of course a significant part of the calculus. Densely populated areas and areas of high energy consumption were consistently prioritized, which meant that Jaffa, Tel Aviv, and Haifa were top priorities, followed by the Mediterranean coastal plain with its energy-intense agriculture, and in particular the lucrative citrus trade. These were predominantly Jewish areas and ventures. It is true, then, that there was an economic rationale behind the preference given to predominantly Jewish areas, on account not of being Jewish but of being large energy consumers. By contrast, when the power company was approached by the government development officer with a request for it to extend a power line to Muqaybla village, located only one mile from the Jenin-Afula road, and the medium-tension line that ran along it, the company refused, citing commercial considerations. Extending the line, building an outdoor transformer station and low-tension distribution would cost £P750. Demand in the village, meanwhile, would come primarily from a single water pump that would consume power at a value of £P30 annually. Consequently, the company would not recover its initial capital outlay for twenty-five years.[59]

Based on such cases, it is tempting to conclude, as Shamir does, that a purely commercial logic explains why only a very small number of Arab villages and towns were connected to the grid by the time the Jewish Yishuv was almost entirely electrified. Besides Jaffa, the only other major town that was connected to the grid at the beginning of the 1930s was Tulkarm, and over the course of the 1930s, only a handful more were connected, despite numerous pleas from towns, including Gaza, Safed, Lydda, Nablus, Beersheba, and Kfar Younes.[60] But such a conclusion overlooks, first of all, that numerous connections were made (and not made) for explicitly political reasons. In March 1935, the power company concluded an agreement with the Palestinian town of Jenin. Construction of the local grid and extension of the high-tension mains would cost £P5,000, and the company estimated the annual revenue at £P500. "Commercially," Rutenberg commented in internal correspondence, "the proposition is not a brilliant one. Politically, from [the] point of view of security of the plant, it is a great success." The extension of the medium-tension line and the construction of the low-tension grid were completed later in the fall.[61] But instances such as the one in Jenin, where political concerns appeared clearly on the surface, fail to address the more important problem with Shamir's contention, namely, that commercial and political logics were not separable; they intermingled with and implied one another. Even a rationale that appeared commercial on the surface was fundamentally shaped by the political context, a context, moreover, that the power company had been instrumental in creating.

HETEROGENEOUS LOGICS

Controversies involving the Palestinian Arabs in making the electrical grid in the mandate period map onto a wider experience of colonialism and anticolonial resistance. In the eyes of many Palestinians, Zionism was of a piece with Western colonialism. They viewed the electric grid, funded and built by active promoters of the Jewish national home, as a part of Zionism's colonial expansion in Palestine. For instance, in 1932 a writer in the Palestinian newspaper *Filastin* warned that "if Rutenberg electricity lights the city of Nablus and Tulkarm one can say that Rutenberg and his works have conquered the land."[62]

For several reasons, as we have already seen, the expansion of the grid constituted the most concrete and invasive manifestation of Zionist encroachment on Arab Palestine. Palestinians' attitudes to the grid also highlighted the ambivalence often encountered between native populations and colonialism. On the one hand, many Palestinians, like many Jews and Britons, were imbued with a certain understanding of progress and civilization, whose attainment depended on having access to cheap electric power. As we saw earlier, when Arab merchants in Jaffa spoke in favor of taking Rutenberg's current, they did so on the conviction that "the latest technical improvements . . . are the only stimulants for the development of a country."[63] On the other hand, there was a constant and lively debate in Palestinian communities over the extent to which modernity should be invited at the expense of what was seen as one's identity, birthright, and autonomy.

The rationale behind electrification, whether of Arab or of Jewish towns, always involved a form of heterogeneous logic that blended economic, political, and other concerns. In the case of Nazareth, businessmen contacted the power company as early as 1923. General Grant, the director of the Department of Public Works at the time, urged Rutenberg to come to an agreement with Nazareth as a means of breaking the Palestinian opposition to his undertaking. Rutenberg and his associates ultimately decided to target Tiberias instead, due to the strong influence of the Nazareth chapter of the Muslim-Christian Association, which was actively and vocally opposed to Rutenberg. When Rutenberg's efforts in Tiberias proved successful, he lost interest in electrifying Nazareth for the time being. Nevertheless, in Nazareth as elsewhere, he still expended considerable effort guarding against any infraction of his statutory rights under the concession. In other words, as the power company was expanding its system at breakneck speed, it was also devoting considerable efforts to preventing any alternative electrification schemes from being constructed, even in areas that the power company had deemed undesirable for expansion. When, for instance, the Department of Public Works forwarded a request from Nazareth in 1923 for permission to carry out a municipal project devoted to overhauling the existing system of public lighting, Rutenberg replied: "I am in principle opposed to authorize the Municipality of Nazareth any change or renewal of their distribution system. The same applies to ... other municipalities. If we are to proceed in this way we would prejudice the market for the Jordan installation which requires, as you know, the investment of a very large capital."[64]

Representatives of Nazareth sent a steady stream of requests for electrification, either directly or through government intermediaries. As time went on, Nazarene businessmen put growing pressure on the municipality and the mayor, Salim Bishara, to negotiate with the power company for a connection to the grid. The mayor, in turn, increased the pressure on the power company, making repeated calls and even traveling up to Haifa for in-person meetings with the management.[65] Ultimately, the deal that the municipality negotiated with the power company was exceedingly beneficial to the company. Under the agreement, the municipality took out a loan at Barclay's Bank, only to lend the money in its turn to the power company, to be repaid in six-month installments. While the loan from the bank to the municipality came with 6 percent interest, the loan from the municipality to the power company was interest free.[66] The agreement between the municipality and Barclay's explicitly stated that the municipality would "hand over the said sum by way of a loan to the [Palestine Electric] Corporation for the purposes indicated in the Electricity Agreement."[67] There was no other way of bringing electricity to the town, as the representatives of the mandatory government informed the municipal council at a meeting in the fall of 1932,[68] though in more informal settings they questioned the power company's motives for demanding an interest-free loan from the cash-poor municipality.⁶⁹ After all, the loan in question was for £P2,000, at a time when the power company's gross revenue was close to £P150,000.⁷⁰ Unsurprisingly, this caused quite a controversy in the Palestinian Arab community, and it received a fair amount of attention in the press.⁷¹ "How is the municipality to be able to lend it [the PEC] the sum at this time when it needs the sum £P2,700 in order to build a water reservoir?"⁷² asked a writer in *Filastin*. When the PEC secretary Yakutiel Baharaw went to see the mayor later in the fall, the mayor expressed concern about whether he was going to "be able to endure the criticism in town."⁷³ He did, however, and so did the mayors of many other Arab towns, including, around the same time, Acre.⁷⁴ These agreements were a far cry from those of the 1920s, in which the power company had assumed all the financial risks and burdens—a clear indication of the shift in the balance of power that had taken place in the intervening decade.

THE NONELECTRIFICATION OF NABLUS

But of course, the more common story, as far as the Palestinian sector was concerned, was not one of electrification but of nonelectrification. Here I will look closer at one such case, the Palestinian Arab town and nationalist stronghold of Nablus. It illustrates the power company's thinking with regard to electrifying Arab towns, but also what the situation looked like from the point of view of Arab denizens. The power company faced one challenge above all, namely, of organizing supply, technically, legally, and politically, in such a way as to minimize sabotage. From the Arab perspective, meanwhile, two major obstacles had to be overcome. First, every Arab town—none more so than Nablus—contained a significant proportion of residents who opposed inviting "the Zionist Rutenberg company" to commence operations in their town.⁷⁵ By the mid-1930s, the opinion was widespread that, as one article from *al-Jami'a al-Islamiya* put it, "the electricity concession stole British policy from the sons of the country [*abna' al-balad*] and gave it to a Zionist Jew [Rutenberg] who works with all his might to turn this country into a Jewish national home."⁷⁶ Nevertheless, a significant section, especially among the economic elites, was keen to have access to large quantities of cheap electricity. Overcoming the opposition without alienating a large portion of a town's population required a great deal of finesse. Second, the kind of large-scale electrification that the power company was engaged in was not well suited to many Arab towns, which would have been better served by a smaller system, requiring more modest front-end capital expenditure, obviating the need to commit to relatively high minimum levels of consumption, as the standard PEC contract did, with reference to justifying the expense of extending the high-tension lines and constructing a transformer stations and low-tension distribution network. But the exclusivity clause of the power company's concession prevented any alternative undertakings, while the considerable opposition to the power company made it even more difficult to make the kind of full-throated commitment that would have been required to make the power company connection profitable.

As early as 1927, members of Nablus's merchant class approached the power company with a request for the town to be connected to the grid. At this time, Nablus was operating a street lighting system of candles that, despite constant maintenance, often failed.[77] At PEC secretary Baharaw's office in Haifa, the Nabulsi merchants Farah Halaby, 'Aref Ja'un, and 'Issa 'Abd al-Reni laid out their proposal. They dangled the prospect of widespread financial participation in the scheme, in return for the power company's agreeing to form a separate company, of which the Nablus municipality would own 51 percent of the shares, and the power company 49. Baharaw rejected the proposal out of hand, as the power company would continue to turn down any proposal that was not based on an open and formal agreement with it directly.[78] In the coming years, the company was approached by numerous individuals offering to serve as intermediaries between the PEC and the Nablus municipality, as well as other towns. The power company was approached by the municipality of Nablus for the first time in 1933 and again in 1934.[79]

In all these cases, the proposals put forward involved electrification through some third party, such as a locally owned company. The municipality at one point admonished the company, "You should not disclose our name to anybody before the definite settlement of the matter between us, and before you let us know your desire which we shall cause to satisfy the wish of the inhabitants and the municipality."[80] The municipal representatives explained that it would not be possible to submit a formal request on account of the strong opposition against the PEC on political grounds. But, they insisted, all of the town's inhabitants, including all the members of the municipal council, wanted to bring electricity to Nablus.[81] The power company stuck to its policy and rejected all these entreaties. As Baharaw stated in a report from one of the meetings: "The answers given by me to these demands are not necessary to enumerate."[82] The underlying reason, too obvious to even comment on, was, once again, sabotage. From the power company's perspective, if it was so unpopular that it was not possible for the municipality to have any public dealings with it, then there was nothing to ensure that the company's equipment would be safe from sabotage. As the power company explained over and over to those who approached it, whether businessmen hoping to make money or newspapermen hoping to make headlines, it was not interested in undertaking electrification in any way other than "officially and openly": "Under no circumstances will we [electrify] against the wish of the population," it stated in one internal memo.[83]

While the power company kept Nablus and other Arab towns at arm's length for fear of sabotage, the significance of electricity for economic growth increased, a fact not lost on observers. A few years into the 1930s, as the Jewish economy was booming and the Arab economy contracting, many were concerned with the issues confronting commercial life in Palestine, and especially in Nablus—a long-standing commercial hub. The town's main industry, soap production, was facing difficulties because its main export market, Egypt, had just imposed high import tariffs to protect local production, and its second-largest market in Syria and Lebanon was beginning to develop its own local soap production. Nablus was even being pushed out of the Palestinian market by competition from new Jewish-run ventures in Haifa and Tel Aviv. Economic development, the newspaper concluded, was sorely needed, and economic development depended on a proper supply of electricity. But, as *Filastin* concluded in an article on the topic, "The Arabs of Nablus prefer to be poor and faithful over being rich and unfaithful."[84]

This, as we have just seen, was not exactly true, and as the 1930s wore on, the divide between the choices outlined by *Filastin* grew starker. By 1934, when it was clear to those in Nablus favoring PEC electricity—the mayor, Sulayman Tuqan, among them—that the company would not agree to conclude anything but a direct agreement with the municipality, their advocacy turned public. Tensions quickly rose. Tuqan was reportedly physically assaulted during a visit to the mosque. Undeterred, he, along with other members of the local council, merchants, and landowners, continued their efforts to negotiate an agreement with the power company. They advocated for their proposed course of action at public meetings and in mosques, and circulated a petition that received many signatures.[85] The issue also became a magnet for other political concerns, such as the long-standing division between the *majlisi* faction supporting Hajj Amin al-Husayni and the *mu'arada* (opposition) faction.[86]

The Arab press in Palestine followed developments in the "nationalist stronghold" of Nablus with great interest.[87] Most conceived of the developments there as a question of staying faithful to the national cause, in a way that once again reflected the sort of oppositional boundary-work discussed earlier. *Al-Jami'a al-Arabiya* claimed that most Nabulsis "would rather light their homes and stores with Arab oil than to light them with foreign electricity."[88] According to an editorial in *Mirat al-Sharq*, "Lighting Nablus with Rutenberg electricity is considered the first step toward settlement by Jews in Nablus."[89] This, of course, made those in favor of doing so, as *al-Jami'a al-Islamiya* put it, "traitors to their city and their homeland."[90]

Through these debates we see that if in Hebrew *Rutenberg* had become a synonym for "electricity," then in Arabic it was a much more loaded and complex term. In Arabic, *Rutenberg* came to signify centralized large-scale electrification, coupled with the political overtones of Zionist conquest. In other words, to those on the losing end of the PEC's electrification endeavor, technology, economics, and politics had merged into a seamless whole: *Rutenberg* became shorthand for precisely that element of Zionist advancement that this study traces, and that has been absent from the literature on mandatory Palestine so far. For example, in Arabic it was possible to say that someone used "generators instead of Rutenberg," both of course producing electricity,[91] or, like *al-Difa'*, to rail against attempts "to introduce Rutenberg in Nablus," which it equated to having "surrendered to imperialism and Zionism."[92] *Al-Karmel* urged that Arabs in favor of accepting "Rutenberg" should "wake up before it is too late" and realize that the linemen, surveyors, and engineers who had recently visited Nablus were "the soldiers of the future Jewish state."[93] A long piece in *al-Wahda al-Arabiya*, published in Jerusalem by Imil al-Ghuri, a Christian member of the Arab Executive and part of Hajj Amin al-Husayni's *majlisi* faction, elaborated on the significance of accepting Rutenberg's current:

> The harm of Rutenberg entering Nablus is not limited to the exploitation of the inhabitants over the high electricity rates and the fact that the factories making electrical wire, lamps, and electrical light switches will derive benefit from it, estimated at tens of thousands of pounds, but that after electricity has entered Nablus panders and land sellers will follow, along with gun runners and immigrants without visas. It also has to be taken into consideration that Nablus is close to Transjordan and the Jews aspire to build the old *Shkhem* and add the hills and strongholds of Nablus to their National Home.[94]

Unsurprisingly, then, talks between the power company and the municipality went nowhere slow. In 1936, a group of merchants submitted a signed petition to the municipality, urging it to conclude an agreement with the power company. This sparked another round of public debate. The newspaper *al-Liwa*, for instance, described it as a "request to the Jew Rutenberg that he come to Arab Nablus and conquer it!"[95]

Not everyone was incensed by the rumors that Rutenberg (and "Rutenberg") was about to arrive in Nablus, though. T. S. Boutagy & Sons, Palestine's largest chain of general stores, wrote several times to the power company asking when it might be connecting Nablus. Currently, Boutagy was experiencing very low sales on its electrical apparatuses, and people were no longer interested in buying battery-operated radios, expecting electricity soon to be available in the town.[96] In other words, the linkage between Palestinian merchant needs and the electricity undermined unified opposition to the current as a form of nationalist struggle, just as in the earlier episode in Jaffa. In the event, the power company was never able to give a clear reply to Boutagy. As talks continued between the company and representatives of the municipality, the tension rose in the town, and in the country as a whole. This was not entirely electricity's doing.

THE GREAT ARAB REVOLT

The growth of the grid and the growth of the economy overall appeared to everyone in the Jewish community and its supporters as an unmitigated success. Lord Reading, the chairman of the PEC, in his address to the board of directors in September of 1935, stated: "The development of Palestine in practically every field of economic activity has been extraordinary during the past few years, and in this development electricity has played a leading part."[97] But Rutenberg had his doubts about the future. In a letter to James de Rothschild in the spring of 1935, he wrote: "Things here are moving fast. On the surface everything is brilliant and prosperous. I think it however to be a 'Purim prosperity.'"[98]

A year later, almost to the day, the Arab Revolt broke out in Palestine. It began with the announcement of a general strike on April 19, 1936, which turned increasingly violent over the course of the summer.[99] The outbreak was preceded by a year of growing tension, caused, among other things, by the discovery in October 1935 of a large arms shipment destined for the labor Zionist underground militia, the Haganah, and the killing of the rebel leader

Shaykh 'Izz al-Din al-Qassam by British forces.[100] Already in February 1935, the Royal Air Force had reported that the feeling was growing in the Palestinian community that "something must be done about the Jewish menace."[101] The revolt had two distinct phases, the first lasting until October 1936, when the general strike ended. The second phase began with the assassination of the district commissioner of the Galilee, Lewis Andrews, on September 26 the following year. The British authorities responded by outlawing the Arab Higher Committee (AHC), the official political leadership of the revolt, formed in April 1936, and exiling several prominent Palestinian leaders, including the head of the AHC, Grand Mufti Hajj Amin al-Husayni. The British also launched a ferocious counterinsurgency operation, fielding twenty thousand troops who drove the rebels out into the countryside, where they roamed as decentralized bands.[102]

The grid emerged as a major target of the Arab rebels, especially in the second, rural phase of the revolt. Of course, sabotaging the grid had been a tactic of the Palestinian national movement from its inception. For the power company, reckoning with sabotage was a permanent feature of grid maintenance, though the frequency of sabotage varied according to other sociopolitical conditions. In 1931, five years before the outbreak of the revolt and after a spate of grid sabotage in the northern regions around the Palestinian towns of al-Tira and Athlit, the PEC prevailed on the Palestine Police to enact collective fines on villages located close to the sites of sabotage. This succeeded in reducing the instances of sabotage considerably, just as the expanded, arguably draconian, protocol of collective punishment enacted after the Wailing Wall Riots of 1929 more generally were effective in keeping the peace. For almost a year thereafter, not a single act of sabotage occurred.[103]

In 1936, there were repeated attacks on the coastal high-tension lines between Haifa and Jaffa, and twice the whole south lost power.[104] When the revolt entered its second stage in the fall of 1937, sabotage of the power lines intensified. Security expenses for the PEC ran to some £P3,000 a month.[105] All work related to the power company's operations became vastly more difficult and took on an increasingly military character. For several months, it was not possible to collect payments from consumers in Arab towns.[106] On one occasion an Arab collector in Tulkarm was robbed and told never to return. Some of the Jewish agricultural settlements were inaccessible except under armed escort.

As noted, the sabotage did cause occasional service interruptions.[107] The most vulnerable section of the grid was located between Hadera and Kfar

Saba, which ran past the Palestinian towns of Tulkarem and Qalkilia, a stretch of about twenty-five miles that was part of the area the British nicknamed "the triangle of terror."[108] At the height of the revolt in 1938, the company had to increase its staff of supernumerary policemen accompanying the repair squads. At first, these supernumeraries lacked proper arms, causing several of them to be killed and wounded on the job. But after appeals from Rutenberg to the government, the company was allowed to acquire its own armored cars and Lewis guns, and the British even provided military training for the company's employees.[109] The controversial counterinsurgency officer Captain Orde Wingate was commissioned for the task of training a special company unit in the tactics from which he and his men had recently garnered such notoriety. Two "Night Squads" of twenty-five men and two officers were taught how to prepare ambushes along the power lines from Zichron Ya'akov to Rosh Ha'ayin.[110]

To be sure, the sabotage was not trivial: the cost of a single damaged pole, including repair and loss of income, was about £P75. To that must be added the cost of the repairmen's armed escort of as many as ten soldiers in some places. On the whole, however, the grid prevailed. Even on the far greater scale of the Great Revolt, sabotage was not as much of a problem in 1936–39 as it had been in 1920–23. Service interruptions were mostly brief and local. At no point during the years of the revolt was there a real threat of widespread system failure, mostly because the system was designed to be able to absorb Palestinian violence. The power company had learned its lesson and put several contingencies in place to handle the risk of interrupted circuits.

It was fastidious about maintaining the integrity of the grid. In one instance in the fall of 1937, at the start of the revolt's second phase, a pole was burned down in the vicinity of Tantura village at 10:00 a.m. The Zichron Ya'akov police informed the Hadera office of the power company shortly thereafter, and by 1:00 p.m., a repairman set out from Athlit, escorted by three British soldiers. They were joined at the site by another five soldiers of the Loyal Regiment. At 2:30 p.m., they had located the burned-down pole, and by 6:27 p.m., less than nine hours after the pole had been sabotaged, the repairs were completed and the power back on.[111] Most important, the past decade of electrification work had taught the company about the importance of a thick electric grid. Indeed, by the mid-1930s the lesson had been absorbed with such fervor that the areas most densely populated by Arabs were clearly discernible on the power company's technical blueprints because of their densely engridded borders. Indeed, the "triangle of terror," appeared on the maps of the PEC as an area demarcated by thick borders of wire along each side, and white within—wires and violence implying each other.

In the annual report for the revolt's first year, 1936, the chairman stated: "Notwithstanding the abnormal conditions which prevailed in Palestine last year, a satisfactory service has been maintained throughout the operating area of the Corporation."[112] This remained the case throughout the revolt.[113] Efforts continued through the rebellion to thicken the grid, making it more resistant to sabotage. So successful was this work that in 1937 it boasted that its supply suffered "a minimum of disruption and compares ... to the most organized stations in the world."[114] During 1937, the company extended the grid by 153 miles, making a total at the end of the year of 706 miles of high-tension transmission lines and cables and 659 miles of low-tension distribution lines.[115] Despite a marked slowdown in consumption, the number of new consumers still grew by 14 percent during 1937 (from 67,000 to 76,000), the slowest growth year for the company in the mandate period.[116]

Overall, however, efforts to expand the grid accelerated during the years of the revolt. In 1936, the power company began making preparations to build a new powerhouse in Tel Aviv. In January 1937, the British approved the lease of a plot of land for the station, just north of the Auja River, and the power company took possession of it on May 1, 1937. The powerhouse, which was named after the founding chairman of the PEC, Lord Reading, was completed in August 1938 and housed two turbo generators.[117] During the time of its construction, the power company employed an additional one thousand workers, providing a significant relief for the economic hardship that the Great Revolt imposed on the country. Indeed, Rutenberg's construction site had been the single largest employer in Tel Aviv during this time of crisis.[118]

These developments did not take shape in spite of the Arab Revolt but directly in response to it. The choice to build another coal-fired plant in Tel Aviv, instead of proceeding with the plans for a second hydroelectric station on the Jordan River, was motivated by the security conditions around the site of the projected works, north of the existing station at Jisr al-Mujami'. More important, the area between the Jordan River on Palestine's eastern border, where the power would be generated, and the areas densely populated by Jews on Palestine's western coastal plain, for which the power was intended, constituted the heartland of the Arab Revolt, especially the strip of land between Hadera and Kfar Saba running east, past Tulkarm and Qalkilya, and later Jabal Nablus, that is, the "triangle of terror." As this area sat between Naharayim and the most intense points of energy consumption on the coast, including Tel Aviv and several of the major agricultural colonies, such as Petah Tikvah and Rishon Letzion, it became necessary to recalibrate not just the geography of the wires but also the geography of power generation. Upon completion, the Reading powerhouse in Tel Aviv was able to step into the breach whenever transmission from the north was interrupted.[119]

Moreover, the revolt compelled the PEC to begin looking south, another initiative that linked the expansion of the grid to the reinforcement of a distinctly Zionist national space in Palestine. In 1937, the company acquired 175 dunams through a subsidiary company near the town of Yabne, located midway between Tel Aviv and Ashkelon. There, it planned to build a high-tension transformer station. Only a small part of this land was required for the works. The rest would be given out to Jewish settlers, in keeping with the company's long-standing policy of establishing buffers of Jewish settlements around its installations.[120] The company also began to prepare plans to build a standby powerhouse in the Negev. The most immediate motivation was the revolt, and the fear that sabotage would risk cutting off the area south of Tel Aviv from the grid. This was a justified fear. On a few occasions, several southern towns and villages were left without light and power, the most important problem of which was the risk of losing their crops (as well as increasing the risk of attacks).[121]

As the political situation evolved, new considerations entered into the expansion of the system. In the fall of 1937, for example, the PEC initiated negotiations with Gaza for a connection. The move was motivated by the work of the British Royal Commission, appointed in 1936 to investigate the causes of unrest in Palestine. The commission, which came to be known by the name of its chairman, Lord Peel, recommended partitioning the land into an Arab and a Jewish part (and a small internationally governed area). Under the Peel plan, Gaza would fall on the Arab side. As Rutenberg explained to his board in the fall of 1937: "The more there will be existing supply in the future Arab territory, the better." The same rationale motivated his intensified efforts to expand the grid throughout the district of Beisan, where the Jordan powerhouse, along with a number of large Jewish settlements, were located.[122] The power company initiated negotiations with Safed and Beisan in 1937, hoping also to benefit from the military camp nearby.[123] In December 1937, the transmission line to Beisan was completed.[124] The final agreement with Gaza was concluded in the spring of 1938, giving the company six months to complete the installation of the low-tension grid.[125] By May 23, 1938, the transmission line to Gaza was completed.[126] But the line had to be abandoned in the fall; as the revolt reached its violent climax, there was no feasible way of defending it.

By May 1939, the revolt had been suppressed, having drained treasuries and ruined lives. The overall defense expenditures in Palestine rose ten times with the outbreak of the revolt.[127] The expenditures for military and police in total amounted to some £P1.5 million, borne by the Palestinian treasury. In addition, it is estimated that the Jewish community in Palestine paid out another £P500,000 for various defensive measures. The attacks and sabotage of citrus and other trees and the destruction of crops covering an area of more than seventeen thousand dunams amounted to additional losses of several hundred thousand pounds.[128] The biggest loss, however, was sustained by the Arab community, counted in Palestine pounds, but especially counted in lives: five thousand Palestinians killed and ten thousand wounded.[129]

The fortune of the power company stands in sharp relief to these grim facts. For the PEC, the revolt was probably a windfall, primarily due to the way it deepened and broadened the gap between the Jewish and Arab economies, especially in the agricultural domain. Except for land sales, foodstuffs trade was the biggest economic interaction between Arabs and Jews by far.[130] Before the revolt, the Jewish sector consumed Arab produce at an estimated value of £P600,000 to £P700,000 a year.[131] But as Arabs ceased selling produce to Jews, Jewish agriculture had to make up the shortfall. As a result, the price of the produce increased, both on account of the higher wage levels of Jewish workers and as a result of the reduced supply. Of course, this strained the urban quarters of the Jewish community. But, in practice, the increased cost of living that resulted in the cities merely amounted to a wealth redistribution from the Jewish city to the Jewish countryside.[132] "As time goes on," one report assessing the impact on the Yishuv predicted, "Jewish agriculture will be increasingly more able to supply the Jewish requirements of those products formerly purchased from the Arabs, namely, fruits and vegetables, eggs and poultry, milk, cattle, fish, etc."[133] And because the increasingly mechanized Jewish agriculture was a far larger consumer of electricity—Arab farms by and large not being connected to the grid—increased productivity in the Jewish agricultural sector meant a further rise in electricity consumption. From the end of 1935 until the end of the Great Revolt in 1939, the yearly amount of electricity consumed for irrigation almost doubled, from about 16,000 megawatt-hours to 28,500 megawatt-hours (see table 3).

Clearly, the power company had grown in the revolt years, and so too had the economic gap between Arabs and Jews, a dynamic reinforced by and reinforcing of the differential impact of the grid. By the end of the 1930s, the most intense period of industrialization and grid expansion, the PEC generated and sold more than 100 million kilowatts annually, producing a value of about £P1 million. The government-produced statistical abstract of Palestine for 1940 stated that "the development of electric power production in Palestine in the past decade has been remarkable."[134]

Regionally, these were significant numbers. Turkey, with a population more than ten times larger than Palestine's, produced only four times as much electricity.[135] Relative to other large commercial ventures in Palestine, the value generated in the energy market was not very considerable. The industry producing the largest value—food—generated a value ten times greater. The sums involved in electricity consumption were also dwarfed by the textile industry, chemical production, metal works, and much else besides.[136] None of these commercial enterprises, however, could have operated at those levels without access to electrical light, heat, and power.[137]

In other words, the response of the power company to the revolt reinforced the effect of the ethnonational differentiation that the grid had already been instrumental in producing. In the starkest indication of the uneven development that had taken place over the course of the decade, more than 90 percent of the electricity was now consumed by Jews.[138] Per capita consumption among Jews had grown from 164 kilowatt-hours in 1936 to 232.4 kilowatt-hours in 1942. By contrast, the Arab per capita consumption of electricity was 11.5 kilowatt-hours in 1942.[139]

THE NONELECTRIFICATION OF NABLUS CONTINUES

A few months after the end of the Great Revolt, interests to connect Nablus to the grid again announced themselves and rekindled the tension between commerce and nation. In January 1940, Robert Newton, the assistant district commissioner in Nablus, approached the power company with a request for a connection to Nablus. Shortly after receiving Newton's letter, the power company received a message from a group calling itself "the Arab Association" of Nablus. The letter stated that "the traitors" who had been interested in bringing the company to the city were "afraid" and could be "counted on one hand." The letter went on: "*Ya*, Director Rutenberg, we are not interested in you or your people or in the power of your rule or your distinction or your money, for our organization is able to set upon you and your workers and your works regardless of your strength. To not consider entering Nablus would be safer for your life and the lives of your workers."[140]

The Arab Association of Nablus could have spared themselves the effort. A new conflict hung over this latest initiative, in the form of world war, and the PEC was loath to commit to further extensions.[141] And in any event, Nablus, Rutenberg noted, was a "peculiar case," best avoided.[142] The terms of the concession prevented the company from outright refusal of service. Instead, the company went through the motions of negotiating and surveying with no serious intention of seeing them through to an actual connection. On the Palestinian side, tension between business and politics reappeared in public discourse. In March, the mayor of Nablus, Sulayman Tuqan, stated in an interview with the newspaper *al-Sirat al-Mustaqim* that he was not prepared to make any decision regarding an electrical connection without the full backing of the municipality, lest he be accused of treason.[143] The statement provoked a response from Nimr al-Nabulsi, a prominent merchant from Nablus. In an open letter published a few days later in the same newspaper, Nabulsi suggested that rather than worrying about what people think of him, the mayor should worry about the "economic and political backwardness" of Nablus. "How is it possible," he demanded, "for you, only out of fear that someone will say what they say, to deprive the city of the economic and political benefits of electricity?" The town's inhabitants needed electricity for their houses and factories, Nabulsi insisted; a connection would immediately double production in the town.[144]

As the power company came closer to an agreement in Nablus, it took special care to protect its future project from sabotage in a novel way. The impetus for this action was the municipal council's passing a resolution to approach the power company with a request for connection.[145] In May 1940, personnel from the power company visited Nablus to survey the town and its equipment.[146] The company presented the municipality with a condition early on. If it was going to electrify Nablus, the town would have to agree to install an electric motor for its water supply system. In meetings with Nablus, the power company presented the condition as part of the consideration of making the connection profitable. In internal correspondence, however, it noted: "Experience has shown that less sabotage occurred in towns dependent on the Corporation's supply for their water."[147] In other words, sabotaging the distribution system would be far less frequent if it meant cutting off one's own fresh-water supply; electrifying the water supply, in other words, was another way of fortifying the grid against sabotage. This round of negotiations, however, also stalled, and as the wartime privations worsened, the power company became increasingly reluctant to take on new extension projects, especially ones of dubious profitability and inflammatory politics.

Toward the end of the war, negotiations resumed, and the particularities of Nablus's politics remained prominent. At a meeting on January 28, 1945, representatives of the Nablus municipality began negotiations with the power company. They again proposed that they would form a company, 51 percent of which would be owned by Nablus and 49 by the PEC. Instead of connecting to the country grid, the new company would install a small diesel generator to power the water supply system, streetlights, and a few houses. They justified the arrangements in two ways. First, the scheme would be compatible with the existing generator, whose sole function it was to pump water from a nearby spring, and in which there were strong "vested interests" by certain residents of the town. The second reason was that "the special psychology of the people of Nablus has to be taken into consideration, which has caused it to remain, unlike all other Arab towns, without an organized supply of electricity for nationalist reasons." The separate company could function as a "device" that little by little would make the nationalists used to the idea of accepting the arrangement, so that ultimately the PEC could take full ownership of the works.

Third, the Nabulsis marshaled technical arguments about costs and benefits to justify their place outside of the countrywide grid. During a follow-up meeting after Baharaw's rejection on February 11, 1945, the Nabulsis made use of the nous of the electrical engineer and Nablus resident Dawud 'Arafat to make this case. There was not enough demand in the town to meet the minimum consumption normally stipulated in the municipal contracts, 'Arafat explained. Extending the high-tension line from Tulkarm to Nablus and connecting the town to the countrywide system would be uneconomical. Only a ways into the future, when the town had undergone considerable economic development, would such a big scheme make sense. For the moment, however, a local solution that could be incorporated into the existing local system and be run on diesel would be preferable both politically and economically. The power company was reluctant, as always, to agree to any scheme that did not involve an open and direct agreement between the municipality and the company. And mismatch between demand and supply had been a feature of the electrification endeavor from the start. Baharaw ultimately rejected the proposal. Nevertheless, the young engineer seems to have left an impression on him; 'Arafat, he scribbled in his notebook, "was clearly very smart."[148]

Once again, then, we see an instance of a complicated interrelation between commercial and political logics. The power supply system and the Jewish economy had created and molded each other; power supply spurred the growth of energy-intensive irrigation, and energy-intensive irrigation spurred the growth of power supply. But these technologies remained largely absent from the Arab economy, and thus too the synergies of technology and economic growth that obtained in the Yishuv. Especially in the wake of the devastation wrought by the Arab Revolt, Arab communities had neither the power infrastructure nor the mechanized agriculture to justify committing to the high levels of energy consumption that the standard PEC contract required.

Negotiations lumbered on for another two years, and the same political tensions as before were manifest in the Palestinian press. They also reflected the commercial logic of large-scale systems electrification versus a small-scale stand-alone system and provoked new threats against the power company and the managing director personally.[149] One letter signed by the Arab League Committee to Boycott Zionist Goods asserted that Mayor Tuqan, in his negotiations with the PEC, "does not represent anybody but himself." The letter promised that "any worker, engineer or surveyor entering Nablus will die. Each pole will be torn down." The letter ended by addressing the director directly: "You should know that you too shall die in your own house. . . . You have been warned."[150]

The negotiations continued nonetheless, though their protractedness ultimately doomed them. By the fall of 1947, Nablus and the PEC made an agreement whereby the town would borrow £P60,000 from the mandatory government and, as in Nazareth, lend that money in turn to the power company to cover the cost of the power connection. The justification for this arrangement was that Nablus was incapable of giving adequate guarantees for minimum consumption levels. Of course, the power company's invocation of this logic was selective; on numerous occasions, as we have seen, the aspirational dimension of power supply had overridden concerns about actual consumption levels. The scheme was approved both by Government House in Jerusalem and by the Colonial Office in London. On February 10, 1948, the supply agreement between the PEC and Nablus was signed.[151] The agreement, however, was never implemented. The civil war between Jews and

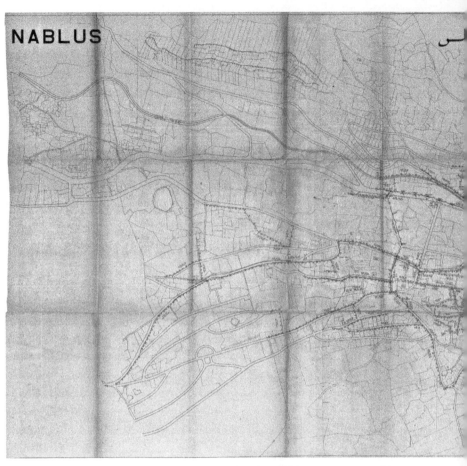

FIGURE 9. Blueprint of Nablus distribution network, 1947. Courtesy of the Israel Electric Corporation Archives.

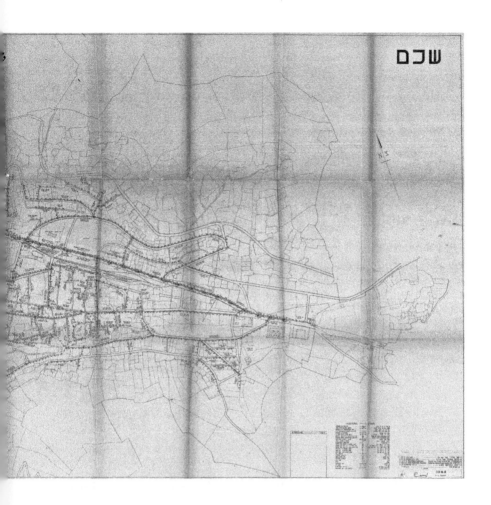

Arabs in Palestine had broken out shortly after the adoption of the UN General Assembly Resolution 181 for the partition of Palestine (see chapter 7), and by May it would expand into a full-blown regional war.

Nablus was never electrified by the PEC. By the end of the 1948 War, Nablus found itself in the territory occupied by Jordan. In the 1950s, no longer hamstrung by the exclusivity clause of Rutenberg's concession, the town built its own electric supply system. When, in June 1967, Israel captured the West Bank in the Six-Day War, the then mayor of Nablus immediately brought a suit against what had by then become the Israel Electric Corporation for breach of contract and demanded repayment of the £P60,000 that the municipality had paid the company in late 1947 for a connection, with 9 percent annual interest.[152]

MAKING JEWISH THINGS: A NUMERICAL APPROACH TO REALITY

The exponential growth of the power supply system throughout the 1930s facilitated the imposition of a new conceptual grid on Palestine. According to the economic historian Jacob Metzer, Palestine's "statistical age" began in 1922, with the British census.[153] Indeed, a new form of knowledge production attended the industrialization of Palestine in the 1920s and 1930s. Beginning in 1921, the Palestine Economic Society started publishing bulletins and articles and compiling tables containing a wealth of statistical data concerning the economic activities of the Jewish part of the population. The Jewish Agency began publishing surveys and analyses in 1923. About the same time, Keren Hayesod began publishing its *Statistical Abstract of Palestine*. In 1925, the Jewish Industrial Association began producing regular publications on the commercial activities of the Yishuv. The following year, the British began publishing the *Palestine Blue Book*. Comprehending Palestine "in numbers," to quote the title of Yehoushoua Ziman's book from the same time, was becoming a pressing imperative.[154]

A significant shift took place during the period of the Yishuv's economic boom in the 1930s. In the government reports from the 1920s, the demographic breakdowns had been on the basis of religion: Muslims, Christians, Jews. In 1928, the mandatory government published the first industrial census, covering the year 1927. It had initially been divided by Jewish and Arab ventures, but the division was discarded for fear of causing further friction. By the mid-1930s, however, the British were forced to accept these sorts of statistics, because it was no longer the government that was producing the most reliable quantitative data on Palestine. That distinction belonged the Jewish Agency, which had no qualms about ethnic categories of quantitative analysis.[155]

Although the mandatory government had originally balked at producing statistical information broken down by ethnicity, it soon found itself reproducing not only the Jewish Agency's numbers but also its worldview. According to this worldview, Palestine contained two separate economies, one Arab and one Jewish. These economies, in turn, housed a number of "Jewish" and "Arab" things, including, of course, "Arab Markets" and "Jewish Markets."[156] Most notably for our purposes, there was suddenly an object called "Jewish industry," which, according to the *Statistical Handbook of Middle Eastern Countries* from 1945, put out by the Jewish Agency, consisted of exactly 2,250 distinct establishments, employing 45,850 people, generating an annual production value (in 1943) of £37,500,000.[157] The second statistical abstract ever published, covering the years 1937–38, reported that "the growth of industry in recent years has been one of the most important features of the post-war Palestinian economy."[158] This growth, as numerous reports pointed out, had taken place almost exclusively in the Jewish sector.

The remaining, non-Jewish segment, meanwhile—the "balance of the population,"[159] as one report put it—was a nebulous collection of varying designation. One British report from 1938 felt the need to clarify that "the word Arab is here used in its strict sense and not as equivalent to 'Non-Jew.'"[160] The government's own statistical abstracts most commonly referred to them as "Arab and other non-Jewish establishments,"[161] reproducing both the language and the spirit of the Balfour Declaration, now close to three decades old and, for all *official* intents and purposes, long since discarded.[162] The declaration lived on, however, in wooden poles, steel earth wires, copper conductors, and the statistical tables with which they had coevolved. Just as the designation of the group itself, the non-Jewish figures were arrived at by way of detraction: the Jewish figures subtracted from the countrywide total. The upshot was an image of an emerging Jewish state and a lagging Arab remainder.

The figures produced in the various statistical reports dating from the 1920s had the effect of reinforcing a Zionist worldview: technocapitalist, modernist, statist. The statistical abstract for 1944, for instance, showed that the gross output of the Palestinian Arab sector was some £P1.5 million, as compared with the Jewish gross output of just over £P6 million. The same report's "Fuel and Light Index," was subdivided into an "Arab Index" and a

"Jewish Index." The Jewish index showed that the sources of power and light among Jews were 30.3 percent electricity, 51.5 percent kerosene, 9.1 percent methylated spirits, and 9.1 percent firewood. The Arab index, meanwhile, listed usage thus: 7.5 percent electricity, 55 percent kerosene, 5 percent firewood, 7.5 percent methylated spirits, and 25 percent charcoal. Many other things had such indices, subdivided into Arabs and Jews.[163] The disparities between "Arab" and "Jewish" figures are obvious, and clearly suggestive of a more "modern" consumption pattern among Jews. So, too, is the greater degree of precision in the Jewish figures, an important element in the performance of modernity.[164]

Interestingly, the wider significance of the statistical abstracts was recognized by the authors of the abstracts themselves. The Palestine government abstract for the year 1937–38 identified the power of statistics to remake sociopolitical worlds: "It is, in general, true that the publication of statistics leads gradually to a conscious need for accurate statistics based on precise definitions completely comprehensible to those whose duty it is to collect the data or to analyse their characters when collected. In this respect the publication of the Abstract has an importance which is not confined to the material uses to which it can be put."[165]

Statistics was thus an important means by which the Zionist movement produced a distinction between the Jewish community in Palestine—producing, consuming, and performing modernity, and with the modern bookkeeping to prove it—and the Palestinian Arab community, possessed of "a mentality," in the words of one Jewish Agency publication, "which does not always view with favour the exact and numerical approach to reality."[166] Of course, this is not to suggest that none of these statistics have any meaning, but rather to point to the way that numbers purporting to represent the world also played a part in making it. The division of the world into Arab and Jewish spheres surely had some basis in reality, but it also no doubt helped to catalyze transformations that made these categories of analysis into modes of existence.

CONCLUSION

"Palestine," wrote the liberal journalist Herbert Sidebotham in 1937, "jumped from a medieval into a modern economy with one leap."[167] Like so many observers at the time, the first concrete expression of that leap that

Sidebotham went on to name was electrification, which he referred to as "one of the most significant aspects of Palestine's industrial development."[168] Before World War I, Sidebotham remarked, the poor majority was lucky if it had an oil-soaked rag to light; the wealthy few if they had kerosene. Now, he noted, electricity was used for lighting and powering household appliances and industrial machinery, and much else besides, and therefore, "the current consumed is a very useful barometer of the modernisation of its industries and its social life."[169] Indeed, virtually every time Palestine's industry was discussed in any document produced by the British governments in Jerusalem or London, or by any other Western observer, the work of the Palestine Electric Corporation was the first concrete aspect of Palestine's development that was mentioned. The report of the UN commission that was formed in 1947 and tasked with finding a solution to the Palestine problem put the matter in the following way: "Palestine is not very favorably endowed for industrial production apart from its geographical location, which is of considerable importance in relation to the whole Middle East. It has no raw materials of any consequence apart from the Dead Sea minerals. Nevertheless, the influx of immigrants with developed tastes for a variety of consumer goods, and the development of electric power by the hydroelectric installation on the Jordan and by oil driven plants at Haifa and Tel-Aviv, provided an important foundation for the industrial development of the last fifteen years."[170]

This, then, constituted another form of momentum, beyond the technological momentum with which this chapter began, and which served the interest of the power company. Earlier, we saw the Conservative MP and leading electrical engineer Philip Dawson say that electricity consumption in Palestine was held back by "primitive methods of agriculture" and "inefficient labour."[171] It is not hard to imagine how the fact that more than 90 percent of electricity was consumed by Jews, or that four-fifths of Arab lighting needs were met with charcoal and kerosene, was construed by someone applying Sidebotham and Dawson's metrics. As we know, they were by no means alone in using electricity consumption as an index of civilizational standing.

Moreover, it is telling that Sidebotham's comments were focused on industry, and not on agriculture. By the time Sidebotham was writing, the industrial sector was twice the size of the agricultural sector, though this was not yet reflected in electricity consumption. In 1937, electricity consumed for irrigation purposes was 25 percent larger than for industry. But only three years later, in 1940, did industry overtake irrigation, and from then on put an ever-growing distance between it and other uses with every passing year (table 3). Since no industry had existed in Palestine before the arrival of the Zionists, according to the claims of numerous reports and statements, building one could never be a matter of dispossession. As Sidebotham assured his readers, "The industrial development represents a net addition to the national income and has clearly displaced nothing."[172] This, of course, depends on how you look at it. The goal here is not to assess the accuracy of Sidebotham's claim but to point out that there was no neutral way of crunching the numbers, and indeed no neutral numbers to begin with.

The rapid expansion of the electric supply system was critical to allowing the continued growth of the Yishuv, enabling it to go on admitting large numbers of immigrants every year. The population growth further increased demand, which allowed the company to continue growing, and so on in a virtuous circle of exponential growth of the system and the Jewish population. In November 1931, a British-conducted survey recorded 174,610 Jews in Palestine; by December 1936, that figure had jumped to 384,000; three years later, in December 1939, the figure stood at 474,000. With the immigrants came capital; an estimated £P50 million arrived with the immigrants of those years, four-fifths of which was private.[173] During the same period, the size of the grid more than doubled, and electricity consumption grew by a factor of ten (table 1).

The Palestinian Arab experience was starkly different. To begin with, the mandate period witnessed a mass exodus of Arabs from the Jewish labor market. In 1922, Arabs made up 13.6 percent of the labor in Jewish-run enterprises; by 1935, that figure had declined to 8.5, and the following year, with the outbreak of the Great Revolt, it underwent a precipitous further drop to 3.7 percent. The trend held through the mandate period, and by 1945 the figure was down to 1.7 percent. Thus, the decline predated the revolt, which lasted until 1939, and continued steadily after it.

During the period of the Yishuv's exponential growth, the Arab economy raised almost no new capital at all.[174] Thus, as Amos Nadan has argued, the different strategies of Arabs and Jews—in which, to most outside observers, the Arabs compared unfavorably—was a function on each side of rational responses to widely differing circumstances, especially with respect to the relative cost of labor and capital. Whereas in the Jewish sector, capital was easily come by and labor dear, the opposite obtained in the Arab sector. For the Jews, it made sense to invest in capital-intensive cultivation methods and industry. In the Arab sector, by contrast, where labor was cheap and credit largely unavailable, it made sense to employ people for the tasks that fell on machines in "modern" agriculture.[175] (Of course, conditions in the Arab sector were no less products of "modernity" than in the Jewish sector.) The same conditions explain why Arab agriculture did not transition into specialization; the sizable investment required in the short to medium term made such a move prohibitive. The absence of credit thus prevented investments in Palestinian businesses, trapping them in a vicious circle.[176]

The same dynamic obtained in the domain of electrification. In the Arab community, the benefits of electrification did not, by and large, justify the initial capital outlay required. But this study has also added a component to this familiar dynamic, namely, that the electricity market was keyed to the Jewish economy, thus making less capital-intensive and thus slower and more gradual growth impossible, while the exclusive rights of the power company under the concession prevented the Arabs from undertaking any alternative electrification schemes. In economies similar to the Palestinian in other parts of the world, the gradual growth of many small grids was usually the order of things.[177]

Ultimately, this means that the unprecedented economic growth of the Yishuv, as presented in the standard works of Palestinian economic history, is artificial in one critical sense: by the late 1930s, after a long process that in effect made socioeconomic differentiation coextensive with ethnic difference, to analytically separate out the economy of the Jewish sector from the Arab sector practically amounts to a kind of creative bookkeeping in the growth equation whereby the lower socioeconomic stratum (i.e., the Arabs) who provided the menial labor, and whose economic progress was de facto blocked by the same process that enabled the economic development of the higher economic stratum (the Yishuv), is eliminated from the calculus. This not only inflates the apparent growth miracle of the "Jewish" economy; it also elides the fact that the growth, at least to a degree, happened at the expense of growth in the Arab sector and, finally, that it was through this selfsame process that "Jewish" and "Arab" became relevant as economic categories in the first place.

SIX

Electrical Jerusalem

IF IN THE 1930S, AS ONE advertisement had it, people believed that "electricity keeps the wheels of industry turning," then by the 1940s, in the words of another ad, they believed that "electricity solves everything."[1] The shift from "industry" to "everything" had, as one might expect, profound consequences for daily life in Palestine. A technology that once was a luxury item associated mostly with lighting—in the home for the middle and upper classes, in public for all—came increasingly to be regarded as a household necessity not just to illuminate but also to iron one's clothes, cool perishables, warm the body, and much else besides: in Tel Aviv, Café Weintraub boasted of "home-made ice creams electrically refrigerated" in the same month that the town got its first "electro-matic" traffic light.[2]

Here I will explore the social ramifications of electricity at the time of its domestication. Over the course of the 1940s, electricity became integral to a wealth of activities, such as maintaining security, law, and order; religious practice and observance; food conservation; meal preparation; as well as public and private health. Moreover, it introduced new categories of crime, such as "electricity theft," and new public and domestic hazards, such as "electrocution." Electricity inserted itself into the fundamentals of private and public life. It appeared on the body and within it; it flowed through some collectives, uniting them, and between others, marking their separation.

Before the arrival of the British and the Zionists, as so many boosters' economic reports asserted, Palestine was almost wholly unelectrified. Yet the appearance of electricity caused a new kind of problem: its absence. As the matron of a leper hospital complained of the difficulties of negotiating dark stairs during blackouts; as a Palestinian father lamented the damage done to the eyes of his studying children by the absence of electric light; and as an ultra-Orthodox community bristled at an unexpected blackout on the Sabbath, they revealed how many agendas—medical, educational, religious—had become bound up with electricity, and then to its absence. That so many of these complaints took the form of accusations of ethnic discrimination, moreover, was a clear sign of how dominant the conflict between Arabs and Jews had become. While these political dimensions of the allegedly apolitical power system became more visible, the power company intensified its efforts to minimize its political profile, an endeavor with its own messy politics.

To illustrate these changes and continuities, I will look more closely at the one major town in Palestine that so far has received hardly any attention at all, Jerusalem. The reason why Jerusalem has largely been left out of the account is that the Palestine Electric Corporation's concession did not apply there. As mentioned earlier, a different agreement was in force in Jerusalem—originally concluded on January 27, 1914—between the president of the Jerusalem municipality, the Mutesarrif of the Sanjak of Jerusalem, and the Ottoman businessman Euripides Mavrommatis for the generation and distribution of electrical energy and the construction and operation of a system of electric tramways in and around the town. The concession area was defined as "the area contained within an imaginary circle of a radius of twenty kilometres having as its centre the central dome of the Church of the Holy Sepulcher in Jerusalem."[3]

Due to the outbreak of World War I, Mavrommatis was not able to act on the concession, whether his intention was to sell it on to a contractor (a common practice in the late Ottoman Empire, among so-called concession-hunters) or develop it himself.[4] When Mavrommatis resurfaced after the war, now a Greek citizen, to claim his statutory rights in Jerusalem, the initial impulse among British officials was to simply dismiss him. But under Article 314 of the Treaty of Sèvres, all contracting parties were bound to respect commercial contracts that had been granted before the outbreak of the war.[5] This clause was carried over into the Treaty of Lausanne, signed on July 24, 1923, which enjoined the contracting parties (on one side, the British Empire, France, Italy, Japan, Greece, Romania, and the Serb-Croat-Slovene State, and on the other Turkey) to respect prewar commercial agreements. Prevented from rejecting Mavrommatis's claims outright, the British instead stalled. A briefing from 1923, prepared by the Middle East Department for William Ormsby-Gore, undersecretary for the colonies, put the matter in the following way: "I have personally no sympathy whatever with Mr Mavromattis [sic]. Unfortunately he has acquired some rights which appear to be valid and is making the most of their nuisance value. He and his friends are quite shrewd enough to see that the anti-Rutenberg agitation gives them a strong card."[6]

After getting nowhere with the British, Mavrommatis turned to the Greek government and succeeded in persuading it to bring a suit against Britain at the newly instituted Permanent Court of International Justice at the Hague. The court's ruling, handed down on March 26, 1925, affirmed Mavrommatis's rights under the Ottoman concession, and the British had no choice but to negotiate with him for a separate electrification concession.[7] A year later, the parties had worked out an agreement, to which a supplementary deed was added two years later, in 1928. It granted a concession for forty-four years and stipulated maximum charges (twenty-five mils per kilowatt-hour) and established it as a public utility body.[8]

In 1928, two years after signing the concession, Mavrommatis sold his venture, without having made any serious effort to build a system himself. It was bought by a British company, which formed a subsidiary by the name of the Jerusalem Electric & Public Service Corporation, Ltd. The company was incorporated in Palestine on November 4, 1930.[9] An ordinance to regulate its activities was promulgated earlier in the year with the Electricity Concession (Jerusalem) Bill.[10] William Campbell-Brown, with forty years' experience in central station work, became the general manager in March 1930. At this point, many of Palestine's larger cities had been electrified, and Jerusalemites of all stripes lamented the fact that they still lingered in relative darkness. One newspaper decried the fact that from a nearby hill Jerusalem looked like "a few scattered and twinkling public incandescent lamps, dimly illuminating the abode of almost one hundred thousand people."[11] Another problem caused by the delay was that for the past six years another, unauthorized company by the name of Hassolel had been operating a small distribution system in the town. Thus, one of Campbell-Brown's first orders of business was to bring an action against the company and its chairman to secure an injunction preventing the sale of electricity by anyone other than the rightful concessionaire. The judge ordered Hassolel to cease and desist, and on July 11, Hassolel turned off the power, leaving all its customers "in total darkness," as one newspaper report put it.[12]

In its first year of operation, the company installed three diesel generators of 275 kilowatts each, installed 140 street lamps (turned on in October), and electrified 120 shops and more than a thousand homes. An engineer at the company held out a vision of a particularly bright future for the city's women: "Every woman in Jerusalem should, in the near future, be able to do all her cooking in electric ovens.... A still greater boon for many a middle-class woman will be the electric iron." Along with the not-so-subtle socioeconomic and gender differentiation came other kinds of distinction. The engineer also pointed out that "Churches and Synagogues use our current, but no Mosque has yet been electrified. Whether this be due to lack of money or to conservatism, I cannot say."[13] The company gradually expanded capacity, and by the close of the decade the station housed nine generators, with seven in operation and two in reserve, for a total generating capacity of 6,600 kilowatts. The distribution system was also significantly expanded during this time. Subsidiary stations were built around the city, the first of which was on Chancellor Street, behind the Stern building, and held five, and later six, lines. A little later, another substation was built on Mount Scopus in eastern Jerusalem.[14]

Consumption in Jerusalem consisted mostly of commercial and domestic use, whereas industrial consumption was negligible. As a result, while the peak load for the PEC occurred in summer, because of the irrigation needs of the hot and dry summer months, the peak load in Jerusalem occurred in winter, because of heating needs.[15] Both the Rutenberg and Jerusalem companies experienced peak loads of about 70 percent over the minimum load.[16] Even though the Jerusalem company experienced remarkable growth in its first years of operation, the PEC still outgrew it, not only in absolute but also in relative terms. In the period from 1931 to 1945, sales grew by a factor of 22.9 for the PEC and 21.2 for the Jerusalem company. The PEC grew from just under 10 million kilowatt-hours a year in 1931 to 200 million kilowatt-hours in 1945. The PEC's single largest proportional increase—70 percent—occurred during the year 1932–33, when Naharayim was put into commission. In the years following, the company's sales grew by about 15 million kilowatt-hours per year through 1936. Growth slowed during the years of the Arab Revolt, after which it rose again to its prerevolt growth rate of 15 million kilowatt-hours a year until 1941, when another period of exponential growth saw an increase until 1945 of 200 million kilowatt-hours sold. The Jerusalem company, by contrast, grew from close to zero in 1931 to 26 million kilowatt-hours sold in 1945. Its single biggest growth year was 1933–34, when it expanded by some 56 percent (see table 1).[17]

THE GRID AND THE CONFLICT

From the start, the company was forced to battle accusations of ethnic discrimination from all sides. By the 1930s, the population was thoroughly steeped in ethnonational antagonisms, which became a prism through which most other things were seen and understood. The charge of ethnic discrimination was never far off. It lurked behind the corner of every dispute, as a repertoire ready to be drawn from and mobilized in the service of a variety of aims.

Already a year into operations, the power company had become the target of a steady stream of accusations of discrimination. Jewish residents were aggrieved by the fact that no Jewish workers had been hired by the power company, sparking a boycott campaign within the Jewish municipal council. It quickly spread to many more companies and parts of the city, including suburbs. The Hebrew press published regular lists of the growing number of firms and individuals who had joined the boycott movement.[18]

Curiously, the same charge was leveled at the company from the opposite direction. One such incident transpired between the power company and the Arab Chamber of Commerce in Jerusalem. Chambers of commerce were established all over Palestine beginning in the late Ottoman period, the earliest dating back to 1912 and the establishment of chapters in Nazareth and Haifa. The first chambers of commerce were ethnically mixed, but in the course of the mandate period, a growing number separated into exclusively Arab and Jewish organizations. The Haifa Chamber of Commerce, for instance, split along ethnic lines in 1921. The Jerusalem Chamber of Commerce went through its ethnic split in 1936, on the initiative of its Arab members, as a result of the revolt. The work of the chambers of commerce intensified in the late 1930s, with the end of the revolt. The trend was reinforced by the economic prosperity brought on by World War II. The chambers, according to Sherene Seikaly, "became sites for building institutions as well as shaping new ideas about the economy as a specifically national domain."[19]

The ethnic split of the Jerusalem Chamber of Commerce coincided with the growing concerns about the ethnic makeup of the Jerusalem power company. George As'ad Khadir, the chairman of the Arab Chamber of Commerce in Jerusalem, addressed these concerns to Campbell-Brown with a request for detailed information on the company's staff and operations. His questions concerned the proportions of Jews and Arabs among the company's employees and consumers. How many Jewish, Arab, and other consumers, Khadir wanted to know, did the company have? How many Jews, Arabs, and others were employed by the company and in what positions? Who worked what hours, and who received overtime compensation? "Are salaries paid on an equal basis to the above employees in the categories mentioned? or is [there] any discrimination between the three categories in so far as the position is concerned? i.e. is an Arab clerk paid as a Jewish clerk—a Lineman, a Meter Reader etc?" The final question was perhaps the most suggestive of the chamber of commerce's concerns: "Do you consider that the present Arab staff represented on your pay roll is a fair percentage to the total Arab consumers and to the amount of electricity consumed?"

In a move that we have become intimately acquainted with through this study, the power company opted to respond with boundary-work. The reply to Khadir began with the statement that the company possessed "no political bias or inclinations whatsoever." The company did not classify consumers or employees by ethnicity or religion, according to the letter. As a result, it was not able to provide the information sought by the chamber. The letter added that the company had had similar requests from Jewish interests in the past, but that it had always "steadfastly refused to supply them or even to discuss these matters."

The reply was not well received. Khadir, echoing earlier oppositional boundary-work of Palestinians working against the grid, wrote back: "From observation, I must frankly state that the Arabic element represented on the staff of the corporation is a negligible item and for that reason the chamber wished to find out what proportion of the Arab employees to the non-Arabs, with a view to granting a fair representation of Arab employment in this public utility."[20]

This was not the last time that the chamber of commerce and the Jerusalem power company would butt heads over the issue of discrimination. In September, Khadir again wrote to the company, complaining about the unfairness of charging Arab shopkeepers the minimum rate, even though they, on account of the Palestinian nationalist strike, had been keeping their shops closed for months. Khadir claimed to have information that Jewish shopkeepers, whose business was interrupted by the strike and attendant disturbances, had been granted relief. It would be only right, he continued, that the Palestinians should receive the same courtesy. The power company rejected Khadir's request and vociferously denied having offered any reductions of rates to Jewish businessmen. "You may rest fully assured," its letter stated, "that there has not been, nor will there at any time be any discrimination whatsoever by this Corporation between its Arab and Jewish consumers."

In the end, the company offered a partial rebate, which it publicized widely in all major newspapers and languages.[21]

Despite the power company's efforts to prove that the grid and the company did not discriminate, charges to the contrary issued in a steady stream from both communities. One example involved the chamber of commerce and J. Gordon Boutagy of Boutagy's Stores, the same company that had been so keen on introducing electricity to Nablus. Complaining to Campbell-Brown (and copying the Arab Chamber of Commerce), Boutagy explained that earlier in the month, a Mr. Z. Dawid of King George Avenue had bought a refrigerator from Boutagy. The purchase came with an installation service, and Boutagy dispatched his electrician to Dawid's home to carry out the work. Dawid then turned to the power company with a request for a connection. An inspector from the company showed up and failed the installation for not being up to standard, providing a list of instructions for required improvements. Yet when they had been made and the inspector came back, charging another 250 mils for his troubles, he failed the installation again. Both Boutagy and Dawid turned to the company asking for an explanation, but none was forthcoming. They also asked for written regulations. Finally, a clerk at the electricity company offices told Dawid to seek out the assistance of an electrician on Princess Mary Avenue to carry out the work. This was the street where the Jewish electricians were housed, and that, Boutagy implied, had been the game all along. Dawid did engage a Jewish electrician, who, according to Boutagy, did not improve the installation but, on the contrary, "made [it] worse," after which the company inspector passed it.

Boutagy was convinced that the power company was engaged in the work of empowering one ethnic community at the expense of another. To avoid similar occurrences in the future, Boutagy sent a clerk to the power company offices in an attempt to get written regulations for electrical installations. The clerk at the company informed Boutagy's clerk that no such document was available, but that if they wanted any information they should consult the engineering firm next door to the company: "(who are Jews)," Boutagy pointedly noted. "Under these circumstances I feel compelled to complain of this most unjust treatment from your staff. It is quite obvious that your Jewish inspector will not facilitate matters if the electricians or consumers are Arabs." He insisted on being given a chance to prove that the initial installation had been carried out correctly, and that the reasons for failing it twice were based on ethnic discrimination. He also requested written regulations for electric power installations, "so that dictatorship of inspectors be avoided."[22] Far from considering the poles of the electrical wires as the "gallows of the nation"—the anti-Rutenberg chant heard on Jaffa's streets in 1923—and thus aiming to sabotage them, Boutagy was eager to conform to the rules of the power company, not least because his store stood to benefit from it. Still, what he perceived to be the unfair economic practices enabled by the power company prompted his protests, not against the grid in general but against its practicalities.

ELECTRICITY THEFT

As electricity entered all spheres of human life, it engendered a range of new phenomena and practices. One was criminal: the new transgression of electricity theft. By the late 1930s, theft of electricity had become a serious problem to both the Rutenberg and the Jerusalem power companies. The Jerusalem company wrote to the attorney general, W. J. Fitzgerald, to complain about the growing frequency of the offense. With the outbreak of World War II, and the strain on existing resources and equipment, the strictures of air-raid blackout routines, and increased fuel costs, electricity theft was one more thing making life difficult for the power company.[23]

There were several methods of manipulating the electrical meter, or circumventing it altogether by hooking one's house, shop, or factory up to the grid in secret. The first case of electricity theft in Jerusalem that went before the Jerusalem Magistrate Court involved the shopkeeper Jamal Namour. Namour had inserted a straw into the meter, which prevented it from recording electricity usage. He had been caught because the company noted a decrease in his electricity consumption over the winter months, when it should have gone up because of the colder weather and fewer hours of daylight. Namour was charged with the theft of electricity and fined £P2 by the magistrate.[24] Similarly, the Armenian proprietor of a barbershop on Mamilla Road was fined £P25 and his employees were ordered five "strokes of the cat" for stealing electricity. Their power was shut off until the fine was paid.[25]

Relatedly, a worker at a tile factory owned by the Palestinian businessman Khalil Bandak was charged before the court with having deliberately sabotaged the electrical meter. The worker claimed to have acted on orders from his manager without realizing the illegality of the act.[26] In 1938, Attorney General Fitzgerald brought a suit against Michael Lorenzo, the proprietor of a bookshop on Jaffa Road, near the Old City, based on a claim by the

Jerusalem power company. Lorenzo stood accused before the District Court of Jerusalem of "fraudulent conversion of power" in violation of Section 285(1) of the Criminal Code Ordinance, 1937. According to the suit, Lorenzo had been able to manipulate his usage rates by inserting "a piece of cinema film in the meter," a common means of manipulation since the early 1930s. The suit charged that Lorenzo, who had been a customer for the past seven years, had been deceiving the company in this way for several years.[27]

Perhaps the most notable case of this new electrical crime was uncovered in February 1938 by the police at Mea Shearim Police Station. It was notable not only because it concerned an unusually large case of electricity theft but also because it took place at the house of Hajj Amin al-Husayni, the Grand Mufti of Jerusalem and leader of the Palestinian national movement. Hajj Amin himself was no longer living in the house by that time, having gone into exile the previous year for his role in the Great Revolt, and it is not clear who, exactly, was responsible for the jerry-rigging. The electricity account was under the name of Fakhri al-Husayni, Hajj Amin's deceased brother, a lawyer and member of the Supreme Muslim Council.[28] When the police uncovered the electricity theft, members of the Husayni family discretely approached the power company, offering to make full restitution in return for keeping the case out of the courts and, presumably, the papers. The power company agreed and wrote to the attorney general, offering its heartfelt gratitude for the diligence of the police in clamping down on "a crime which is becoming all too prevalent." But, the letter continued, "we think that in this case any action may be construed as political bias on the part of our Corporation, who claim to be entirely outside politics in our dealings with consumers."[29] It is of course a great irony that the company was making a blatantly political move in order to appear "outside politics," but it should not surprise us, given Rutenberg's diligent work to appear above the fray precisely by manipulating it to the utmost. The case made waves among British officials in Palestine as well as the Criminal Investigation Department of the Palestine Police Force, which recommended that the case be dropped to avoid "undesirable publicity."[30]

STRUGGLING TO KEEP UP

By the outbreak of World War II, the Jerusalem power company had seen a long period of considerable system growth, and the generating capacity of the system was expected to be sufficient for the coming four or five years. In 1941, the peak load on the system was 2,575 kilowatts; this grew to 3,200 kilowatts the following year, well within the capacity of the station's maximum output of 6,600 kilowatts. But demand for electricity continued to grow precipitously during the war years, both because new consumers were added but, more important, as a result of the increased per capita consumption, mostly as people developed a taste for high-powered lightbulbs and electrical heaters. (The soaring demand for electrical heaters was chiefly the result of electricity's being the only source of energy whose cost did not grow several times over the course of the war.) Overall, consumption increased during World War II by as much as 80 percent, and average consumption per consumer nearly doubled. An added difficulty of the increase was that despite repeated calls from the power company that consumers should give notice of any increase in consumption, most did not. The peak load on the system in 1939 was 2,500 kilowatts, but the demand by the end of the war had more than doubled to 6,000 kilowatts, creeping up on the maximum output capacity of the plant.[31]

The war ushered in a period of prosperity in mandate Palestine. Although the war sent prices soaring, the cost-of-living allowances paid out by most companies more than covered the difference.[32] Campbell-Brown, the general manager of the Jerusalem power company, noted with a mixture of enthusiasm and concern that even regular wage earners were becoming much better off, and "wishful of the 'luxuries' they could not previously afford." The reason for his concern was that such luxuries included electricity. "There seems to be no let up in the number of applications for supply," he noted with alarm in a letter to his London supplier.[33]

The power company became increasingly concerned over its ability to meet demands on the system. Starting midway through the war, it put out repeated calls on consumers to restrict electricity usage during peak hours, from 5:00 to 9:00 p.m., by reducing the number of lamps in use, especially the new, high-powered ones that had just begun to appear on the market, and avoiding using electric radiators altogether. As late as the end of 1942, the power company still believed that its current capacity would carry it through the war years. But 1943 proved to be a critical year, which saw "a sudden and abnormal rise in consumption of electricity." Further increases of generating capacity at the plant became an acute necessity, as it became clear that the power company would not be able to meet demand for much longer. Moreover, according to the company's estimates, demand was likely to continue to grow rapidly after the war, as wartime privations caused people to save more than usual. "The general standard of living will go, and remain, up," predicted Campbell-Brown.[34]

The peak load the previous winter (1942–43) was 4,100 kilowatts, and the power company anticipated that the load for the coming winter would be far greater, upwards of 5,000 kilowatts. Among other things, the company expected a run on electrical heaters. The maximum generating capacity of the plant, we recall, was 6,600 kilowatts, distributed over nine engines. As a result, if a single generator failed, the power company would not be able to meet demand. This caused an additional problem because the company had to run its engines much harder than advisable, significantly shortening their life span, which under normal circumstances would have been fifteen to twenty years. Demand was likely to grow even more by the following winter, 1944–45, for which the company estimated a peak load of 6,000 kilowatts. The war's end was expected to generate an immediate surge in demand of 500 kilowatts. One major concern associated with insufficient reserve capabilities is that it becomes difficult to comfortably meet fluctuations in demand. For instance, a temperature drop of five degrees centigrade resulted in an average increase in demand of 200 kilowatts. In other words, by next winter, even a slight malfunction or brief cold spell would cause blackouts.[35]

An expansion of generating capacity was urgently needed at the Jerusalem powerhouse, but wartime privations posed obstacles to the delivery of the necessary equipment. The general manager wrote to his suppliers in Britain asking them to begin making arrangements for two 2,000-kilowatt generators to be manufactured and shipped as soon as possible. In March 1944, the company formally put in the order for the first generator. The manufacturers promised to deliver the first set in December 1944 to meet the winter load. But there were severe restrictions in effect in Britain on the manufacture of heavy electrical machinery for civilian purposes, as all available manpower and materials (particularly steel) was directed primarily toward the war effort. To get an order approved, the company had to request a Certificate of Essentiality and an import license. The Palestine Government in turn had to forward such requests for consideration by the Middle East Supply Board in Cairo. Once those permits were in hand, the company had to turn to the Colonial Office to obtain the necessary permits from the Inter-Departmental Committee of Heavy Industries. Meanwhile, the company's suppliers in London estimated that it would take about twelve months for the first engine to arrive in Palestine.[36]

After much effort, including intense pressure put on various government bodies in Britain and the Middle East, the power company was able to cut through some of the red tape (although the efforts to expedite the order itself involved a fair amount of red tape).[37] When the matter was finally settled in April 1944, the engine on order—a Crossley-Premier of 2,400 horsepower—would not be ready to ship until February or March the following year, a result, among other things, of labor shortages having grown more severe.[38]

Meanwhile, the company had been proved correct in its assessment of future growth. This is ironic, because Campbell-Brown had in fact inflated his future projections to convince his directors in London to spend the necessary £50,000 on a new engine and also to persuade the director of public works in Palestine to grant the Certificate of Essentiality. The rapid growth of electricity consumption in Jerusalem had rendered an exaggeration an understatement.[39] Peak load during the winter of 1943–44 increased over the previous year by 50 percent, to 4,860 kilowatts, and the following winter, 1944–45, it rose to 6,390 kilowatts. The company made it through the winter months by the skin of its teeth, barely managing to avoid major service interruptions. With no hope of getting a new engine on time, considerable disruptions to the service looked inevitable during the coming winter peak of 1945–46. By February 1945, the power company raised its assessment of future demand, predicting a peak load in November 1945 of 7,800 kilowatts, while the plant's capacity remained at 6,600 kilowatts. Even with all engines running at 110 percent, the company could not hope to cope with demand.[40] The company put in the order for the second engine in March, scheduled for delivery in another twelve months.[41]

In the event, the first engine was not brought into commission in Jerusalem until July 1946, at which point it was already insufficient to meet demand, which was then approaching 7,000 kilowatts. The power company hoped to finally catch up with demand with the arrival of the second engine. It did not arrive until late 1946 and early 1947 (in several shipments; the engine in its entirety weighed 222 tons and was shipped in fifty crates) and put into commission in early 1947. Even before the arrival of the first engine, the company had put in orders for two more 2,000-kilowatt generators. They too were greatly delayed as a result of the shortage of materials and labor that persisted into the postwar period, as well as a dockworkers' strike in Britain, involving forty thousand dockworkers who struck for five weeks in the fall.[42]

PRESENCE OF ABSENCE

As the skyrocketing demand for electricity in Jerusalem suggests, by the early 1940s, electricity was fast becoming an integral part of everyday life and commerce, and a substantial part of daily expenses. The lawyer 'Issa Nahla, whose electricity bills were typical in every respect, spent £P12,600 on electricity every month, about one-tenth of what he spent on rent.[43] The power company's inability to keep up with demand therefore seriously impacted all areas of life in Jerusalem. The rolling blackouts affected all parts of the city, though some worse than others. If electricity had become the commodity that "solves everything," its absence was now a universal obstacle. To make matters worse, the power company was prevented by the authorities from announcing power outages in advance out of safety concerns.[44] The blackouts thus help clarify electricity's significance at the time.

One such instance of a newfound reliance on electricity came from the matron of Jerusalem's leper hospital in the fall of 1946. Writing to the district commissioner, J. Tuqan, she complained that the daily nuisance of blackouts made work difficult. Tuqan wrote back with assurances that the Jerusalem power company was working to fix the situation. The matron soon wrote again, however, telling Tuqan that two recent power outages had caused accidents at the leper home. The first accident happened to one of the sisters, who had fallen down and hurt her head and arm, injuries from which, three weeks later, she had still not recovered. And only the day before, in an event that had prompted the matron to write the letter, a helper had tumbled down a staircase when the lights went out and broken her arm. The matron reminded the district commissioner that the hospital was doing very important work, and that in the absence of proper lighting the old building, with its many stairs and steps, could be "very dangerous for patients and staff." In consideration of this, the matron asked that the leper home be exempt from the rolling blackouts. Tuqan was naturally compelled to let her know that it was not technically possible to exempt a single building from the sector-wide blackouts.[45] But the broader significance of the matron's claims has to do with the degree to which people had come to rely on electricity in Jerusalem. The old building that housed the leper hospital had of course been used for years before electricity had been the source of illumination. Yet the sudden absence of electric light had made the place dangerous in a way that it had not been before its introduction, when perhaps people were more familiar with darkness and its dangers or relied on alternative technologies for lighting the structure's darkened interior.

Electricity—including the equally new and important phenomenon of its absence—also had consequences beyond tumbles down stairs. In the fall of 1946, Harry Gottlieb, a discharged soldier recently returned to Jerusalem, wrote to the power company with an urgent plea. On his return in the spring after four years of service, Gottlieb began to set up a woodworking workshop. He was able to get a loan to buy the lathe, and the district commissioner requisitioned a workshop for him on Ratisbone Street. Gottlieb then applied to the power company for a connection, but the application was denied—"to my great chagrin," Gottlieb noted in his letter. He described the rejection of his application as a "death sentence." Without electricity, he explained, "my motor cannot work and I cannot start to work and earn a living for myself and my family, nor can I pay the large debt contracted by me in order to be able to buy the machinery for my workshop." He also turned to the district commissioner, describing his access to electricity as a "question of life or death." The district commissioner took pity on Gottlieb and intervened on his behalf with the power company. Soon thereafter, Gottlieb had electricity at his workshop.[46] As in the case of the leper hospital, Gottlieb's expectation of access to electricity had made him more vulnerable to its absence.

At the same time, another ex-serviceman, Raji Qubayn, a resident of Beit Jala, was experiencing the presence of the same absence. Upon his discharge, he moved with his family into a house, rented from Salim Katimeh. In the early fall, in anticipation of the cold and dark of the winter months, Qubayn paid Katimeh £P7 to arrange for the installation of electricity in the house. Katimeh took the request and the money to the offices of the Jerusalem power company. Three months later, however, the house was still not connected. For Qubayn, the greatest damage of the absence was done to his children's schooling. "I have four children," he told the power company, "who all go to school, and have to study their lessons during night time, and the light of the lamps we are using at present, besides being not very helpful to them, hurts their vision." Three Sundays in a row—Qubayn's only day off—he had gone to the company offices to seek redress, but without success.[47]

One of the most common uses of electricity mentioned in the context of its absence was crime prevention. During the blackouts of the winter of 1946–47, according to a report in the *Palestine Post*, a "sneak thief" cloaked by the darkness had climbed into a ground-floor bedroom on Abyssinia Street and stolen £P2 from the nightstand.[48] At the end of the long winter, the *mukhtars* of Wadi al-Joz and Aqabat al-Suwana, both neighborhoods in eastern Jerusalem, addressed the district commissioner with a request for the installation of electric light in their neighborhoods. They wrote, "We are in sore need of the illumination of this area especially during winter. Burglary and larceny are committed frequently in this area." Ultimately, the claim was dismissed by the municipality, with the motivation that Aqabat al-Suwana was outside the municipal boundary and Wadi al-Joz was already "adequately lit."[49]

Questions of electricity also had important implications for religious observance among Jerusalem's many confessional groups. On January 4, 1947, Joseph Weinberg, the chairman of the neighborhood committee of Beth Israel, a quarter of ten thousand residents, almost all ultra-Orthodox Jews, wrote in rage to the district commissioner. The day before, the neighborhood had had to endure a blackout that lasted for three hours and fifteen minutes, from 5:00 to 8:15 p.m. This happened to be the Sabbath, and because of the religious proscriptions against lighting candles on the day of rest, the unannounced blackout had forced the community to "stay in darkness on Friday night, both at home and at the synagogue." The chairman pointed out that if the community had been given advance notice of the power outage, it would have been able to make other arrangements. There was general agreement among the neighborhood residents that the power outage was deliberate and discriminatory. The power company, the letter stated, "dared not stop the supply of electricity on Christian Festivals in Christian quarters or Bethlehem." Yet not only had residents of Beth Israel had to spend this recent Sabbath in darkness; the previous Rosh Hashanah the power had also been cut, despite repeated requests that the neighborhood would be spared during the holidays. The malice was all the more evident, according to the chairman, given that "no shops, factories or bakeries are open on Friday night," and a blackout at that time therefore saved precious little electricity. The behavior of the company, the letter asserted, "contravenes all the articles of its concession and utterly disregards the Jewish public." The district commissioner forwarded the letter to the power company and urged it to avoid cutting power to communities of observant Jews on the Sabbath. In its reply, the power company assured the district commissioner that the blackout was the result of "extreme emergency and not willful." The letter claimed that the company did its utmost "to observe the religious susceptibilities of all communities when arranging our switching programme." The district commissioner sought to assuage Beth Israel by echoing the power company's assurances regarding the reasons for the power outage.[50]

If what brought all these cases together was a lack of electricity, there was also the opposite problem: too much electricity, that is to say, electrocution.

This new kind of hazard ensured that the technology could be, as Chris Otter terms it, "a death force as well as a life force."⁵¹ In the earliest days of the Jerusalem grid, Ibrahim 'Abd al-Ahad of Bethlehem lost his life while employed on a maintenance crew at a local construction site. He accidentally came in contact with a low-hanging 6,600-volt high-tension electrical wire. With the help of Awni Abd al-Hadi, the prominent Palestinian lawyer (and political activist in the Istiqlal Party), al-Ahad's widow, Maryam, was able to receive compensation from the company in the amount of £P250, a significant sum, but a pittance in the face of the burden of providing for four children, the oldest of whom was only eleven.⁵²

BERNARD JOSEPH V. JERUSALEM ELECTRIC AND PUBLIC SERVICE CORPORATION

The most spectacular consequence of the power outages took the form of a much-discussed court case that soon took on countrywide significance. On the morning of December 3, 1945, at the main power station in the German Colony, a piston crown in engine number 8 cracked, forcing the supervisor to shut the engine down for the day. A replacement piston was fitted during the night, and the engine was up and running again the following day. In order to carry out the repairs over the next few days, however, it was necessary to cut off several feeders in Jerusalem, which meant blacking out large sections of the city. This was done shortly before 5:00 p.m., and a rolling blackout was enforced in the city. From 5:00 to 7:00 p.m. it hit the central area of Jerusalem, as three of the six lines on the Chancellor Street substation were disconnected. This was not the first time the company had such troubles. In fact, working with machinery in dire need of repair and spare parts, it had been facing similar issues since late 1944. The station's mechanical engineer sent numerous reports warning about the poor state of the equipment. The company tried to remedy the situation by sending spare parts to Egypt for repair. Usually, however, this was a drawn-out process, and the welded cylinders held up for only a short period, up to six months at best, but sometimes they lasted for as little as a couple of days. By November 1945, the company had started to run its system with restricted capacity due to the poor state of the equipment.⁵³ It publicized warnings in at least eight newspapers, in the three official languages—English, Arabic, and Hebrew—as well as in German and Polish. The notice urged consumers to adhere to five rules, which all served to reduce nonessential electricity use.[54] The public did not heed the company's call to restrict electricity use. On the contrary, one survey of three hundred consumers found that the vast majority had actually increased their use.[55]

The service disruptions were becoming a real nuisance in the town, and the Jerusalem power company was under considerable pressure to remedy the situation. The company, especially its general manager, Campbell-Brown, was acutely aware of the delicate position in which the company found itself vis-à-vis the public, a position all the more precarious given that the PEC had successfully avoided similar problems in the rest of the country, even in the face of wartime shortages. Apparently deciding that offense is the best defense, the company blamed consumers for the outages. In a statement to the Public Information Office, Campbell-Brown noted that for fourteen years he had run the power company with almost no interruptions. The difficulty that it experienced in meeting peak demand in the winter of 1943–44 was entirely the result of people's irresponsible consumption, in particular, their desire to offset the war-induced shortages of coal and fuel oil by means of electrical generators.[56]

In December, the *Palestine Post* interviewed Campbell-Brown regarding the issue. He assured the *Post* that the company was not insensitive to the inconvenience and financial burden that blackouts caused consumers, or to the fact that after urging people for years to use electrical equipment, the company "suddenly had to ask them to stop." He promised that there would be no need for restrictions next winter, as another generating set was on its way and was expected to arrive in June 1946. When the interviewer pointed out that the PEC had been able to meet demand in the rest of the country throughout the war, despite laboring under the same conditions as the Jerusalem company, Campbell-Brown denied that there could be any comparison. When pressed further, he shot back in an obvious reference to the PEC that "certain interests in this country want to get a stranglehold of the country," by obtaining an electricity monopoly that they could then use for "political purposes."

In response, the managing director of the PEC, Abraham Rutenberg, who had taken over after his brother Pinhas's death in 1942, stated to the *Post* that his company had no designs on the Jerusalem market. The PEC already supplied more than 90 percent of the electricity market in Palestine, and Jerusalem would make "very little difference." He reminded the *Post*'s readers that "apart from the very large sums of money, it had cost the company the lives of a number of its employees to maintain an uninterrupted service during the riots [i.e., the Arab Revolt]." In general, he went on, "the PEC's part in the building up of this country is so substantial and so self-evident that it needs no comment. This country's considerable contribution to the war effort would have been impossible but for the large-scale conception of my late brother, Pinhas Rutenberg." "The PEC had never failed Palestine, even in the most difficult times," Rutenberg concluded, dismissing Campbell-Brown's "insinuations" as "utter and unworthy rubbish." He added that "a breakdown of service need not be followed by mental and moral breakdown as well." Yet, he said, the PEC would still be happy to provide the Jerusalem power company with assistance if it requested it, which it had not yet done.[57]

Early the following year, the law firm Messrs. Bernard Joseph & Co., on Princess Mary Avenue, brought a case against the power company before the Jerusalem Magistrate Court for breach of the concession and the customer agreement. The amount of the suit was £P20 for loss of income and working capacity during two major blackouts. The case garnered considerable attention in the Palestinian press because it highlighted the question of whether the company was deliberately discriminating against Jews. The liberal Hebrew daily *Ha'aretz* published an article giving the details of the case, where it also voiced other protests against the company, which perceived a "growing lawlessness" as a result of its poor service. One letter published in the paper decried it as "abuse" of the Jewish community. That it was the result of ill will and discrimination was "proven," according to the letter, by the fact that all parts of the town were not affected evenly. Only the Jewish neighborhoods suffered this "'malfunction,'" a word put in scare quotes in the letter, which concluded: "The Jews of Jerusalem wonder how long the company will go on abusing them?"[58]

The sums involved in the case were negligible. Bernard Joseph, after all, had sued the company for £P20, a pittance by any measure. By comparison, the lawyer retained by the power company charged a £P100 retainer fee, and £P25 for each court sitting. In the end, the trial ran over ten hearings, and the bill ended up amounting to several hundred pounds. As of May 1, 1946, the defense had spent almost as much on photocopies (£P18) as the suit sought in the first place.[59] In his ruling, the judge pointed to the principal importance of the case, noting that the "proceedings were not instituted only with a view to obtaining payment of LP.1 or LP.50 but for the purpose of causing defendants to perform their duties towards the public." As was clear to all parties involved, what was at stake was the question of responsibility more broadly. According to a report from the proceedings published in *Davar* in the spring of 1946, both sides considered the matter as a "test case."[60] It was clear, in other words, that more than money was involved, and much of the proceedings took on an emotionally charged tone.[61]

The first court date was set for March 19, and the proceedings continued through the spring and summer. In October, the Magistrate Court handed down the verdict and ordered the power company to pay damages to Bernard Joseph for "non-fulfillment of their duty." In addition to the £P20, the power company was ordered to pay court fees of another £P34.[62] According to the judge's ruling, the machinery in the plant was "old and weak," and thus the breakdown could not be considered "unforeseen, fortuitous or unexpected." Instead, the power cut was the result of the defendants' failure to maintain the plant properly. The power company, which, the judge noted, had made considerable profits during the war years, could have—and should have—obtained spare parts locally. (The power company denied this, claiming that locally produced spare parts were three times as expensive and half as good.)

Another critical consideration in the judge's ruling was the failure of the Jerusalem power company to seek assistance from the PEC. Regardless of who was to blame for the shortages, he wrote, "it was not beyond the power of the Defendants to try and alleviate the situation by approaching the other undertaking of electric supply in Palestine, the P.E.C., with a request for cooperation as was, in fact, done in the United Kingdom by the various similar undertakings in order to maintain an uninterrupted supply of electric current to the consumers." Indeed, the PEC had publicly expressed its willingness to assist numerous times. In the proceedings, it transpired that Campbell-Brown had looked into such assistance but in the end had decided against it because it would have taken longer to hash out the terms and construct the connection between the systems than it was expected to take for the ordered equipment to arrive.[63] The closest point between the two systems was the PEC network in Ramleh and the Jerusalem company's line to Kiryat Anavin, eight miles west of Jerusalem, near the Arab town of Abu Ghosh. According to Campbell-Brown's testimony at trial, it would have taken three to four months to complete the intercommunicating high-tension mains between these two points, separated by about twelve miles. When Wilson Blaze, the Jerusalem company station manager, was asked during the trial whether it would have been possible to receive assistance from the PEC, he replied that he did not know, but added that they did operate similar diesel engines.[64]

The episode demonstrated how complicated the links between electricity and political power had become. The blackouts that prompted the case in the first place had likely been the result of a straightforward business failure. But the seemingly straightforward failure soon overlapped with the ongoing ethnonational conflict within Palestine. In an attempt to deflect criticism, Campbell-Brown had invoked the specter of Jewish control of the grid; meanwhile, they themselves were targets of accusations of discrimination on ethnic grounds.

Continued difficulties to obtain new equipment and spare parts, along with the loss in court, compelled the Jerusalem power company to abandon its earlier opposition to forge a PEC interconnect. In anticipation of another winter of widespread service interruptions, the power company initiated negotiations with the PEC in the late summer of 1947. The agreement, which was finally concluded in December 1947, stipulated that the PEC would build and operate a high-tension line and a receiving step-down transformer substation in Jerusalem on the old Jaffa-to-Jerusalem road. The Jerusalem company would buy the energy in bulk and transmit it over the local grid. The terms of the agreement reflected the balance of power between the two companies. The Jerusalem power company was to pay the PEC £P400,000 for the capital expenditure, of which the PEC demanded £P100,000 to be paid as a cash advance before signing the agreement in December. The rest of the money was to be paid upon signing the agreement. Despite this, ownership of all the materials would still be in the hands of the PEC.[65]

CONCLUSION

The stories collected in this chapter give an indication of the social, economic, and political reverberations of electricity at street level and, more precisely, how electricity, as it was domesticated, came to interact with the society around it in a variety of ways. As a new technology is introduced and becomes ubiquitous, it transforms material and mental landscapes: its absence becomes an event. The arrival of a new technology creates a presence, and thus also the possibility of the presence of an absence. Moreover, as several of the incidents discussed in this chapter indicate, most notably the Bernard Joseph trial, once electricity became ubiquitous, the disputes became a window onto, and were interpreted through, existing friction points, most notably, of course, the Jewish-Arab antagonisms that by the 1940s had grown both deep and wide. What the Palestinian businessman Boutagy, the ultra-Orthodox residents of Beth Israel, and the lawyers at Bernard Joseph all had in common was that they read electrical service issues through the prism of ethnonational antagonisms.

Nevertheless, one obvious consequence of the atmosphere that prevailed in 1940s Jerusalem was that the power company was extremely wary of appearing to discriminate against any group in particular. It was, for instance, willing to overlook a significant instance of electricity theft, because it had taken place at a house belonging to a prominent Palestinian family. This points to a corollary, and more significant point, about how the heterogeneous engineers of this electrical enterprise negotiated the question of politics. They were extremely anxious to avoid the impression of political bias and had no qualms about employing decidedly political means to appear apolitical. No one seemed to reflect on the blatant contradiction inherent in doing so.

SEVEN

Statehood and Statelessness

FOR THE FIRST TWO DECADES OF British rule in Palestine, economic considerations dominated British policy. The white paper of 1922 established a procedure for determining Jewish immigration quotas based solely on the economic capacity of the country to absorb them. "Leave politics alone," Herbert Sidebotham summarized the policy laid down by the white paper with respect to the Jews. "The sole test of your rights of immigration are [*sic*] economic; so long as your immigration does not exceed the economic capacity of the country to absorb it, all will be well."[1] While the picture is perhaps not as one-sided as Sidebotham would have it, it is true that every major instance of violence engendered a commission of inquiry, all of whose reports identified economic issues as being at the heart the problem. After the so-called Wailing Wall Riots of 1929, which killed 234, the Shaw Commission diagnosed the violence as a consequence of immigration having exceeded economic absorptive capacity, though it also identified several political grievances. Nevertheless, the subsequent follow-up studies, which laid the foundation of Palestine policy in the 1930s, focused their recommendations primarily on economic initiatives.[2]

Only in 1937, with the report of the Royal Commission, commonly known as the Peel Report, presented in the fall, did the British abandon their economistic approach to Palestine and begin to reckon with the *political* situation not as a symptom but as the primary cause of what the report took to calling "the disease."[3] The moment that the British concluded that the conflict in Palestine could not be solved by economic means, they seemingly abandoned any hope of solving it at all. The widespread sense among the British that economic problems were solvable, whereas political problems were not, may itself have delayed the realization that politics, not economics, lay at the heart of Arab-Jewish relations in Palestine. Be that as it may, the conclusion that the problem was political caused the British to abandon hopes for harmonious coexistence and instead embrace the idea of partition, "the only method we are able to propose for dealing with the root of the trouble." With the idea of partition, borders once again became a central concern. The Peel Report advised dividing Palestine into three parts. Some 20 percent would be allotted to a Jewish state; most of the rest would become an Arab state, with a small third area designated as an internationally governed *corpus separatum* that included Jerusalem and Bethlehem, as well as an east-west land corridor affording access to the Mediterranean coast.[4]

In the area around Naharayim, the report called for borders that left only the western half of the Sea of Galilee and none of the land east of the Jordan River in the future Jewish state. In other words, large portions of the Jordan hydroelectric works, including the powerhouse itself, a significant stretch of the company's new high-tension line, and a number of Jewish settlements in the Samakh Triangle (including Degania A and Degania B, Kfar Gun, Afiqim, and Dalhamiya), would fall outside the borders of the Jewish state. Rutenberg faced down this challenge in typically unrelenting fashion. Before the report was even published, he and his associates at the power company had begun lobbying the authorities to change the border. In one note to the Colonial Office, Rutenberg stated: "There seems no reason to have the boundaries traced in such a way as to have sections of our main trunk transmission lines in different states under different police control, thus endangering the normal service of these lines."[5] In July, Colonial Secretary William Ormsby-Gore complained that Rutenberg had been "sitting" on him for the last week and a half.[6] Rutenberg's concerns, however, were premature. As his protestations traveled from the Middle East Department to the colonial secretary, an official scribbled in the margin, "The Royal Commission sketched only a very approximate boundary and it is doubtful whether they ever considered in detail the relative advantages or disadvantages of excluding from the Jewish State the small triangle between the Yarmuk and Jordan."[7] To Gideon Biger, the Peel Report's exclusion of the area around Naharayim from the Jewish state "hinted that desolate Jewish settlements couldn't influence the determination of the boundary delimitation."[8] But contrary to Biger's claim that the overriding concern of all partition proposals was settlement disbursement patterns, every single partition proposal following Peel included Naharayim and the whole of the sparsely populated Samakh Triangle in the future Jewish state.

The Peel Report itself noted that its border proposal was not final and called for the creation of a special border commission that would prepare a more considered partition plan. That commission was formed in 1938. Officially named the Palestine Partition Commission, it was more commonly referred to by the name of its chairman, the weathered India hand John Woodhead. The commission's final proposal was indeed more in line with Rutenberg's wishes, identifying the location of the powerhouse and the presence of the Jewish settlements as strong reasons to adjust the borders proposed by the Peel Plan. Since "only a small part of the electric energy produced by the Palestine Electric Corporation is sold to consumers in the Arab State outlined," and by far the majority was sold to consumers in the projected Jewish state, this boundary "should be modified so as to include the power-station within the Jewish State."[9] Indeed, the report went on to suggest that not only the area of the works should be included in the Jewish state but also the surrounding 6,000 dunams that the PEC had bought from Emir Abdullah in 1927. The report also adjusted the borders so as to keep all high-tension mains on Jewish territory, quite in accordance with Rutenberg's counterproposal (See map 2).[10]

With the renewed efforts to solve the Palestine question after World War II, new border proposals were floated. They all recommended the inclusion of the Samakh Triangle in the Jewish state, even though, as Major Young had griped in 1925, it amounted to "throwing a part of Trans-Jordan into Palestine," given its location on the eastern bank of the Jordan River.[11] The Morrison-Grady plan from 1946, which recommended allocating only 17 percent of the land to the Jewish state, nevertheless supported the inclusion of the area around the Jordan works, as well as all high-tension transmission lines.[12]

In February 1947, Britain announced its decision to withdraw from Palestine and referred the matter to the United Nations, which formed the United Nations Special Committee on Palestine (UNSCOP) on May 15, 1947. According to UNSCOP's proposal, presented early in the fall, 56 percent of Palestine would be included in the Jewish state, containing a population of 900,000 people, of whom 500,000 were Jews and 400,000 were Palestinian Arabs. The entire area of northern Palestine, including all the land around Naharayim, was designated for the Jewish state. The Arab state, under the proposal, was to be allocated 33 percent of the land, home to

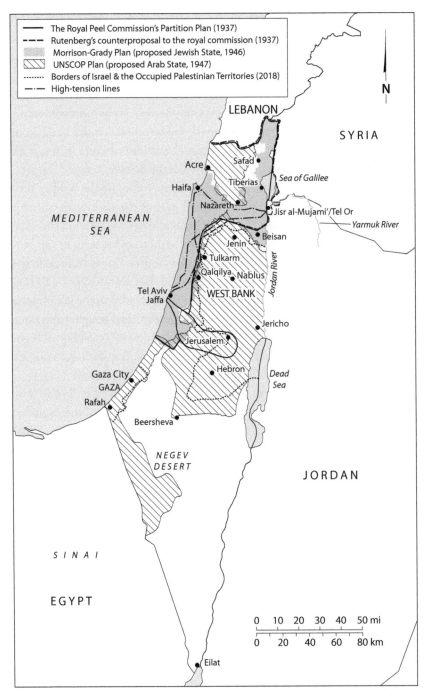

MAP 2. Partition proposals and the grid at the end of the mandate.

700,000 people, 10,000 of whom were Jews. The international zone centered on Jerusalem, and again including Bethlehem, contained 200,000 people, half of whom were Jews. The UNSCOP proposal was then passed on to an ad hoc committee, which made several smaller changes, including moving the border of the Jewish state to include more of the Dead Sea at the insistent urging of Moshe Novomeysky, the concessionaire and engineer who had been operating a mineral extraction factory there since 1930.[13] This further underscored the strong influence on the partition deliberations of large industrial ventures.

Of course, in the end, none of the numerous partition proposals that were presented in the decade from 1937 to 1947 ended up deciding the borders. Ultimately, it was troop placement and armistice negotiations that decided the shape of the future Jewish state. Nevertheless, the obvious importance of industry and infrastructure in considering the border proposals demonstrates the central role they occupied in the minds of colonial officials at the time. And although Naharayim, as we will see in the next section, would soon be occupied by Arab forces, the transmission network remained wholly within the borders of Israel by the end of hostilities in 1949. In fact, the most recently constructed high-tension line, which ran between Naharayim and Tel Aviv and had been built more than a decade before the 1948 War, clearly traces out the border between Israel and the West Bank, which since 1967 has been under Israeli occupation. Moreover, the proposal for the new high-tension line contains a list of place-names, meant to indicate the route of the wires. Almost all of the Arab villages listed, Kawkab al-Hawa, Danna, Kafra, Sirin, Lajjun, and so forth, were depopulated in 1948.[14]

THE 1948 WAR AND THE DESTRUCTION OF NAHARAYIM

The war for Palestine broke out almost immediately after UNSCOP's partition proposal was adopted by the UN General Assembly in November 1947. Until May of the following year, the fighting was limited to Palestinian Arabs and Jews. Amid a civil war whose fault lines the power company had been instrumental in creating over the past few decades, the PEC engaged in now-familiar boundary-work that presented the company as outside the realm of politics. In March 1948, the company drafted a letter to the mayors of all Arab towns in the country, expressing regret that "irresponsible people have caused damage to the Corporation's lines where same pass Arab areas." According to the letter, high- and medium-tension lines were "blown up and damaged" between Jisr al-Mujami' and Jenin and between Afula and Tulkarm, and also near Qalqilya. The company decried these acts, which it characterized as "theft and robbery," and chastised those elements of the press that had construed them as "acts of warfare": "We feel that our Corporation which has always stood and must of necessity, as a public utility body, stand outside political and racial strife, and which has always served all sections of the populations impartially and cannot be accused of having accorded any particular section of the population preferential treatment, should be kept out of inter community imbroglios and that it and its installations should be considered strictly 'neutral,' if one may use the term." The statement perfectly captures the politics of non-politics that the company had practiced since the early 1920s.[15]

On the afternoon of May 14, 1948, in the main hall of Beit Dizengoff in Tel Aviv, David Ben-Gurion read out Israel's declaration of independence. Except in Jerusalem, where the electricity had been cut, the Jewish population of Palestine heard the prime minister's words in real time, huddled in front of their electrical radios.[16] At precisely the same time, Iraqi troops were spotted approaching the Jordan powerhouse. Having declared war on the new state along with a number of other Arab countries, the Iraqis sacked the plant that night and detained the thirty-eight workers still at the site. The power company sent numerous letters pleading with the king of the newly independent Hashemite Kingdom of Jordan to secure their release. "Some of these people," one letter reminded the monarch, "carried out various technical jobs at different times in your palaces and were privileged to become acquainted with your noble personality."[17] The pleas fell on deaf ears, and the workers were not released until February 1949, after the conclusion of the Israeli-Jordanian armistice agreement.

Under that agreement, most of Naharayim remained within Israel's borders, and members of the nearby kibbutz Ashdot Ya'cov continued to farm the land surrounding it. The plant itself, however, sat precariously on the border between two states that would remain in a formal state of war for another half century, leaving the station inoperable. Following the armistice agreement, PEC officials met with Abdullah in Amman in hopes of convincing the king to allow the plant to be reopened, but he proved unwilling.[18] As the so-called Clapp Report of the United Nations Economic Survey Mission for the Middle East was issued later in the year, recommending a large-scale development project that would divert the waters of the Jordan and Yarmuk Rivers in a multinational cooperative venture between Israel, Jordan, Syria, and Lebanon, the PEC continued, fruitlessly, to argue its case before the Jordanian court and government.[19] Early in 1953, Jordan and Syria concluded a series of agreements to develop the resources of the Jordan and Yarmuk through a system of irrigation and electricity generation.[20] Later that year, Jordan signed an agreement with the United Nations Relief and Works Agency for a $40 million development scheme called the Yarmuk-Jordan Valley Project.[21] The PEC, meanwhile, prepared a long letter addressed to the Jordanian government. The letter called attention to the company's concessionary rights, but the real stress lay on the mutual benefits that Naharayim stood to generate. "Viewing the situation from a practical standpoint," the letter argued, "the fact is that the hydro-electric development which could be of immense benefit to the peoples of both Israel and of Jordan has remained idle for five years."[22] The letter went on to suggest that allowing operations to recommence would be a first important step toward the rehabilitation of the Palestinian refugees in the area. The letter then presented a vision for an "integrated power system ... under common and unified control," which would not only bring great economic benefits but also promote "stable neighborly relations, economic and otherwise."[23] The PEC never received a reply. In March 1954, Jordan formally canceled the PEC concession.[24]

For four decades, Naharayim remained out of reach to Israelis due to its proximity to the armistice lines, but with the peace agreement signed between Jordan and Israel in 1995, Israel formally ceded the land to its eastern neighbor in return for a lease agreement allowing the members the kibbutz Ya'acov Ashdod to continue farming the land. The site of the powerhouse itself was remade into a binational park, called the Island of Peace.

NATIONALIZATION

By the time King Abdullah of Jordan canceled the concession, ownership of the Palestine Electric Corporation had passed into the hands of the young Israeli state. In the years after independence in 1948, the power company experienced considerable challenges, including war and several large immigration waves that doubled the population of the country in the first three years of its existence. Furthermore, shortly before leaving, the British expelled

Palestine from the sterling area, making for a rocky start for the Israeli pound (£I), introduced shortly thereafter.

In 1948, the company installed another generating set in the Reading powerhouse in Tel Aviv, which later took a direct hit from aerial bombardment. In addition, seven employees perished during maintenance work before the end of the war in early 1949. The year 1948 was the first in the company's history in which it experienced a decrease in the number of units sold. It was also the first year in which the company failed to meet demand; the loss of Naharayim had reduced its generating capacity by one-third, and in the summer of 1948 it was forced, for the first time, to impose restrictions on consumption.

Unsurprisingly, perhaps, the flight and expulsion of some 750,000 Palestinian Arabs in the course of the war did not affect the company significantly. As the chairman summarized the matter at the annual meeting: "The evacuation of a great part of the Arab population necessarily involved a loss of consumers; nevertheless the total number of consumers increased by 4,241, making the total at the end of the year, 128,561."[25] Indeed, Israel in its infancy was not a country at risk of depopulation, and the consumption of electricity in 1950 rose by 40 percent over the previous year and by six times over its level in 1938. As the chairman stated at the annual meeting of 1951: "All the towns are surrounded by larger areas of new buildings; hundreds of new villages have been founded; factories and workshops are springing up in all the industrial district and in many of the larger settlements; the main roads are thronged with lorries; the harbours are working to full capacity and large extensions are beginning."[26]

The company was faced with many serious challenges in the early years of Israeli statehood. In 1950, the British Treasury refused to sanction the issue of new capital for the company, which prevented it from making its planned expansions. As a result, in the summer of 1951, the company again proved incapable of meeting demand and was forced to impose "severe restrictions" on electricity use, by both staggering industrial working hours and cutting off supply for domestic consumption entirely during certain hours of the day. Although the arrival of a 30,000-kilowatt turbo-alternator set and boiler plant in December 1951 put the company back ahead of demand,[27] the PEC was soon hit by the biggest crisis of its existence. This unprecedented crisis, ironically, was inflicted by the Israeli government. In February 1952, the government introduced its so-called New Economic Policy (NEP), whose measures included a readjustment of the exchange rate of the Israeli pound. In an effort to encourage production and reduce consumption, Israel devalued its currency to about one-third of its former value in relation to the sterling. As a result, labor and commodity prices rose sharply.

In the short run, this had two consequences for the PEC. First, it scuttled the advanced merger negotiations between the PEC and the Jerusalem power company, when the latter, British-registered company, effectively rose in value by 300 percent. As a result, the Jerusalem company remained independent, transforming with time into today's Jerusalem District Electric Company, serving East Jerusalem, Ramallah, Bethlehem, and Jericho.[28] Second, the Palestine Electric Corporation was forced to raise its rates considerably, first in the summer of 1952, and then again the following summer. Such raises now required the approval of the Israeli government. In another demonstration of the kind of social engineering that is possible by adjusting electricity rates, the Israeli government instructed the company to shift the majority of the rate increase onto domestic consumers. This encouraged industry and depressed domestic consumption somewhat, in line with the stated aim of the Israeli government's new economic policy and its focus on large-scale industrialization.[29]

The most important long-term impact of the NEP on the power company was nationalization. Immediately after the enactment of the NEP, the company and the Israeli government initiated talks about how the government could support the company. In October 1953, the government proposed nationalization, and after prolonged negotiations, continuing into the following year, the company accepted. It was the end of an era. At the last annual general meeting of the power company before nationalization, on September 30, 1954, Herbert Samuel, Palestine's first high commissioner and since 1936 the chairman of the PEC, gave his final statement. It was a curious coincidence, though hardly random, that the man who had granted the first electrification concession to Pinhas Rutenberg on September 12, 1921, would be the one to give the company's eulogy as chairman thirty-three years later: "Your Directors will be handing over to the State of Israel what is now a large electrical enterprise, served by a fully competent technical and administrative staff. As the principal public utility service, it has provided the economic foundation indispensable to the energetic progressive community that is reviving in the modern world an ancient land illustrious in history."[30]

Conclusion

ELECTRICAL PALESTINE

> It is difficult to imagine present-day Palestine without the racial acrimony that has unfortunately been so prominent a feature of the public life of the country, but the skeptic might find some consolation in the fact that it would be equally difficult to imagine modern Palestine without those far-reaching works such as the one of the Palestine Electric Corporation, which will leave their impress forever on the development and future of the Holy Land.
>
> —*Alfred Bonne Hebrew University (1931)*[1]

PAST HISTORIANS HAVE RELATED THE story of Pinhas Rutenberg's work in Palestine much the way Mary Shelley did the story of Dr. Frankenstein: electrifying dead matter, he brought it back to life. To most observers, Ottoman Palestine was all but deceased, a casualty of neglect by a degenerate Islamic civilization. The renowned French interwar economist Charles Gide described the thirteen centuries of Islamic rule in Palestine as "one long dark night."[2] According to Gide's widely shared view, Zionist electrification brought a new dawn, the "resurrection," as Rutenberg put it at the inaugural ceremony of his hydroelectrical powerhouse, of a "barren and scorched country, abandoned by God and man."[3] It was the great self-described virtue of the Jewish state-building project that, in lifting the Jews out of their wretched diasporic state, it also injected new life into the land and its non-Jewish inhabitants.

British officials, meanwhile, regarded Palestine and its inhabitants in colonial terms that were eminently compatible with the Zionists' worldview.

European Jews would go to Palestine and serve as a developmental vanguard; they would spur economic growth and provide a model for imitation by the Palestinian Arabs, who, like natives elsewhere in the empire, were assumed to lack the ability to undertake their own development.

Palestine's supposed degeneration and the prospects for its regeneration along certain lines, therefore, led naturally into an Anglo-Zionist alliance. The imagined power of the Jewish immigrants to transform Palestine for the benefit of all its citizens allowed the British to square the circle of the dual mandate, that is, their commitment, on the one hand, to Jewish nationalism and, on the other, to the well-being of the Palestinian Arab majority. As one would expect from such a framing, the relationship between Jews and Arabs was not one between equals. The inequality, powerfully expressed in the Balfour Declaration's negative definition of Palestinians as "non-Jews," gained a material expression in Rutenberg's power system, which then deepened and fortified it. It also carried this unequal relation through the entire mandatory period, even long after the Balfour Declaration and the inequality it implied had been officially abandoned by the British rulers. In short, it was a critical aspect of Palestine's modern history that the Zionists arrived when they did, espousing a technocapitalist program whose root metaphors were of a piece with Britain's colonial vision. They came in the guise of a modern Prometheus and won the land for their effort.

The assumptions undergirding this worldview have also found their way into much of the scholarship on mandatory Palestine, even among Zionism's harshest critics. We see it in the tendency to narrate Zionism as an economically productive force, on the assumption that economic growth is apolitical, even if its consequences are not. The same applies to the technologies that facilitated that growth. But the proclivity to ignore the mutual influences running between technology, economics, and politics amounts to a form of question-begging that fails to account for the way technological objects function as both causes and effects of social change. Technical objects, in Thomas Hughes's words, are cultural artifacts, "both socially constructed and society-shaping."[4] Mistaking a political object for a natural one elides and thus reproduces the politics built into it. In effect, it naturalizes an aspect of political claims-making and thus the prevailing relations of power.

My argument, simply put, is that Palestine's electric grid produced what it illuminated: a Jewish national space, a frame for Jewish-Arab relations, and a powerful model of economic development. By looking more closely at the mutual influences of technology and politics, this book has limned the importance of technological processes for how power and capital circulate and accumulate. The electric supply system in Palestine contributed certain necessary conditions for the making of a Jewish state, without which such a state might not have been, and without which it would surely not have been the same. Conversely, the history of Palestine's electrification is deeply political: the means of electricity generation and the design and growth of the grid were influenced by political considerations. As a result, the electric grid inscribed itself on the conflict, as the conflict in turn inscribed itself on the grid.

Insofar as Rutenberg's venture serves more generally to explain how the Zionists were able, in a few short decades, to gain control of Palestine and build a state in this Arab-majority area, this book calls for a reimagination of the way we understand the operations of political power in the context of the conflict and beyond. Rather than an independent quantity wielded by geniuses and heroes, political power was and remains a function of the local and global associations of many varied agents. Rutenberg's vision spoke to certain preexisting ideas and preferences linked to the prevailing agenda for colonial development, including technological preferences (waterpower), scalar preferences (the nation-state, global trade), and organizational preferences (centralization). And while Rutenberg's plan therefore satisfied a need that flowed from these preferences, it also shaped them in turn.

The story also testifies to the importance of boundary-work in modern politics. We live in a world of proliferating connections that bring all facets of the human condition into a network that gets denser by the day. At the same time, on the rhetorical level the modern world has been characterized by the increasingly hardening boundaries between domains, such that politics, the economy, science, and culture are all supposedly distinct and incommensurable. This modern paradox generates a new source of political power and a new form of politics. The most important factor in Rutenberg's success, as in the success of systems builders worldwide, was the ability to move nimbly between supposedly distinct domains, while rhetorically reaffirming the inviolability of the borders separating them.

This is not, however, an invitation to regard electrification in Palestine as a Machiavellian power play. There was no master plan according to which the power system would deliver Palestine to Zionism. Electrification proceeded as a series of contingent responses to immediate needs. What nevertheless gives this story a structured logic is the inherent directionality of technocapitalism's root metaphors together with the specific properties of the technological system. To borrow a phrase from Richard White, the movers and shakers of the Palestine Electric Corporation were "more opportunistic than calculating."[5]

Nevertheless, we still want to know about Rutenberg's end goal. Was he a capitalist seeking profit, or a Zionist seeking Jewish statehood? Which, we want to know, was the means and which the end? This, as it turns out, is the wrong question. He was both equally, which brings us to the real point: the tight coupling of the "rational" ordering of economics, society, and politics went hand in hand with the Jewish state-building project in Palestine. To distinguish between the two is anachronistic, an analytic imposition that Rutenberg himself would not have recognized. He was attracted to industrial capitalism and to Zionism precisely because, to his way of seeing, they were both expressions of an underlying rational organization of life. And so, all his political goals could be achieved without ever stooping to the base wrangling of "politics." Pure reason, as he and his contemporaries understood it, did his political bidding. The result was *electrical Palestine*. The power system coevolved with a kind of economy of rule, a manner of management that organized individuals, knowledge, capital, and the land and its resources in a new administrative apparatus. That is, the power system emerged together with a larger sociopolitical order, in the context of which the system and its needs became common sense, not least because it had been intimately involved in the making of that order.

This created new sources of and pathways for political power. Throughout, we have seen the importance of producing new knowledge, especially in numerical form, and the way in which the electrification project spurred the project to acquire such knowledge, which then became increasingly powerful. As the quantitative and scientific claims obtained epistemological power, the people who wielded them became powerful too. By going up to the Beisan district in late 1919 and taking measurements, Rutenberg was staking a new kind of claim to the land, one with which other land practices, so powerful under prior regimes, could not compete: claims on land based on hundreds of years of grazing cattle, fishing in the Sea of Galilee, or farming the alluvial plains of the Jordan Valley were swept aside by flow rate graphs, soil composition charts, and annual rainfall tables. Politics had become a numbers game.

The Palestinian Arabs may have thought they were fighting Rutenberg and his power company and, by extension, Zionism. But in fact they were fighting a vast social movement, tied to a particular understanding of science, human difference, and the demands of human development. In opposing electrification, then, the Palestinians were confronting not just Rutenberg, or Zionism, or even the British colonial state, but the new world order of the interwar period. Edward Said sketched the beginnings of this argument already in 1979, when he wrote that Zionism's success stemmed from its "policy of detail":

> Palestine was not only the Promised Land, a concept as elusive and as abstract as any that one could encounter. It was a specific territory with specific characteristics, that was surveyed down to the last millimeter, settled on, planned for, built on, and so forth—*in detail*. From the beginning of the Zionist colonization this was something the Arabs had no answer to, no equally detailed counter-proposal. They assumed that since they lived on the land and legally owned it, that it was therefore theirs. They did not understand that what they were encountering was a discipline of detail—indeed, a very culture of discipline by detail—by which a hitherto imaginary realm could be constructed on Palestine, inch by inch and step by step.[6]

This, moreover, was a process critically informed by what this study has referred to as *technocapitalism*. Capitalism and technology are mutually constituted and share a characteristic in their ability to distinguish themselves from any alternatives so as ultimately to appear to have no viable alternatives. The history of Palestine's electrification therefore lends credence to Fredric Jameson's quip that we could sooner imagine the end of the world than the end of capitalism.[7] Zionists and British officials were no more capable of imagining an alternative to centralized, large-scale electrification, in part because of its critical role in scaling Palestine in a way suitable to world capitalist trade.

There is no doubt that the Zionists were far more successful than the Palestinian Arabs in terms of developing their community and growing their economy. When asking why this is, we have to keep in mind that the Zionists were instrumental in designing the criteria by which such success was measured, and thus setting Palestine's political, economic, and social order. The historian of technology David Edgerton has observed that the assumption when evaluating the significance of a given technology is often that there was no alternative to it. In the case of Palestine's electrification, the assumed "hidden counterfactual" of the Palestine Electric Corporation was darkness. But of course that was not the real alternative. The alternative was a collection of several, less capital-intensive local or regional systems. Edgerton also points out that for a technology to become ubiquitous it does not have to be significantly better than the alternatives, only "*marginally* better."[8] While he acknowledges that "better" is not always the decisive factor in determining use, he leaves out another question, which hovers over this study, but which it cannot answer, namely, the question of how to determine what is "better." As Said notes, "Israel was and is the culmination of a politics of a *certain kind* of effectiveness."[9] In certain important respects, large-scale electrification truly was better: quite possibly, it was the most efficient means of providing abundant cheap power to spur the Palestinian economy. By elevating that goal into a socioeconomic—indeed a moral—imperative, it became a powerful political tool. If nothing else, then, it was a hugely effective means of Zionist state building. The Palestinians were not able to partake in that development and ended up disenfranchised. From their perspective, centralized, large-scale electrification was hardly "better." It cost them the land.

NOTES

INTRODUCTION

1. Scott Huler, *On the Grid: A Plot of Land, an Average Neighborhood, and the Systems That Make Our World Work* (Emmaus, PA: Rodale, 2010).
2. "Quasi-conference" [*Shibh mu'atammar*], *Filastin*, June 1, 1923.
3. Najwa al-Qattan, "When Mothers Ate Their Children: Wartime Memory and the Language of Food in Syria and Lebanon," *International Journal of Middle East Studies* 46, no. 4 (November 2014): 730.
4. "Blow to Rutenberg: Jaffa Refuses His Electricity," *Daily Mail,* May 28, 1923, National Archives of the United Kingdom (hereafter TNA) CAOG 14/108.
5. 'Abd al-Wahhāb Kayyālī, *Wathā'iq al-muqāwamah al-Filasṭīnīyah al-'Arabīyah ḍidda al-iḥtilāl al-Barīṭānī wa-al-ṣihyūnīyah, 1918–1939* [Documents of Palestinian Arab resistance against the British and Zionist occupation] (Beirut: Institute for Palestine Studies, 1968), 73.
6. Stuart Hall, "The Toad in the Garden: Thatcherism amongst the Theorists," in *Marxism and the Interpretation of Culture,* ed. Cary Nelson and Lawrence Grossberg (Urbana: University of Illinois Press, 1988), 44.
7. Draft letter from the Palestine Electric Corporation to Sulayman Tuqan, Mayor of Nablus, March 29, 1948, Israel Electric Corporation Archives (hereafter IECA) 0429-1131.
8. George Orwell, "Politics and the English Language," in *All Art Is Propaganda: Critical Essays* (New York: Mariner Books/Houghton Mifflin Harcourt, 2009), 270–87.
9. Cf. Herbert Marcuse, "Some Social Implications of Modern Technology," in *The Essential Frankfurt School Reader,* ed. Andrew Arato and Eike Gebhardt (New York: Continuum, 1982), 138–62. Marcuse distinguishes between *technics*, defined as "the technical apparatus of industry, transportation, communication," and *technology*, which also encompasses the "social process."
10. Beshara Doumani, *Rediscovering Palestine: Merchants and Peasants in Jabal Nablus, 1700–1900* (Berkeley: University of California Press, 1995).

11. These are rough numbers. For more exact specifications, see *Interim Report on the Civil Administration of Palestine*, League of Nations, July 31, 1921, accessed February 17, 2015, http://unispal.un.org/UNISPAL.NSF/0/349B02280A93081305 2565E90048ED1C; *Report and General Abstracts of the Census of 1922*, October 23, 1922, compiled by J. B. Barron, Superintendent of the Census; for 1948, see "Official Records of the Second Session of the General Assembly," Supplement No. 11, United Nations Special Committee on Palestine, Report to the General Assembly, vol. 1, September 3, 1947, accessed February 17, 2015, http://domino.un.org/unispal.nsf /9a798adbf322aff38525617b006d88d7/07175de9fa2de563852568d3006e10f3?Open Document. The first census in Palestine carried out by the British, on October 23, 1922, followed the Ottoman precedent of counting people according to religious affiliation. In the second major census carried out by the British, in 1931, respondents were asked to identify both their religious affinity and their ethnicity. For urban-to-rural proportions at the end of the mandate, see Rashid Khalidi, *The Iron Cage: The Story of the Palestinian Struggle for Statehood* (Boston: Beacon Press, 2006), 17. The changes described were the result of urbanization among the Palestinian Arabs, and the preference from the start among Jewish immigrants to live in urban environments. Jewish rural dwellers never composed more than 27 percent of the Yishuv. See Jacob Metzer, *The Divided Economy of Mandatory Palestine* (New York: Cambridge University Press, 1998), 8.

12. By comparison, the second-highest growth rate in the region was Turkey's, at 4.1 percent, while Egypt's was 0.2 percent. Metzer, *Divided Economy*, 18–19.

13. Yehoshua Ziman, *Hakalkala ha'eretzyisra'elit bemisparim* [The economy of the land of Israel in numbers] (Tel Aviv: Davar, 1929).

14. *Statistical Abstract of Palestine 1944-45*, table 4, p. 53; Fuel and Light Index, p. 117, Israel State Archives (hereafter ISA) M-5314/11.

15. Jewish Agency for Palestine Economic Research Institute, *Statistical Handbook of Middle Eastern Countries, Palestine, Cyprus, Egypt, Iraq, the Lebanon, Syria, Transjordan, Turkey* (Jerusalem, 1945).

16. Metzer, *Divided Economy*, 104.

17. David E. Nye, *Electrifying America: Social Meanings of a New Technology* (Cambridge, MA: MIT Press, 1990), 26.

18. "Marquis of Reading: Rutenberg Scheme Director," *Daily Telegraph*, September 28, 1926, TNA CAOG 14/109.

19. Letter from Middle East Department Director John Shuckburgh to Secretary of State for the Colonies (S/SC) Winston Churchill, September 7, 1921, TNA CO 733/15.

20. Letter presented to Parliament in June 1922, ISA M-4381.

21. IECA 2369-4.

22. I am not the first historian to point this out, or to attempt a corrective. More than two decades ago, Ken Alder did for the French Revolution what this book attempts for the Arab-Israeli conflict: "Rather than limit political activity to the realm of symbols and representations," he writes, "this book seeks to *expand* our understanding of politics to include contests over the terms of the material life." Ken Alder,

Engineering the Revolution: Arms and Enlightenment in France, 1763–1815 (Princeton: Princeton University Press, 1997), xiii–xiv. Another book, also about France and published around the same time, is Gabrielle Hecht's, *The Radiance of France: Nuclear Power and National Identity After World War II* (Cambridge, MA: MIT Press, 1998). Hecht, from whom I borrowed the title of my fourth chapter, is a prominent exponent of the insights to be gained from bringing technology studies to political and cultural history. Her articulation of the concept of "technopolitics" as "the strategic practice of designing or using technology to constitute, embody, or enact political goals," in *The Radiance of France*, 15, guides my usage of the term. Numerous other Science and Technology Scholars have guided me in the writing of this book. Their work appears in later footnotes and in the bibliography. In the interest of readability, however, this books seeks to *show* its indebtedness to STS, rather than tell about it. Discussions of theory and method are therefore kept to a minimum. Nevertheless, the richness of this historical subfield makes the absence of its perspectives in the historical mainstream all the more notable. Recently, a small but growing number of Middle East scholars have begun to draw on STS. They include: On Barak, *On Time: Technology and Temporality in Modern Egypt* (Berkeley: University of California Press, 2013); Katayoun Shafiee, "A Petro-Formula and Its World: Calculating Profits, Labour and Production in the Assembling of Anglo-Iranian Oil," *Economy and Society* 41, no. 4 (November 2012): 585–614; Katayoun Shafiee, *Machineries of Oil: An Infrastructural History of BP in Iran* (Cambridge, MA: MIT Press, 2018); Ronen Shamir, *Current Flow: The Electrification of Palestine* (Stanford, CA: Stanford University Press, 2013); Timothy Mitchell, *Rule of Experts: Egypt, Techno-Politics, Modernity* (Berkeley: University of California Press, 2002); Timothy Mitchell, *Carbon Democracy: Political Power in the Age of Oil* (London: Verso, 2011); Toby Craig Jones, *Desert Kingdom: How Oil and Water Forged Modern Saudi Arabia* (Cambridge, MA: Harvard University Press, 2010). Finally, Derek Penslar's *Zionism and Technocracy* stands out in the literature for its rich exploration of the "symbiosis" of technical expertise and political leadership within the Zionist movement. It tracks the emergence of technocracy as a result of the influence of German and German-educated Zionist leaders between 1896 and 1914. *Electrical Palestine* carries forward Penslar's critical insight that technocracy is not "an ideologically neutral administration by apolitical technicians," but it construes the relationship between the two somewhat differently. Whereas, on Penslar's view, politicians set the agenda and experts realize it, this book identifies a dialectical relationship between different actors and agendas, where influence runs both ways. Derek Penslar, *Zionism and Technocracy: The Engineering of Jewish Settlement in Palestine, 1870–1918* (Bloomington: Indiana University Press, 1991), 151.

23. Ian Hacking, *The Social Construction of What?* (Cambridge, MA: Harvard University Press, 1999).

24. Thomas P. Hughes, "The Evolution of Large Technological Systems," in *The Social Construction of Technological Systems: New Directions in the Sociology and History of Technology*, ed. Wiebe E. Bijker, Thomas P. Hughes, and Trevor Pinch, anniversary edition (Cambridge, MA: MIT Press, 2012), 45; Langdon Winner, "Do Artifacts Have Politics?," *Daedalus* 109, no. 1 (Winter 1980): 121–36; John Law,

"Notes on the Theory of the Actor-Network: Ordering, Strategy and Heterogeneity," *Systems Practice* 5, no. 4 (August 1992): 379–93.

25. This is true not least for Sherene Seikaly's stimulating excavation of Palestine's "men of capital," and specifically her observation that "economics functions as a metonym for science." Sherene Seikaly, *Men of Capital: Scarcity and Economy in Mandate Palestine* (Stanford, CA: Stanford University Press, 2015), 34.

26. As numerous studies have noted, the concept of the "free market" signified an ideological commitment to the unregulated market, rather than an actual state of affairs. Although, as Giovanni Arrighi has shown, ensuring the freedom of trade served as the British Empire's chief legitimating device from the mid-1700s to the early 1900s, the British consistently manipulated markets in numerous ways, not least in Palestine. Giovanni Arrighi, *The Long Twentieth Century: Money, Power and the Origins of Our Times* (New York: Verso, 2010), 53–55. See also John Gallagher and Ronald Robinson, "The Imperialism of Free Trade," *Economic History Review* 6, no. 1 (1953): 1–15. For British market manipulation in Palestine, see Barbara J. Smith, *The Roots of Separatism in Palestine: British Economic Policy, 1920–1929* (Syracuse, NY: Syracuse University Press, 1993).

27. Joseph Morgan Hodge, *Triumph of the Expert: Agrarian Doctrines of Development and the Legacies of British Colonialism* (Athens: Ohio University Press, 2007), esp. 7.

28. Ernest Gellner, *Nations and Nationalism* (Ithaca, NY: Cornell University Press, 1983); Sven Beckert, *Empire of Cotton: A Global History* (New York: Alfred A. Knopf, 2015).

29. Benedict Anderson, *Imagined Communities* (New York: Verso, 1991), esp. 175, 184.

30. Shamir, *Current Flow*, 27.

31. Shamir, *Current Flow*, 17, 27.

32. Khalidi, *Iron Cage*, 31–34. See also Benny Morris, *Righteous Victims: A History of the Zionist-Arab Conflict, 1881–1999* (New York: Alfred A. Knopf, 1999); Avi Shlaim, *The Iron Wall: Israel and the Arab World* (New York: W. W. Norton, 2000), esp. 7–10; Tom Segev, *One Palestine, Complete: Jews and Arabs under the Mandate* (New York: Metropolitan Books, 2000), esp. 49.

33. James Renton, *The Zionist Masquerade: The Birth of the Anglo-Zionist Alliance, 1914–1918* (New York: Palgrave Macmillan, 2007), 7, 10. Jonathan Schneer, whose book on the Balfour Declaration is held by some scholars as superior to Renton's, is equally convinced of the Balfour Declaration's singular importance. Jonathan Schneer, *The Balfour Declaration: The Origins of the Arab-Israeli Conflict* (New York: Random House, 2010).

34. E.g., Anita Shapira, *Israel: A History* (Waltham, MA: Brandeis University Press, 2012), 74–75.

35. Religion: Renton, *Zionist Masquerade*, 12, 19–20, 149–50; Bernard Wasserstein, *The British in Palestine: The Mandatory Government and the Arab-Jewish Conflict, 1917–1929* (London: Royal Historical Society, 1978), 11; Segev, *One Palestine*, 32, 36-43. Realpolitik: William M. Mathew, "The Balfour Declaration and the

Palestine Mandate, 1917–1923: British Imperialist Imperatives," *British Journal of Middle Eastern Studies* 40, no. 3 (2013): 231–50; David Engel, *Zionism* (Harlow, UK: Pearson/Longman, 2009), 80–81. Racialization: Renton, *Zionist Masquerade*, 14; Jacob Norris, *Land of Progress: Palestine in the Age of Colonial Development, 1905–1948* (Oxford: Oxford University Press, 2013), 84–85; Segev, *One Palestine*, 32. Indeed, Shapira makes all these points together (*Israel*, 71–73).

36. Shapira, *Israel*, 73.

37. Khalidi, *Iron Cage*.

38. Segev, *One Palestine*, 43. See also Gudrun Krämer, *A History of Palestine: From the Ottoman Conquest to the Founding of the State of Israel* (Princeton, NJ: Princeton University Press, 2008), 151–55; Martin Bunton, *Colonial Land Policies in Palestine, 1917–1936* (Oxford: Oxford University Press, 2007).

39. The question of nonhuman agency is controversial, not least among historians. Resistance to the idea can be attenuated, I think, if it is understood not as an attempt to elevate the status of nonhumans to the level of intentional agency, but as a means of highlighting the way all human intentions are waylaid by evolving contingencies outside our control. See Linda Nash, "The Agency of Nature or the Nature of Agency?," *Environmental History* 10, no. 1 (January 2005): 67–69.

40. Graham Burchell, Colin Gordon, and Peter Miller, eds., *The Foucault Effect: Studies in Governmentality* (Chicago: University of Chicago Press, 1991), 92–96; Pierre Bourdieu, *In Other Words: Essays towards a Reflexive Sociology* (Stanford, CA: Stanford University Press, 1990), esp. 11.

41. Nye, *Electrifying America*, 11, 28; Thomas P. Hughes, *Networks of Power: Electrification in Western Society, 1880–1930* (Baltimore: Johns Hopkins University Press, 1983), 370; Jack Casazza and Frank Delea, *Understanding Electric Power Systems: An Overview of the Technology and the Marketplace* (Hoboken, NJ: John Wiley and Sons, 2010).

42. William J. Hausman, Peter Hertner, and Mira Wilkins, *Global Electrification: Multinational Enterprise and International Finance in the History of Light and Power, 1878–2007* (Cambridge: Cambridge University Press, 2008, Kindle Edition), Kindle Location 592; Hughes, *Networks of Power*, 365–66.

43. Indeed, electric power systems, along with other large infrastructures such as telegraph networks and railroads, are often referred to in the literature as "natural monopolies." A natural monopoly occurs in sectors of dramatically decreasing average unit costs as production rises; Hausman, Hertner, and Wilkins, *Global Electrification*, Kindle Locations 5459–60. Historically, the industries counted as natural monopolies have included telephone, telegraph, trolley companies, and electricity utilities; Nye, *Electrifying America*, 175.

44. Hughes, *Networks of Power*, 2.

45. This slippage is also found in the work of many historians of technology, including Thomas Hughes's deservedly celebrated history of electrification in Western societies. Hughes, *Networks of Power*, 260; see Chris Otter's critique, *The Victorian Eye: A Political History of Light and Vision in Britain, 1800–1910* (Chicago: University of Chicago Press, Kindle Edition, 2008), 250–51.

46. Hughes, *Networks of Power*, 360–61.

47. Hughes, *Networks of Power*, 361. In Britain, the electricity rates fell by 70 percent after the introduction of the National Grid.

48. Michael Adas, *Machines as the Measure of Men: Science, Technology, and Ideologies of Western Dominance* (Ithaca, NY: Cornell University Press, 1989), 144–45.

49. Leo Marx, "Technology: The Emergence of a Hazardous Concept," *Technology and Culture* 51, no. 3 (July 2010): 561–77.

50. Bruno Latour, *Pandora's Hope: Essays on the Reality of Science Studies* (Cambridge, MA: Harvard University Press, 1999), 182.

51. These things listed in Thomas F. Gieryn, *Cultural Boundaries of Science: Credibility on the Line* (Chicago: University of Chicago Press, 1999), 1.

52. Sa'd al-Din, "Rutenberg's Zionist Current," *al-Jami'a al-Islamiya*, December 20, 1934 (emphasis added). Alder, *Engineering the Revolution*, ch 8, and his discussion of Revolutionary engineers in the context of "the technocratic pose."

53. Barry, *Political Machines*, 9. This point is made well in Bernward Joerges, "Do Politics Have Artifacts?" *Social Studies of Science* 29, no. 3 (1999): 411-431. The classic study in this regard is Shapin and Schaffer, *Leviathan*.

CHAPTER ONE

1. Thomas Carlyle, *Critical and Miscellaneous Essays* (New York, 1896), vol. 2, 60; quoted in Adas, *Machines as the Measure of Men*, 213.

2. Eli Shaltiel, *Pinhas Rutenberg: 'Aliyato unefilato shel "ish chazak" beeretz yisra'el, 1879–1942* (Tel Aviv: Am Over, 1990), 41.

3. Parliamentary Debates, House of Commons, Motion on Rutenberg Plan, July 4, 1922, Central Zionist Archives (hereafter CZA) A342/3.

4. December 4, 1923, M. Elsasser to Flexner, CZA A342/2.

5. Chaim Weizmann, *Trial and Error: The Autobiography of Chaim Weizmann* (New York: Harper, 1949), 149.

6. "Wealth from the Dead Sea," *Popular Mechanics*, November 1930, 798.

7. "Attempt to Introduce Rutenberg in Nablus," *al-Difa'*, December 17, 1934.

8. Shaltiel, *Rutenberg*, 21–22; Salo Baron, *The Russian Jew under Tsars and Soviets* (New York: Macmillan, 1964).

9. Gregory Guroff, "The Legacy of Pre-revolutionary Economic Education: St. Petersburg Polytechnic Institute," *Russian Review* 31, no. 3 (July 1972): 272–81; Jonathan Coopersmith, *The Electrification of Russia, 1880–1926* (Ithaca, NY: Cornell University Press, 1992), 39.

10. Coopersmith, *Electrification of Russia*, 26.

11. Coopersmith, *Electrification of Russia*, 151–53, 155, 157–63; Sergei Zhuravlev, "'Little People' and 'Big History': Foreigners at the Moscow Electric Factory and Soviet Society, 1920s–1930s," *Russian Studies in History* 44, no. 1 (Summer 2005): 10–14.

12. G. G. Lapin, "70 Years of Gidroproekt and Hydroelectric Power in Russia," *Hydrotechnical Construction* 34, nos. 8–9 (August 2000): 374.

13. Many of the ideas, as well as much of the language in this paragraph, echo Thomas Hughes's magisterial study of electrification in Western society. See Hughes, *Networks of Power*, esp. 18, 33–34, 201, 207, 216–17, 334–35, 339. The concept of the "systems entrepreneur," however, is my own.

14. Gregory Guroff, "The Legacy of Pre-revolutionary Economic Education: St. Petersburg Polytechnic Institute," *Russian Review* 31, no. 3 (July 1972): 272–81.

15. T. Clayton Black, "Icon of the Revolution: The Putilov Factory and the Dynamics of the Bolshevik Master Fiction," in *Russia's Century of Revolutions: Parties, People, Places: Studies Presented in Honor of Alexander Rabinowitch*, ed. Michael S. Melancon and Donald J. Raleigh (Bloomington, IN: Slavica, 2012), 89–110.

16. Shaltiel, *Rutenberg*, 28.

17. Daniel Headrick, *The Tentacles of Progress: Technology Transfer in the Age of Imperialism, 1850–1940* (Oxford: Oxford University Press, 1988), 177. It was to Italy, for instance, that the British administration in Egypt sent its engineers in preparation for the construction of the first Aswan Dam around the turn of the last century. Jennifer L. Derr, "Drafting a Map of Colonial Egypt: The 1902 Aswan Dam, Historical Imagination, and the Production of Agricultural Geography," in *Environmental Imaginaries of the Middle East and North Africa*, ed. Diana K. Davis and Edmund Burke (Athens: Ohio University Press, 2011), 140.

18. According to Yitzhak Ben-Zvi, PR had been exiled by Jamal Pasha. Article published on the occasion of PR's death in 1942, a version of the commemorative speech he held on January 11, 1942, ISA P-1968/3; see also Weizmann House Archives (hereafter WHA) 2-299.

19. ISA P-2027 contains multiple copies of the pamphlet in Yiddish and Hebrew.

20. IECA 2375-11; ISA P-2027/2; Rafael Medoff and Chaim I. Waxman, *The A to Z of Zionism* (Lanham, MD: Scarecrow Press, 2009).

21. Pamphlet dated 1 Av 5679: (July 28, 1919), by Nachman Syrkin, who met PR in Paris at a dinner party, IECA 2377-3.

22. Reports from surveys in Nahr el Jalud in the Beisan Valley, January 19, 1920; the Jisr al-Shaykh Husayn, January 18; footbridge January 13, IECA A348-28.

23. Edward Said, "Zionism from the Standpoint of Its Victims," *Social Text*, no. 1 (Winter 1979): 36.

24. Tzadok Eshel, *Naharayim: Sipura Shel Tachanat Ko'ach* (Haifa: Israel Electric Corporation, 1990), 22.

25. Parliamentary Debates, House of Commons, Motion on Rutenberg Plan: London Press Report, July 4, 1922, CZA A342/3. For another example, see report published by Sir Charles Metcalfe & Selves on PR's project, commissioned by Lord Mond, IECA 0359-230, IECA 2377-7.

26. *The Values of Precision* (Princeton, NJ: Princeton University Press, 1997), ed. M. Norton Wise. For a discussion of the evolution and politics of precision in the context of the Middle East, see Mitchell, *Rule of Experts*, 82.

27. "Electrische Centrale in Nazareth," October 1923, IECA 0036-14.

28. Report by PR, 1923, TNA CO 733/44.
29. Acting Chairman of ZC, E. W. Lewin-Epstein to the Zionist Organization (London), February 19, 1919, CZA L4\606-1-6.
30. "Representatives of the Jewish colonists in the district" to ZC, Jaffa, March 8, 1919, CZA L4\606-10.
31. Weizmann to Louis Mallet, June 18, 1919, in Shaltiel, *Rutenberg*, 43.
32. In the United States, in 1927, for instance, the systems of Philadelphia Electric, Public Service Electric & Gas of New Jersey, and Pennsylvania Power and Light were interconnected, through the establishment of a new entity, the Pennsylvania–New Jersey Interconnection. At the time, this was the largest integrated, centrally controlled pool of electric power (generating a total of 1.5 million kW). And in connection with this move, the companies built a dam on the Susquehanna River, near Conowingo, Maryland. The agreement also called for a ring of 220,000-volt trunk lines, with transmission lines from each of the participant's systems to the ring. The system began service in 1930. Hughes, *Networks of Power*, 331. Leonard DeGraaf, "Corporate Liberalism and Electric System Planning in the 1920s," *Business History Review* 64 (Spring 1990): 1–31.
33. Jonathan Coopersmith, "The Electrification of Russia, 1880 to 1925" (PhD diss., Oxford University, 1985), 144, 168–76. Also quoted in Ronald R. Kline, *Consumers in the Country: Technology and Social Change in Rural America* (Baltimore: Johns Hopkins University Press, 2000), 3.
34. Kline, *Consumers in the Country*, 3.
35. The British National Archives CO 733/17b.
36. Report published by Sir Charles Metcalfe & Selves on PR's project, commissioned by Lord Mond, August 10, 1920, p. 1, IECA 0359-230, 2377-7.
37. CZA L3\82\6-85.
38. White, *The Organic Machine*; David E. Nye, *America as Second Creation: Technology and Narratives of New Beginnings* (Cambridge, MA: MIT Press, 2003).
39. Jennifer Alexander, "The Concept of Efficiency: An Historical Analysis," in *Handbook of the Philosophy of Science*, vol. 8, *Philosophy of Technology and Engineering Sciences*, ed. Anthony Meijers (London: Elsevier, 2009), 1007–30; Jennifer Pitts, *A Turn to Empire: The Rise of Imperial Liberalism in Britain and France* (Princeton, NJ: Princeton University Press, 2005), 2. As Mitchell points out, these and other key thinkers of the era spent parts or all of their working lives in the service of the East India Company and, after nationalization, the India Office. Mitchell *Rule of Experts*, 96. See also Adas, *Machines as the Measure of Men*, 209–12.
40. Richard Harry Drayton, *Nature's Government: Science, Imperial Britain, and the "Improvement" of the World* (New Haven, CT: Yale University Press, 2000), 128; Pitts, *Turn to Empire*, 11–16; Daniel Headrick, "The Tools of Imperialism: Technology and the Expansion of European Colonial Empires in the Nineteenth Century," *Journal of Modern History* 51 (June 1979): 263; Adas, *Machines as the Measure of Men*, 212; see also Robert H. MacDonald, *The Language of Empire: Myths and Metaphors of Popular Imperialism, 1880–1918* (Manchester, UK: Manchester University Press, 1994); Thomas R. Metcalf, *Ideologies of the Raj* (Cambridge:

Cambridge University Press, 1994); Alice L. Conklin, *A Mission to Civilize: The Republican Idea of Empire in France and West Africa, 1895–1930* (Stanford, CA: Stanford University Press, 1997).

41. In Richard Drayton's words, what emerged was an "imperialism of 'improvement.'" Drayton, *Nature's Government*, xv. Joseph Hodge similarly characterizes it as an "imperialism of science and knowledge." Hodge, *Triumph of the Expert*, 11. See also James C. Scott's famous definition of what he terms "high modernism." James C. Scott, *Seeing Like a State: How Certain Schemes to Improve the Human Condition Have Failed* (New Haven, CT: Yale University Press, 1998), 89–90.

42. Hodge, *Triumph of the Expert*, 12, 24; S.B. Saul, "The Economic Significance of 'Constructive Imperialism,'" *Journal of Economic History* 17, no. 2 (June 1957): 173–92.

43. Adas, *Machines as the Measure of Men*, 218–20.

44. Stephen Constantine, *The Making of British Colonial Development Policy 1914–1940* (London: F. Cass, 1984), 11.

45. Ibrahim A. Abu-Lughod, *The Transformation of Palestine; Essays on the Origin and Development of the Arab-Israeli Conflict* (Evanston, IL: Northwestern University Press, 1971), 35–37.

46. Robert G. Weisbord, *African Zion; The Attempt to Establish a Jewish Colony in the East Africa Protectorate, 1903–1905* (Philadelphia: Jewish Publication Society of America, 1968).

47. Hodge, *Triumph of the Expert*, 42–43; Conklin, *Mission to Civilize*; Suzanne Moon, "Empirical Knowledge, Scientific Authority, and Native Development: The Controversy over Sugar/Rice Ecology in the Netherlands East Indies, 1905–1914," *Environment and History* 10, no. 1 (2004): 64; Juhani Koponen, *Development for Exploitation: German Colonial Policies in Mainland Tanzania, 1884–1914* (Helsinki: Finnish Historical Society, 1994); Andrew Zimmerman, "'What Do you Really Want in German East Africa, Herr Professor?': Counterinsurgency and the Science Effect in Colonial Tanzania," *Comparative Studies in Society and History* 48, no. 2 (April 2006): 419–61; Penelope Hetherington, *British Paternalism and Africa 1920–1940* (Totowa, NJ: Frank Cass, 1978), 4.

48. F.C. Linfield, "Empire Development," *Contemporary Review* 128 (1926): 324; quoted in Hetherington, *British Paternalism and Africa*, 93–94.

49. Charles D. Marx, "Social and Economic Aspects of Hydro-Electric Power," *Proceedings of the American Society of Civil Engineers* 48, no. 9 (November 1922): 1781–82.

50. Adas, *Machines as the Measure of Men*, 225 (emphasis added); also quoted in Hodge, *Triumph of the Expert*, 32.

51. Adas, *Machines as the Measure of Men*, 228.

52. In 1898, Benjamin Kidd, a prominent British sociologist and civil servant, published a best-selling book, first serialized in the *London Times*, with the title *The Control of the Tropics* in which he estimated that Africa had "some of the richest territories on the earth's surface." Hodge, *Triumph of the Expert*, 40–41.

53. Adas, *Machines as the Measure of Men*, 229.

54. Adas, *Machines as the Measure of Men*, 229. Theodeophilus French called them "the great civilizer of modern times." Quoted in Richard White, *Railroaded: The Transcontinentals and the Making of Modern America* (New York: W. W. Norton, 2012), xxii.

55. Between 1860 and 1920, the Indian railway system grew by an average of 594 miles annually. Manu Goswami, *Producing India: From Colonial Economy to National Space* (Chicago: University of Chicago Press, 2004), 47–51; Casper Andersen, *British Engineers and Africa, 1875–1914* (London: Pickering and Chatto, 2011), 1, 19–20.

56. Rudolf Mrázek, *Engineers of Happy Land: Technology and Nationalism in a Colony* (Princeton, NJ: Princeton University Press, 2002), 7.

57. The exact years were 1871 to 1914. Adas, *Machines as the Measure of Men*, 143.

58. Donald Denoon, *Settler Capitalism: The Dynamics of Dependent Development in the Southern Hemisphere* (Oxford: Clarendon Press, 1983), 8; see also Hodge, *Triumph of the Expert*, 24; Drayton, *Nature's Government*, 229; P. J. Cain and A. G. Hopkins, "Gentlemanly Capitalism and British Expansion Overseas II: New Imperialism, 1850–1945," *Economic History Review*, n.s., 40, no. 1 (February 1987): 1–26. According to E. A. Brett, the British were "obsessed with the need to create an export economy which would draw [their colonies] directly and profitably into the British system of international trade." E. A. Brett, *Colonialism and Underdevelopment in East Africa: The Politics of Economic Change 1919–1939* (London: Heinemann, 1973), 54.

59. Antony Anghie, *Imperialism, Sovereignty, and the Making of International Law* (Cambridge: Cambridge University Press, 2005), 157; Antony Anghie, "Civilization and Commerce: The Concept of Governance in Historical Perspective," *Villanova Law Review* 45, no. 5 (2000): 887; see also Hetherington, *British Paternalism and Africa*, 94.

60. Jason W. Moore, *Capitalism in the Web of Life: Ecology and the Accumulation of Capital* (New York: Verso, 2015), 20.

61. Valeska Huber, *Channelling Mobilities: Migration and Globalisation in the Suez Canal Region and Beyond, 1869–1914* (New York: Cambridge University Press, 2013).

62. David Arnold, *Science, Technology, and Medicine in Colonial India* (New York: Cambridge University Press, 2000), 121.

63. Susan Pedersen, *The Guardians: The League of Nations and the Crisis of Empire* (New York: Oxford University Press, 2015), 109. See also Karen E. Fields, *Revival and Rebellion in Colonial Central Africa* (Princeton, NJ: Princeton University Press, 1985); Bruce Berman and John Lonsdale, *Unhappy Valley: Conflict in Kenya and Africa* (London: John Currey, 1992). Timothy Mitchell makes an argument along similar lines with respect to Egypt. Mitchell, *Rule of Experts*, 41.

64. Mitchell, *Rule of Experts*, 41. See also Claire Jean Cookson-Hills, "Engineering the Nile: Irrigation and the British Empire in Egypt, 1882–1914" (PhD diss., Queen's University, 2013); Headrick, *Tentacles of Progress*, 201. See also Jennifer L. Derr, "Drafting a Map of Colonial Egypt: The 1902 Aswan Dam, Historical Imagination, and the Production of Agricultural Geography," in *Environmental Imagi-*

naries of the Middle East and North Africa, Diana K. Davis and Edmund Burke (Athens: Ohio University Press, 2011), 136–57.

65. Coopersmith, *Electrification of Russia*, 10–15, 152–53.

66. Headrick, *Tentacles of Progress*, 180. See also David Ludden, "Patronage and Irrigation in Tamil Nadu: A Long-Term View," *Indian Economic and Social History Review* 16, no. 3 (September 1979): 358–60; Nicholas B. Dirks, "From Little King to Landlord: Colonial Discourse and Colonial Rule," in *Colonialism and Culture*, ed. Nicholas B. Dirks (Ann Arbor: University of Michigan Press, 1992), 175–209.

67. Constantine, *British Colonial Development Policy*, 19.

68. Drayton, *Nature's Government*, 227–29.

69. Hodge, *Triumph of the Expert*, 28–29.

70. Hodge, *Triumph of the Expert*, 30–31; Zaheer Baber, *The Science of Empire: Scientific Knowledge, Civilization, and Colonial Rule in India* (Albany: State University of New York Press, 1996), 205.

71. Helen Tilley, *Africa as a Living Laboratory: Empire, Development, and the Problem of Scientific Knowledge, 1870–1950* (Chicago: University of Chicago Press, 2011), 76–82.

72. Frederick D. Lugard, *The Dual Mandate in British Tropical Africa* (London: William Blackwood and Sons, 1922). Hetherington refers to *The Dual Mandate* as "probably the most influential book on colonial affairs in the whole inter-war period." Hetherington, *British Paternalism and Africa*, 4.

73. Brett, *Colonialism and Underdevelopment*, 47.

74. Rogers Brubaker, "Aftermaths of Empire and the Unmixing of Peoples: Historical and Comparative Perspectives," in Rogers Brubaker, *Nationalism Reframed: Nationhood and the National Question in the New Europe* (New York: Cambridge University Press, 1996), 148–79; Dawn Chatty, *Displacement and Dispossession in the Modern Middle East* (Cambridge: Cambridge University Press, 2010), 289; Carole Fink, *Defending the Rights of Others* (Cambridge: Cambridge University Press, 2006); Pedersen, *The Guardians*, 8–9.

75. Patricia Clavin, *Securing the World Economy: The Reinvention of the League of Nations, 1920–1946* (Oxford: Oxford University Press, 2013); Pedersen, *The Guardians*.

76. Covenant of the League of Nations, Article 22, http://avalon.law.yale.edu/20th_century/leagcov.asp.

77. The rough correspondence to the contemporary map is as follows: Palestine: Israel, the West Bank, the Gaza Strip, and Jordan; Mesopotamia: Iraq; Greater Syria: Syria and Lebanon.

78. Covenant of the League of Nations, Article 22. http://avalon.law.yale.edu/20th_century/leagcov.asp; Central African territories were Class B; South-West Africa and the South Pacific Islands constituted Class C. Quincy Wright, *Mandates under the League of Nations* (Chicago: University of Chicago Press, 1930).

79. San Remo Convention, available online through UNISPAL: http://unispal.un.org/UNISPAL.NSF/0/DB662E3B80797A9685257A130073F02E; Gideon Biger, *The Boundaries of Modern Palestine, 1840–1947* (London: Routledge, 2004), 68.

80. The Balfour Declaration, available online through UNISPAL: http://unispal.un.org/UNISPAL.NSF/9a798adbf322aff38525617b006d88d7/e210ca73e38d9e1d052565fa00705c61?OpenDocument; the Palestine Mandate Charter, article 2, available online through UNISPAL: http://unispal.un.org/UNISPAL.NSF/0/2FCA2C68106F11AB05256BCF007BF3CB.

81. See, for instance, the Permanent Mandates Commission in 1925; the report of the Royal Commission, 1937. Chaim Weizmann describes it in his autobiography from 1949 as "the duality of the mandate." Weizmann, *Trial and Error*, 325.

82. "Zionist Rejoicings: British Mandate for Palestine Welcomed," *The Times*, April 26, 1920.

83. Parliamentary Debates, House of Commons, Motion on Rutenberg Plan, July 4, 1922, CZA A342/3; "Mr. Churchill's Reply," *The Times*, July 5, 1922, TNA CAOG 14/109.

84. Theodor Herzl to Yusuf Diya' al-Khalidi, March 19, 1899; http://cosmos.ucc.ie/cs1064/jabowen/IPSC/articles/article0025759.html (accessed December 4, 2015).

85. Weizmann, *Trial and Error*, 149.

86. Weizmann, *Trial and Error*, 463.

87. Benjamin Beit-Hallahmi, *Original Sins: Reflections on the History of Zionism and Israel* (New York: Olive Branch, 1993), 16; quoted in William M. Mathew, "War-Time Contingency and the Balfour Declaration of 1917: An Improbable Regression," *Journal of Palestine Studies* 40, no. 2 (Winter 2011): 38.

88. Norris, *Land of Progress*, 65, persuasively makes this argument.

89. "Fifth Meeting of the Advisory Committee to the Palestine Office," May 10, 1919, TNA FO 608/100; cited in Norris, *Land of Progress*, 89

90. For instance, on meeting Jan Smuts for the first time, Weizmann found "a sort of warmth of understanding radiated from him, and he assured me heartily that something would be done in connection with Palestine and the Jewish people.... He treated the problem with eager interest, one might say with affection." Weizmann, *Trial and Error*, 159. Lord Cecil, according to Weizmann, "saw [Zionism] in its true perspective as an integral part of world stabilization. To him the re-establishment of a Jewish Homeland in Palestine and the organization of the world in a great federation were complementary features of the next step in the management of human affairs." See Weizmann, *Trial and Error*, 191.

91. Smith, *Roots of Separatism in Palestine*, 52.

92. Herbert Samuel, "The Future of Palestine," presented to the British cabinet in January 1915, TNA CAB 37/123/43.

93. "Sir Herbert Samuel," *Daily Herald*, March 11, 1931.

94. Weizmann, *Trial and Error*, 289.

95. Brett, *Colonialism and Underdevelopment*, 71.

96. Hetherington, *British Paternalism and Africa*, 45–48.

97. Palestine Royal Commission Report, Cmd. 5479, London: HMSO, July 1937, 58; UNSCOP document A/AC.13/69, Special Committee on Palestine, Working Documentation prepared by the secretariat, Volume IV.

98. Royal Commission Report, 370.
99. Marx, "Social and Economic Aspects of Hydro-Electric Power," 1784.
100. Brett, *Colonialism and Underdevelopment*, 46
101. Quoted in Drayton, *Nature's Government*, 232–33.
102. Jan C. Smuts, "African Settlement," in *Africa and Some World Problems* (Oxford: Oxford Clarendon Press, 1930), 50–51.
103. Smuts, "African Settlement," 49.
104. Pedersen, *The Guardians*, 71.
105. House of Commons Debate, July 4, 1922, vol. 156, col. 335. See also TNA CAOG 14/109, "Mr. Churchill's Reply," *The Times*, July 5, 1922. Expressing this sentiment, Churchill was echoing ideas of the profligate savage, living in natural abundance but also, for failure of adequate industry, in human squalor. In his pronouncement on the Palestinian Arabs, there are strong echoes of John Locke's writings on the North American Indians in the section "Of Property" in *Second Treatise of Government*. This quotation has been reproduced many times in the literature, including in Binyamin Netanyahu, *A Durable Peace: Israel and Its Place among the Nations* (New York: Warner Books, 2000); David Fromkin, *A Peace to End All Peace: Creating the Modern Middle East, 1914–1922* (New York: H. Holt, 1989), 523; Norris, *Land of Progress*, 85.
106. Churchill, speech to the Liverpool Chamber of Commerce, June 1921.
107. Lugard, *Dual Mandate*, 295; Constantine, *British Colonial Development Policy*, 53–55; Brett, *Colonialism and Underdevelopment*, 44–45.
108. Brett, *Colonialism and Underdevelopment*, 68; Hodge, *Triumph of the Expert*, 31–32.
109. Caroline Elkins and Susan Pedersen, *Settler Colonialism in the Twentieth Century: Projects, Practices, Legacies* (New York: Routledge, 2005), 9–10; Smith, *Roots of Separatism in Palestine*. Another indicator of the Zionists' relatively less privileged position in relation to other settler groups is land possession. In some of the settler colonies with the strongest settler privilege, most of the land ended up in the colonists' hands. For instance, in Australia and the Orange Free State, white settlers owned more than 90 percent of the land by the end of the nineteenth century. In Rhodesia and South West Africa, at least one-third of the land belonged to white settlers. The Zionists, by contrast, owned only some 6 percent of the land by the end of the mandatory period in 1948, and the British launched several initiatives designed to limit land sales from Arabs to Jews. In 1946, the Zionist industrialist E. Schmorak complained, "The fact that the Jewish economic sector comprises to-day only 1,700,000 dunams, that is, no more than 6% of the total area and 10% of the immediately cultivable land, is due, in no small degree, to the Land Policy of the Government, which is an agreement neither with the spirit not the letter of the Mandate." E. Schmorak, *Palestine's Industrial Future* (Jerusalem: Rubin Mass, 1946), 16. See also Kenneth Good, "Settler Colonialism: Economic Development and Class Formation," *Journal of Modern African Studies* 14, no. 4 (1976): 597–620. In South West Africa, as much as 55 percent of the nondesert land in the "Police Zone" was owned by whites. Pedersen, *The Guardians*, 129.

110. Smith, *Roots of Separatism*, 107; Charles Kamen, *Little Common Ground: Arab Agriculture and Jewish Settlement in Palestine, 1920–1948* (Pittsburgh: University of Pittsburgh Press, 1991), 73.

111. Samer Alatout, "Bringing Abundance into Environmental Politics: Constructing a Zionist Network of Water Abundance, Immigration, and Colonization," *Social Studies of Science* 39, no. 3 (June 2009): 369. See also Gabrielle Hecht, "Introduction," in *Entangled Geographies: Empire and Technopolitics in the Global Cold War*, ed. Gabrielle Hecht (Cambridge, MA: MIT Press, 2011), 3: "The allure of technopolitical strategies is the displacement of power onto technical things, a displacement that designers and politicians sometimes hope to make permanent." In a later work, Hecht demonstrates how at a critical juncture technopolitics served to render Gabonese and Nigérien uranium a banal commodity, thereby moving it beyond the reach of political considerations. *Being Nuclear: Africans and the Global Uranium Trade* (Cambridge: MIT Press, 2012), 139.

112. Weizmann, *Trial and Error*, 291.

113. Said, "Zionism from the Standpoint of Its Victims," 8; see also Abdul Latif Tibawi, *Anglo-Arab Relations and the Question of Palestine, 1914–1921* (London: Luzac, 1977), 1–33.

114. Haim Gerber, *Remembering and Imagining Palestine: Identity and Nationalism from the Crusades to the Present* (New York: Palgrave Macmillan, 2008).

115. Krämer, *History of Palestine*, 155. According to the last Ottoman administrative reorganization in 1873, the Palestine that ended up being defined under the British corresponded roughly to the three Ottoman provinces, or *sanjaks*, of Jerusalem, Nablus, and Acre. Most of the Negev Desert belonged to the province of Hijaz. It was the first time that these areas were organized under a single authority. For a more detailed description, see Biger, *Boundaries*, 13–14; Bernard Lewis, "On the History and Geography of a Name," *International Review* 2, no. 1 (January 1980): 1–12; Haim Gerber, "'Palestine' and Other Territorial Concepts in the 17th Century," *International Journal of Middle East Studies* 30, no. 4 (November 1998): 563–72, esp. 565.

116. *Encyclopaedia Britannica*, 11th ed., s.v. "Palestine," vol. 20 (1910–11), 600, https://archive.org/details/encyclopdiabri20chis.

117. Lewis, "History and Geography of a Name," 7. Incidentally, the 1910–11 edition was the first to mention the concept of "capitalism." Jörgen Kocka, *Capitalism: A Short History* (Princeton, NJ: Princeton University Press, 2016).

118. Naval Staff, Intelligence Department, *Handbook of Syria (including Palestine)*, June 1919, 11, ISA M-4409.

119. Biger, *Boundaries*, 15–18.

120. "From Dan to Beersheba" was the phrase that Weizmann initially proposed at the Paris peace conference. Margaret MacMillan, *Paris 1919: Six Months That Changed the World* (New York: Random House, 2002), 415, 423; Asher Kaufman, *Contested Frontiers in the Syria-Lebanon-Israel Region: Cartography, Sovereignty, and Conflict* (Baltimore: Johns Hopkins University Press, 2014), 30.

121. The two major documents produced during the war that are often seen as stages in the emergence of Palestine are the Husayn-McMahon Correspondence and

the Sykes-Picot Agreement. With regard to the former, to this day the argument remains unsettled whether Palestine was included in McMahon's inexpert, or most likely deliberately vague, locution: "west of the districts of Aleppo, Hamma, Homs and Damascus." As for the Sykes-Picot Agreement, the territory that ended up being included in mandatory Palestine was divided over all seven areas sketched in the agreement. An international zone covered most of Palestine; the rest was divided over a British-controlled area, a French-controlled area, Arab "Area A" (under French patronage), Arab "Area B" (under British patronage), the Hijaz, and Egypt. UNISPAL, http://unispal.un.org/UNISPAL.NSF/0/232358BACBEB7B55852571100078477.

122. Balfour Declaration.

123. Recommendations of the King-Crane Commission with regard to Syria-Palestine and Iraq (August 29, 1919), UNISPAL, http://unispal.un.org/UNISPAL.NSF/0/392AD7EB00902A0C852570C000795153#sthash.wvRFlJvV.dpuf; Memorandum Submitted to the Conference of Allied Powers at the House of Commons, Submitted by General Hoddad Pasha (Hejaz Army) and delegated by His Royal Highness Amir Faisal, on March 10, 1921, UNISPAL, http://unispal.un.org/unispal.nsf/9a798adbf322aff38525617b006d88d7/14f06fe1edd50616852570c00058e77e?OpenDocument#sthash.X2laYxlc.dpuf.

124. The British themselves recognized that it was not necessary to control Palestine in order to protect Suez. Segev, *One Palestine*, 35–36.

125. Alexander Schölch, "European Penetration and the Economic Development of Palestine, 1856–82," in *Studies in the Economic and Social History of Palestine in the Nineteenth and Twentieth Centuries*, ed. Roger Owen (London: Macmillan, 1982); Gad G. Gilbar, *Ottoman Palestine, 1800–1914: Studies in Economic and Social History* (Leiden: Brill, 1990); Arnold Blumberg, *Zion before Zionism 1838–1880* (Syracuse, NY: Syracuse University Press, 1985); Paul Cotterell, *The Railways of Palestine and Israel* (Abingdon: Tourret, 1984); Eugene L. Rogan, *Frontiers of the State in the Late Ottoman Empire: Transjordan, 1850–1921* (Cambridge: Cambridge University Press, 1999), 65–66; James L. Gelvin, *The Israel-Palestine Conflict: One Hundred Years of War* (New York: Cambridge University Press, 2005), 30–33; Ilan Pappe, *The Rise and Fall of a Palestinian Dynasty: The Husaynis, 1700–1948* (Berkeley: University of California Press, 2010), 113; Norris, *Land of Progress*, 36; TNA CO 733/28; TNA CO 733/41. On October 27, 1922, Platon G. Franghia submitted an application for a concession at Jaffa port. He originally submitted an application in 1910 to the Ottoman Ministry of Public Works. Full plans were submitted in October 1912, TNA CO 733/7; TNA CO 733/40; Midian Ltd.'s counsel, H. W. Stock, to the US/SC, April 8, 1921, TNA CO 733/10; CZA A153/1/143; also cited in Norris, *Land of Progress*, 40–42.

126. In 1880, the Jews thus made up some 5 percent of Palestine's population. Engel, *Zionism*, 11.

127. Engel writes that "economic rather than sentimental considerations led him to favour" Jewish immigration to Palestine. Engel, *Zionism*, 29–33. Other scholars see Smolenskin as a cultural nationalist.

128. "Resolution on Palestine (July 31, 1905)," Seventh Zionist Congress, July 27–August 2, 1905, in Basel, Switzerland. *Jewish Chronicle*, August 4, 1905, 21. From

the start, the Zionist movement endeavored to place its members in as many influential positions as possible, including the growing number of Palestinian Chambers of Commerce. CZA L3\372-58.

129. See also Penslar, *Zionism and Technocracy*, esp. his discussion of Max Bodenheimer, 44.

130. Gelvin, *Israel-Palestine Conflict*, 30.

131. Already by the late 1870s, the Ottomans had enacted laws restricting Jewish immigration to Palestine. Neville J. Mandel, *The Arabs and Zionism before World War I* (Berkeley: University of California Press, 1976).

132. CZA L3\372-5; Confidential memo to G. S., OET, Mount Carmel, Haifa, September 30, 1919, CZA L3\372-7; letter summarizing Report on Syria, Part II, June 10, 1919, CZA L3\372-8; M. D. Eder to Julius Simon, June 12, 1919, CZA L3\372-9; CZA L3\372-13; Zionist Organization of America to Julius Simon, September 18, 1919, CZA L3\372-18; L3\372-20; a more detailed list of concessions, June 16, 1919, prepared by N. Wilboushevitch, CZA L3\372-46-51.

133. Deputy Acting Chairman of the ZC to the Zionist Organization, Political Dept, Paris, August 20, 1919, CZA L3\372-32,33.

134. CZA L3\372.

135. IECA 0349-42-43; for Nablus, ISA GL-16614/5; for Tulkarem, TNA CO 733/29, TNA CO 733/39; for Bethlehem, TNA CO 733/29; for Haifa and Jaffa, IECA 0349-42, "Two memoranda submitted to The Council & Permanent Mandates Commission of the League of Nations respectively through H. E. the High Commissioner for Palestine by The Executive Committee Palestine Arab Congress, 12 April 1925." For Haifa, see also May Seikaly, *Haifa: Transformation of an Arab Society 1918–1939* (London: I. B. Tauris, 2002), 197. In January 1920, an application was received from Mr. Dabdub and Mr. Handal for a concession related to agricultural undertakings, telephone, electricity, tramways, and other business. There were no specifics here, only a request to be kept in mind once the government was ready to start giving out concessions. Pal Gov to S/SC, July 2, 1922, TNA CO 733/23.

136. On April 8, 1919, Sayer and Colley applied for an exclusive concession for electricity in Jerusalem. They received the reply that "laws and usages of war" made the provisional military administration unable to grant a concession, but it would be kept for future consideration. Pal Gov to S/SC, July 2, 1922, TNA CO 733/23.

137. Norris, *Land of Progress*, 146; Headrick, *Tentacles of Progress*, 171–207.

138. "Two memoranda submitted to The Council & Permanent Mandates Commission of the League of Nations respectively through H. E. the High Commissioner for Palestine by The Executive Committee Palestine Arab Congress, 12 April 1925." In the fall of 1920, a Palestinian businessman in Haifa applied to the British DG for a concession to electrify the city. On November 1 and 5, the Jaffa municipality applied to the governor of the Jaffa District with an electrification scheme of its own. On November 11, 1920, the Chief Secretary (CS) replied to the municipality: "I am to inform you that the <u>Government cannot consider this application before the ratification of the Mandate</u>." IECA 0349-42-43 (underlining in the original).

For Haifa, see also Seikaly, *Haifa*, 197; Deputy Acting Chairman of the ZC to the ZO, Political Dept, Paris, August 20, 1919, CZA L3\372-32,33.

139. David Ben-Gurion and Yitzhak Ben-Zvi, *Eretz Yisra'el be'avar uvahoveh* (Jerusalem: Hotsa'at Yad Yitsḥak Ben-Tsevi, 1979); also quoted in Biger, *Boundaries*, 58–60.

140. Ben-Gurion and Ben-Zvi, *Eretz Yisra'el be'avar uvahoveh*, 227.

141. S. Ilan Troen, *Imagining Zion: Dreams, Designs, and Realities in a Century of Jewish Settlement* (New Haven, CT: Yale University Press, 2003, Kindle Edition), Kindle Location 5523.

142. Norris, *Land of Progress*, 65.

143. John Marlowe, *Milner: Apostle of Empire* (London: Hamish Hamilton, 1976); 210–11, 330–33. Norris gives a rich account of the link between "new imperialism" and the making of the Balfour Declaration. Norris, *Land of Progress*, 8–9, 65.

144. "Palestine and Jewish Nationalism," *The Round Table: The Commonwealth Journal of International Affairs* 8, no. 30 (1918): 327.

145. Leopold Amery, *My Political Life*, vol. 2 (London: Hutchinson, 1953), 115; quoted in Marlowe, *Milner*, 331–32.

146. Biger, *Boundaries*, 59–60; Shaltiel, *Rutenberg*, 53. In the years after World War I, many British officials, politicians, and military personnel presented their visions for Palestine, including William Ormsby-Gore, Sir Earl Richers of the Foreign Office, and Colonel Richard Meinertzhagen. See, for instance, Richard Meinertzhagen, *Middle East Diary, 1917–1956* (New York: Yoseloff, 1960), 63–64.

147. Statement of the Zionist Organization regarding Palestine, February 3, 1919, UNISPAL, http://unispal.un.org/unispal.nsf/9a798adbf322aff38525617b006 d88d7/2d1c045fbc3f12688525704b006f29cc?OpenDocument#sthash.aunezJki. dpuf; Biger, *Boundaries*, 74–76, 109–10.

148. Cf. Neil Smith, *Uneven Development: Nature, Capital, and the Production of Space* (Athens: University of Georgia Press, [1984] 2008), 181, 193–94, 293n15. Smith identifies the three primary scales associated with the production of space under capitalism as the scale of the urban/local, the nation-state, and the global.

149. "Sir Hugo Hirst: British Work in Palestine: An Interview," *Sydney Morning Herald*, December 31, 1929, IECA 2379-16.

150. Shaltiel, *Rutenberg*, 44–46. The Zionist Organization to acting Chairman of ZC, E. W. Lewin-Epstein, February, 19, 1919, CZA L4\606-1-6; Ben-Zvi's eulogy upon PR's death on January 3, 1942, delivered January 11, ISA P-1968/3.

151. In this, Zionism was of a piece with myriad nationalist movements, positing both an unbreakable continuity and a radical break at the core of the movement.

152. TNA CO 733/17b.

153. *Science* 62, no. 1600 (August 28, 1925): xii.

154. "Palæstinas opdyrkning og elektriticering—Et norsk ingeniørsprojekt," *Krig og Fred*; "Die Urbarmachung und Elekrifizierung Palästina durch das Mittelmeer als Kraftquelle" [The reclamation and electrification of Palestine with the Mediterranean as power source], by Ingeniør H. Sørbye. The original article appeared in the Norwegian magazine *Teknisk Ukeblad*, as "Palestinas gjenvinning

og elektriticering ved Middelhavet som strømkilde," August 29, 1919. "Ein plan zur wirtschaftlichen Erschließung Palästinas" [A plan for the economic development of Palestine], Dr. Alfred Grobenwik, CZA K13/50; CZA A130/238.

155. Weizmann, *Trial and Error*, 149.

CHAPTER TWO

1. John Shuckburgh to J. Masterton Smith, January 19, 1922, TNA CO 733/29.
2. Letter from Samuel to London, undated, early fall, 1920, IECA 2375-4.
3. IECA 0357-195.
4. Letter from Julius Simon and Nehamie de Lieme to HC Samuel, dated August 24, 1920, IECA 2384-1, IECA 2371-17, IECA 2370-7.
5. Minutes of a meeting at Government House in Jerusalem between HC Samuel, CS Wyndham Deedes, and Director of the Department of Commerce and Industry Ralph A. Hariri, November 24, 1920, IECA 2371-17.
6. PR's proposal, December 8, 1920, preface, 1, CO 733/9.
7. PR's proposal, December 8, 1920, 16, TNA CO 733/9.
8. Nahum Vilbush, *Haroshet ha-ma'aseh: kovets letoldot ha-ta'asiyah baaretz: lezekher po'olo shel Nahhum Vilbush, haluts hata'asiyah ha'ivrit hahhadishah baaretz* (Tel Aviv: Hotsa'at Milo, 1974); Eshel, *Naharayim*.
9. PR's proposal, December 8, 1920, 48, TNA CO 733/9.
10. The conference was hosted by Britain at its Empire Exhibition and sponsored by the British Electrical and Allied Manufacturers' Association in cooperation with technical and scientific institutions and industrial organizations from around the world. In the five points that made up the instructions to speakers, for instance, waterpower, the only power source that was explicitly specified, was mentioned in four. There was also a section of the conference entirely devoted to waterpower, again being the only power source explicitly mentioned. D. Heineman, "Electricity in the Region of London," in *The Transactions of the First World Power Conference*, vol. 4 (London: Percy Lund, Humphries and Co., 1924), TNA CO 733/46.
11. Marx, "Social and Economic Aspects of Hydro-Electric Power," 1782.
12. Walker, *Toxic Archipelago*, 28.
13. James L. Gelvin, *Divided Loyalties: Nationalism and Mass Politics in Syria at the Close of Empire* (Berkeley: University of California Press, 1998); Wasserstein, *The British in Palestine*, 62; Michael Provence, *The Great Syrian Revolt and the Rise of Arab Nationalism* (Austin: University of Texas Press, 2005), 45.
14. Munir Fakher Eldin, "British Framing of the Frontier in Palestine, 1918–1923: Revisiting Colonial Sources on Tribal Insurrection, Land Tenure, and the Arab Intelligentsia," *Jerusalem Quarterly* 60 (2014): 42–58.
15. PR's proposal, December 8, 1920, 37, TNA CO 733/9. For more on the Bedouin, see Administrative Report, March 1922, TNA CO 733/10; Report of the Administration of the Government of Palestine for the period July 1920–December 1922 for the League of Nations, TNA CO 733/46.

16. General HQ of British army in Cairo to War Office, April 23–26, 1920, TNA FO 371/5118, cited in Fakher Eldin, "British Framing of the Frontier in Palestine," 42.

17. PR's proposal, December 8, 1920, vi, 36, TNA CO 733/9.

18. This was not wholly unprecedented. When the Walchenseewerk, the large German hydroelectric station, with a projected 50,000-hp capacity, was approved in the 1920s, the local government in Bavaria demanded an assessment of future demand from the head consulting engineer on the project, Oskar von Miller. Of course, there was no sense in this case in which von Miller was actually trying to create load centers by sending power to certain points. See Hughes, *Networks of Power*, 345.

19. TNA CO 733/44.

20. "Rutenberg Power-House Damaged by Floods Caused by Torrential Rains," *Jewish Telegraphic Agency*, February 16, 1931.

21. Andrew Needham, *Power Lines: Phoenix and the Making of the Modern Southwest* (Princeton, NJ: Princeton University Press, 2014), 78.

22. Palestine Government, *A Survey of Palestine: Prepared in December 1945 and January 1946 for the Information of the Anglo-American Committee of Inquiry*, vol. 2 (Washington, DC: Institute for Palestine Studies, 1990), 148 (1922 figures) and 12 of supplement (1946 figures); cited in Norris, *Land of Progress*, 101. The number of Jews in Haifa rose from 6,230 to 74,230 during the period. Norris, *Land of Progress*, 123.

23. *Statistical Abstract of Palestine*, 1940, table 10; *Statistical Handbook*, 1947, p. 48; cited in Metzer, *Divided Economy*, 78.

24. PR's proposal, December 8, 1920, 48, TNA CO 733/9. Beer Sheva did indeed quickly develop into the largest town of the Negev. After Israeli statehood, it grew exponentially. Today, it is the fourth-largest city in Israel, often referred to as "the capital of the Negev."

25. PR's proposal, December 8, 1920, 62, TNA CO 733/9.

26. Snell's report, dated March 7, 1921, TNA CO 733/13.

27. Snell's report, dated March 7, 1921, TNA CO 733/13.

28. Snell's report, dated March 7, 1921, TNA CO 733/13.

29. Cf. Steven Shapin, *A Social History of Truth: Civility and Science in Seventeenth-Century England* (Chicago: University of Chicago Press, 1994).

30. IECA 2375-4; Shaltiel, *Pinhas Rutenberg*, 42–43. The British transitioned Palestine from military to civilian rule in the summer of 1920 and appointed Herbert Samuel as the first high commissioner.

31. Letter from Shuckburgh to Churchill, August 19, 1921, TNA CO 733/4.

32. TNA CO 733/17b

33. *HaPoel HaTzair*, "From the Press," quotes the *Daily Chronicle*, November 25, 1921. IECA 2377-4; see also "A Preliminary Outline of the Erection of an Electrical District Station in Palestine," prepared by the engineer I. Archavsky in 1919, CZA L3\82\6-83.

34. Parliamentary Debates, House of Commons, Motion on Rutenberg Plan, July 4, 1922, CZA A342/3.

35. Parliamentary Debates, House of Commons, Motion on Rutenberg Plan, July 4, 1922, CZA A342/3.

36. Handwritten note by Shuckburgh, May 7, 1923, TNA CO 733/44.

37. Summary of the proceedings of the Permanent Mandates Commission session in October 1924, WHA 1-983.

38. US Dept of Commerce booklet "Palestine: Its Commercial Resources with Particular Reference to American Trade," by Addison E. Southard, American Consul at Jerusalem, October 12, 1922, TNA CO 733/46.

39. I. W. Majercik to PR, October 29, 1922, IECA 2375-11.

40. "Sir Hugo Hirst: British Work in Palestine: An Interview," *Sydney Morning Herald*, December 31, 1929, IECA 2379-16.

41. PR to Samuel, March 24, 1921, IECA 2375-4.

42. CAOG 14/109. Until 1927, the currency in Palestine and Transjordan was the Egyptian pound (£E); thereafter it was the Palestine pound (£P).

43. Memorandum by PR on railway electrification, the first stage of the Jordan, "Water Resources of Palestine: Report (Alternative to Section II). First Jordan Hydroelectric Power Installation: 'Jisr-el-Mujamyeh,' dated May 15, 1921, TNA CAOG 14/109; CO 733/3; CO 733/9, IECA 0357-199, IECA 2372-2 (emphasis added); Memorandum by PR on Railway electrification, dated May 15, 1921, TNA CAOG 14/109.

44. June 26, 1921, TNA CO 733/3.

45. Dispatch from HC Samuel to S/SC Churchill, June 7, 1921, CAOG 14/09.

46. Handwritten memo from June 1921, CO 733/3.

47. R. V. Vernon of the Colonial Office described PR's plan as "of overwhelming importance for the future general development of Palestine"; he added, "other less urgent needs of the country might quite properly be postponed to it." August 13, 1921, TNA CO 733/17b; Locomotive Superintendent in Palestine: "I am of the opinion that the hydro-electric scheme would eventually be a great benefit to the country generally and the Railways and would be one of the best means of developing the country but the Government should be prepared to withstand the loss of the first few years." Locomotive Superintendent's Office, Haifa, November 25, 1921, TNA CAOG 14/109; the government's Consulting engineers: "even if it is not possible for us to give now figures showing definite monthly saving with electricity over coal the indirect advantages are considerable." December 23, 1921, TNA CAOG 14/109.

48. G. L. M. Clauson, June 26, 1921, CO 733/3.

49. CO 733/7. In the event, electrification was never actually carried out. A few months into 1922, opinion started to shift among British officials, mostly as a result of a continuous fall in the price of coal. By 1923, the majority view was against electrification. TNA CO 733/58, 733/61.

50. IECA 2375-4-78; Report: PR, "Utilization of River Auja: Part I Hadrah-Jerishe: Water Power and Irrigation for Jaffa District," January 1921, TNA CO 733/17b. PR presented a detailed report on the Auja to the British later that spring, on April 27, 1921. See IECA 2371-17-108. See also Michael B. Oren, "Ha-Masa o-Matan lehasgat Zikayion LeKhevrat Ha-Kheshmal" [The negotiation to obtain a concession to electrify Palestine], *Cathedra* 26, no. 3 (5743; 1982): 143.

51. Undated letter from late fall 1920, IECA 2375-4.

52. Report by Consulting Engineers on the Auja scheme, dated August 30, 1921, TNA CAOG 14/108.

53. US Department of Commerce booklet on "Palestine: Its commercial resources with particular reference to American trade" by Addison E. Southard, American Consul at Jerusalem, published 1922, submitted to DoC October 12, 1922: "As insufficient power will be developed by the small volume of water available, a Diesel engine will be installed to supplement it," 564, TNA CO 733/46.

54. Shamir, *Current Flow*, 37–38.

55. PR to Samuel, March 24, 1921, IECA 2375-4-79.

56. Shuckburgh for Churchill to Lord Curzon, March 1, 1921, TNA CO 733/13. According to Shaltiel, *Pinhas Rutenberg*, 74–75, there was another reason for the FO to agree to Churchill's request, namely, the belief that Jews were influential in the United States and that granting a concession in Palestine that would enable American Jews to invest would appease the United States on the issue of oil concessions. Oren, "Ha-Masa o-Matan," 144.

57. Shamir, *Current Flow*, 37–38.

58. Letter dated July 15, 1921, TNA CAOG 14/109; Crown Agents to S/SC, June 6, 1921, TNA CO 733/9; Letter June 1, 1921, TNA CAOG 14/108; Crown Agents to Vernon, TNA CAOG 14/108.

59. Churchill to PR, July 2, 1921, TNA CAOG 14/108.

60. "Agreement for the granting of a concession in connection with the provision and supply of electrical energy for lighting and power purposes and for irrigation within the District of Jaffa Palestine" [2nd proof], September 12, 1921, IECA 0362-290, IECA 2371-8.

61. PR's Report: "Water Resources of Palestine: Report (Alternative to Section II). First Jordan Hydroelectric Power Installation: 'Jisr-el-Mujamyeh,'" May 15, 1921, CO 733/9, IECA 0357-199, IECA 2372-2.

62. Telegram from Churchill to HC, July 1, 1921, CO 733/3.

63. Shuckburgh to PR, June 23, 1921, notifying PR of the approval, contingent on a few modifications, mainly to do with increasing British influence over the venture. On June 27, 1921, PR wrote back essentially agreeing to the amendments. On July 2, 1921, S/SC Churchill wrote to PR formally giving him provisional approval. TNA CAOG 14/108.

64. Shuckburgh to Churchill, TNA CO 733/15.

65. "Agreement for the granting of a concession for the utilization of the waters of the Rivers Jordan and Yarmuk and their affluents for generating and supplying electrical energy," September 21, 1921, IECA 0362-290.

66. In the first agreement an electric utility in Germany, the Deutsche Edison Gesellschaft and Berlin in 1884, electricity rates were similarly specified, and 10 percent of the utility's gross income and 25 percent of its annual profit would go to the city. The contract period was thirty years. In return, the company was granted a monopoly of the inner-city area. See Hughes, *Networks of Power*, 185.

67. Undated memorandum by Clauson from June 1921, TNA CO 733/3.

68. Note from Vernon to Churchill, August 13, 1921, TNA CO 733/17b.

69. Samuel to PR, November 17, 1921, IECA 2371-8-93.
70. PR's proposal, December 8, 1920, 31, TNA CO 733/9.
71. Biger, *Boundaries*, 129.
72. See, for instance "Report of General Grant and Mr. P. Rutenberg on their expedition to Beirut as experts of the Anglo-French Water Commission," November 22, 1921; see also Clauson's note from November 26, 1921, protesting the way PR, despite being "told to report on a very minor matter... seized the opportunity to make a report which went far beyond their terms of reference, &, worse still, negotiated on this basis with the French." Clauson went on to say that PR's proposals "suit the Rutenberg project & the French; but not the wider interests of Palestine or Transjordania." PR was defended by the HC, in a telegram to the S/SC, who blamed any failure on Newcombe, November 24, 1921, TNA CO 733/7.
73. Eshel, *Naharayim*, 16.
74. Note dated August 16, 1921, TNA CO 733/5.
75. Notes dated August 8, 16, and 28, 1921, TNA CO 733/5; CO 733/7 contains several long discussions, including a memorandum written by Clauson, November 26, 1921; Biger, *Boundaries*, 149.
76. PR to Downie, US/SC, July 29, 1937, TNA CO 733/337/2.
77. Newcombe to CO, personal note, September 19, 1921, TNA CO 733/17b; S/SC to HC, September 8, 1921, CO 733/5.
78. Weizmann and Cohen to Brandeis, January 30, 1920, CZA Z4/40522.
79. "To Incorporate Mettula in British Palestine," *Jewish Telegraphic Agency*, April 25, 1923.
80. Secret dispatch from HC Wauchope to CO, May 24, 1932, containing a "Note on variations between the frontier between Palestine and Syria as laid down in the Anglo-French Convention of the 23rd December, 1920, and the Anglo-French Agreement of the 7th March, 1923, regarding the Boundary between Palestine and Syria," ISA M-107/2.
81. Weizmann to Halpern, October 14, 1920, WHA 6-599.
82. Handwritten note by Major Young, dated January 26, 1922. Young reports back that he has settled a dispute between PR and Newcombe over Lake Huleh in Palestine. Young settled the matter in PR's favor, to Newcombe's evident chagrin, TNA CO 733/39; report by Newcombe from September 19, 1921, in TNA CO 733/17b.
83. Secret dispatch from HC Wauchope to CO, May 24, 1932, containing a "Note on variations between the frontier between Palestine and Syria as laid down in the Anglo-French Convention of the 23rd December, 1920, and the Anglo-French Agreement of the 7th March, 1923, regarding the Boundary between Palestine and Syria," ISA M-107/2; ISA P-920/3.
84. *League of Nations Official Journal*, November 1922, 1188–89, 1390–91; Philip Robins, *A History of Jordan* (Cambridge: Cambridge University Press, 2004), 19–20; for more on borders, see Yitzhak Gil-Har, "British Commitments to the Arabs and Their Application to the Palestine–Trans-Jordan Boundary: The Issue of the Semakh Triangle," *Middle Eastern Studies* 29, no. 4 (1993): 690–91; Yitzhak

Gil-Har, "Boundaries Delimitation: Palestine and Trans-Jordan," *Middle Eastern Studies* 36, no. 1 (2000): 70; Biger, *Boundaries*, 177.

85. Gil-Har, "British Commitments," 693–95; Oren, "Ha-Masa o-Matan."

86. Memorandum by Major Young, dated April 8, 1925, TNA CO 733/92. In the report, Young initially lists three reasons for the border irregularity, but he proceeds to dismiss the two that do not concern PR's scheme.

87. *League of Nations Official Journal*, November 1922, 1390–91.

88. HC's report on meeting with Abdullah, October 25, 1922, TNA CO 733/37.

89. "Power and Irrigation Development in Palestine," *Engineering News-Record* 88, no. 25 (June 22, 1922): 1042.

90. The settlement pattern resembling the letter N was first proposed by Arthur Ruppin in 1907 and referred to a territory along Palestine's northern coast, a diagonal line inland that connected to another north-south line along the Jordan Valley. See Krämer, *History of Palestine*, 306.

91. TNA CO 733/9.

92. "Power and Irrigation Development in Palestine," 1042.

93. The idea has a long colonial precedent. Adas, *Machines as the Measure of Men*.

94. "Palestine Could Easily Support Five Times Present Population If Modern Methods of Farming Are Introduced," *Jewish Telegraphic Agency*, March 27, 1931.

95. "Power and Irrigation Development in Palestine," 1042.

96. "Sir Hugo Hirst: British Work in Palestine: An Interview," *Sydney Morning Herald*, December 31, 1929, IECA 2379-16.

97. "Power and Irrigation Development in Palestine," 1042.

98. PR's Proposal, December 8, 1920, 37, 62, TNA CO 733/9.

99. *Statistical Abstract of Palestine*, 1929, Keren Hayesod, Jerusalem, 1930, ISA M-5316.

100. Owen Tweedy, "Cheap Electricity for Palestine: Harnessing the Jordan: First Stage Nearing Completion," *Financial Times*, May 1930. Tweedy had served in the Palestine campaign during World War I.

101. TNA CO 733/165/3.

102. Cf. Diana K. Davis, *Resurrecting the Granary of Rome: Environmental History and French Colonial Expansion in North Africa* (Athens: Ohio University Press, 2007).

103. For a detailed look at the way this theme played itself out in the social sciences in mandatory Palestine, see Troen, *Imagining Zion*.

104. Hughes, *Networks of Power*, 364, 286.

CHATPER THREE

1. Shuckburgh to S/SC, September 7, 1921, TNA CO 733/15.

2. Addison E. Southard, "Palestine: Its commercial resources with particular reference to American trade," booklet published by the United States Department of Commerce, October 12, 1922, TNA CO 733/46.

3. PR to Samuel, June 11, 1923, ISA M-10/2.

4. IECA 0361-275.

5. Dispatch from Civil Secretary to Legal Secretary reporting on a meeting between the Financial Secretary and the Mayor of Jaffa, ISA GL-16614/5.

6. Shaltiel, *Pinhas Rutenberg*, 41.

7. "Correspondence with the Palestine Arab Delegation and the Zionist Organization, Presented to Parliament June 1922," ISA M-4381.

8. "Blow to Rutenberg: Jaffa Refuses His Electricity," *Daily Mail*, May 28, 1923, TNA CAOG 14/108.

9. Steven Shapin and Simon Schaffer, *Leviathan and the Air-Pump: Hobbes, Boyle, and the Experimental Life* (Princeton, NJ: Princeton University Press, 2011), 332.

10. Thomas F. Gieryn, "Boundary-Work and the Demarcation of Science from Non-science: Strains and Interests in Professional Ideologies of Scientists," *American Sociological Review* 48, no. 6 (1983): 781–95; see also Gieryn, *Cultural Boundaries of Science*, definition on p. 4.

11. Report by Bols, June 6, 1920, FO 731/5114, cited in Fakher Eldin, "British Framing of the Frontier in Palestine, 1918–1923," 47.

12. Wasserstein, *The British in Palestine*, 101–2; Segev, *One Palestine*, 173–90; Sahar Huneidi, *A Broken Trust: Herbert Samuel, Zionism, and the Palestinians, 1920–1925* (London: I. B. Tauris, 2001), 127. For an in-depth examination of Samuel's change of heart, see Evyatar Friesel, "Herbert Samuel's Reassessment of Zionism in 1921," *Studies in Zionism* 5, no. 2 (Autumn 1984): 213–37.

13. Wasserstein, *British in Palestine*, 93.

14. Telegram from Samuel to Churchill, April 24, 1921, TNA CO 733/5.

15. Samuel's fear was not unfounded. See "Political report on disturbances" by W. N. Congrave, Lieutenant-General, Commanding, Egyptian Expeditionary Force, May 30, 1921, CO 733/3.

16. For a fuller view of Samuel's shifting thinking, see Wasserstein, *British in Palestine*, 17 and chap. 5; also Oren, "Ha-Masa o-Matan, 145–48.

17. Annual report to the PEC, December 31, 1923, IECA 0361-275.

18. "Conversations between D. K. and Ibrahim Shammas at the Hotel Cecil, 5 Jan 5 1922," pp. 6–7, IECA 0349-42-6. The person referred to as "D. K." is never identified in the records. He appears to be a spy working for PR.

19. Mitchell, *Carbon Democracy*, 37.

20. Yael Zerubavel, "The Politics of Interpretation: Tel Hai in Israeli Collective Memory," *Association for Jewish Studies Review* 16 (1991): 133–60; Yael Zerubavel, *Recovered Roots: Collective Memory and the Making of Israeli National Tradition* (Chicago: University of Chicago Press, 1995).

21. IECA 0349-42.

22. IECA 2375-4.

23. Dispatch from Civil Secretary to Legal Secretary reporting on a meeting between the Financial Secretary and the Mayor of Jaffa, November 19, 1921, ISA GL-16614/5. The mayor stated, "I have always had before my mind the question of the advance of the town and I told them the people in general were in favour of such

things." Later in the same interview, he stated, "Most of the wealthy people of Jaffa and District are willing to participate [financially in the scheme]. The notables have been to me and have told me that this is their idea." This aligns with Yehoshua Porath's claim that Jaffa businessmen, most of whom were engaged in the citrus trade, were decisively pro-British, as opposed to the supporters of pan-Arabism or Greater Syria. Yehoshua Porath, *The Emergence of the Palestinian-Arab National Movement, 1918–1929* (London: Cass, 1974), 90.

24. Samuel to S/SC, August 25, 1921, TNA 733/5.

25. Samuel to CO, August 25, 1921, IECA 2375-04-31.

26. TNA CO 733/5; Samuel to Churchill, August 25, 1921, IECA 2375-04-31.

27. Another letter to PR notifying him of provisional approval, July 2, 1921, TNA CAOG 14/108.

28. Auja Scheme submitted by the Crown Agents to the Consulting Engineers, August 27, 1921, TNA CO 733/5.

29. Report submitted by the Consulting Engineers to the Colonial Office, August 30, 1921, TNA CAOG 14/108.

30. August 27, 1921, TNA CAOG 14/108.

31. IECA 2371-17.

32. TNA CO 733/6.

33. Handwritten note by Major Young, September 9, 1921, TNA CO 733/6/50.

34. Samuel to Churchill, September 2, 1921, TNA CO 733/17b.

35. Walid Khalidi, *All That Remains: The Palestinian Villages Occupied and Depopulated by Israel in 1948* (Washington, DC: Institute for Palestine Studies, 1992), 259–60.

36. PR to Samuel, October 31, 1921, ISA GL-1664/5; Minutes of meeting on October 15, 1921, IECA 0349-42-9.

37. Report: PR, "Utilization of River Auja: Part I Hadrah-Jerishe: Water Power and Irrigation for Jaffa District," January 1921, TNA CAOG 14/108.

38. PR to Samuel: "financial participation of the Jaffa municipality is not essential as money required can be obtained elsewhere," TNA CAOG 14/108.

39. PR to Herbert Samuel, October 19, 1921, IECA 2371-1.

40. Letter from PR to Mayor of Jaffa, April 8, 1921, IECA 2375-4; "Conversations between D.K. and Ibrahim Shammas at the Hotel Cecil, January 5, 1922," pp. 6–7, IECA 0349-42-6.

41. January 18, 1923, TNA CO 733/61.

42. Unsigned note by government official, November 17, 1923, ISA M-10/2.

43. PR to Mayor of Jaffa, April 8, 1921, IECA 2375-4-67.

44. PR to Samuel, October 19, 1921, ISA GL-16614/5. Agreement to get the government guarantee was given on June 5, 1921. Although Britain's foreign investments during the empire's heyday were unprecedented in history—double the French and three times the German figure at the same time—almost all of the investments in the colonial world (a total of £1.8 billion) went to colonies established before the final decades of the nineteenth century, primarily India and the Americas. Only negligible sums were invested in Africa after the Scramble and in the Middle East:

Niall Ferguson, *Empire: The Rise and Demise of the British World Order and the Lessons for Global Power* (New York: Basic Books, 2003), 202–3. Indeed, in the period following World War I, Britain's policy all over its sphere of influence was that its holdings would have to pay their own way; Roger Owen, *State, Power and Politics in the Making of the Modern Middle East* (New York: Routledge, 2004), 12–13.

45. TNA CO 733/44/215.

46. "Jaffa Municipal Decision No. 560 of the 16th day of November 1921," ISA GL-16614/5; Letter from Jaffa Municipal Council to Samuel, November 16, 1921, IECA 2375-4-23.

47. Following letter from Governor of Jaffa with the Municipal Decision, November 18, 1921, ISA GL-16614/5.

48. Dispatch from the Jaffa DG, November 18, 1921, ISA GL-16614/5.

49. PR from unnamed British official. The quotation is by Colonel Symes, December 8, 1921, ISA GL-16614/5.

50. Late November and early December 1921 saw a flurry of activity in this regard. On December 13, the governor of Jaffa sent PR his final draft, explicitly asking for PR's approval. PR gave it, contingent on a few minor changes. See IECA 2375-4.

51. PR to Legal Secretary Norman Bentwich in Jerusalem, November 24, 1921, IECA 2375-4-22; also available in ISA M-756/4.

52. Harry Sacher to Governor of Jaffa, December 14, 1921, IECA 2375-4-18: "The Company will earmark L.E.25,000 worth of the shares for the Jaffa Municipality should the Jaffa Municipality decide to take up that number and all the inhabitants of Jaffa will have an unrestricted right to apply for shares in the Company on the terms of the the issue on the same footing as other applicants."

53. TNA CO 733/9.

54. TNA CO 733/17b.

55. IECA 0361-275.

56. IECA 2369-15-171.

57. IECA 0361-275-30.

58. IECA 0361-275.

59. Hughes, *Networks of Power*, 355.

60. PR himself often referred to the need of the British to maintain this business framing: "The Colonial Office supports the Concession on the basis that it is a business and not a political undertaking," he wrote to Judge Mack on September 22, 1922, IECA 2371-8-54.

61. One of the first recorded mentions of it in the British records is from a letter from Churchill to Samuel about PR's attempts to negotiate with the Arab delegation in London. Churchill reports, "Rutenberg had discussed the matter with them as a business undertaking and not on political lines and is not dissatisfied with the results." S/SC to HC, September 12, 1921, TNA CO 733/6; TNA CO 733/15.

62. This happened in connection with Herbert Samuel's visit to London in May 1922. TNA CO 733/40.

63. According to an internal British memo, PR informed them that "certain Arab landowners have raised their prices by three or fourhundred [*sic*] percent in

order to stultify the Auja Scheme. They are said to be waiting for a Moslem Christian Delegation to talk sense into them." ISA GL-1664/5.

64. TNA CO 733/6: Political report for August 1921 prepared by Wyndham Deedes.

65. Baruch Kimmerling, *Zionism and Territory: The Socio-territorial Dimensions of Zionist Politics* (Berkeley: University of California Press, 1983), 11. See also Warwick P. N. Tyler, *State Lands and Rural Development in Mandatory Palestine, 1920–1948* (Brighton: Sussex Academic Press, 2001), 9–10.

66. Sacher to PR, September 9, 1922, IECA 2375-11.

67. PR to Sir Alfred Mond, February 11, 1923, IECA 2377-7-265.

68. The Auja landowners were asking for £E25 per dunam, bringing the total cost of the land to £3,750—not an insignificant sum to be sure, but considering that a single diesel set of 500 hp cost about £E5,000 and the first year's profits alone were estimated at £E30,000, the cost of land remained a proportionally small capital investment. What is more, within two years, PR would conclude a transaction with Arab landowners in Haifa for seventy dunams of land at £E64 per dunam, or £E4,500 total. And in 1922, he paid £E8,000 for twenty dunams of (Jewish-owned) land in Tel Aviv, which comes to £E400 per dunam. Report: "Cost of power house and distribution system." The plot was purchased jointly by the Jaffa Electric Company and the PEC, the latter paying £E5,600 for fourteen of the twenty dunams. IECA 0361-275-6.

69. IECA 2371-1-39.

70. According to Shamir, PR did not make the decision to switch technologies until December–January 1921/2, as a result of consultations with German engineers. The record clearly shows, however, that PR had made his decision earlier. November 16, 1921, IECA 2381-5: In this letter from PR's assistant, Singer, to Julian Mack, it is stated that PR will go to Germany "to get final estimates."

71. PR to US/SC, ME Dept, January 31, 1922, TNA CO 733/39; Consulting Engineers' Report on PR's offer of March 8, March 23, 1922, TNA CO 733/29; PR to Judge Mack detailing his plans for electrification in Palestine, February 14, 1922, IECA 2371-8-78.

72. Harry Sacher to PR, January 22, 1922, IECA 2371-1.

73. Harry Sacher to PR, January 22, 1922, IECA 2371-1.

74. Hughes. *Networks of Power*, 80.

75. PR to Judge Mack, February 14, 1922, IECA 2371-8-78.

76. PR to Judge Mack, February 14, 1922, IECA 2371-8-78.

77. ISA GL-16614-5.

78. Note dated July 13, 1923, "handed to Mr. Vernon," IECA 2371-1-43.

79. Zachary Lockman, *Comrades and Enemies: Arab and Jewish Workers in Palestine, 1906–1948* (Berkeley: University of California Press, 1996).

80. Note dated July 13, 1923, "handed to Mr. Vernon," IECA 2371-1-43.

81. PR to Churchill, July 24, 1922, IECA 2371-1. PR's lawyer, Harry Sacher, had prepared the ground with a letter to Shuckburgh in May: Sacher to Shuckburgh, May 14, 1922, ISA GL-16614/5.

82. Harry Sacher to CO, May 14 1922; PR to CO, August 3, 1923; CO report summarizing the various extensions, September 13, 1923; PR to S/SC, February 18, 1924; PR to CO, August 1, 1924; HC to PR, October 7, 1924, approval of another extension, ISA GL-1664/5.

83. Director of the Department of Industry and Commerce to Samuel, December 15, 1922, ISA GL-16614/5.

84. AG to CS, December 19, 1922, ISA GL-16614/5.

85. CS to PR, informing him that he is released of the obligation to build the Yarkon powerhouse but is allowed to keep the irrigation part of the concession, dated October 19, 1926, ISA GL-1664/5; Official Gazette, Promulgation of Ordinance: The Electricity Concessions Ordinance, No. 9 of 1927 was promulgated on March 7, 1927, IECA 0359-232.

86. Indeed, already in 1923 he drew up a plan for the project and allocated £E20,000 to it, of which £E2,000 was allotted to land purchases. Draft of plan for development of the Auja, July 1923, IECA 2371-1; PR to Baharaw, July 30 1923; Telegram from Baharaw to PR, August 8, 1923, IECA 2371-1-35.

87. Director of DPW to CS, February 1931, ISA P-3047/4. The PEC formed a subsidiary company, the Auja Irrigation Company, and on August 6, 1929, it submitted a scheme to the DPW for approval, which envisioned irrigating the high-lying citrus land in Jewish possession; the plan, however, lacked provisions for irrigation of the lands of the Auja basin, which was largely Arab owned. The file also contains a "Memorandum on Irrigation from the Auja River and the Rutenberg Irrigation Concession," by a Mr. Shepherd. An important reason why PR opted not to irrigate the basin was that he deemed it "commercially untenable." PR's reason for not irrigating Auja was that the area was highly developed agriculturally, and there was a wealth of wells, sunk at unsystematic, dense patterns, which had caused the groundwater to be lowered, and in some instances alkaline water to be drawn, which was harmful to the plants. If an above-ground irrigation scheme was carried out, at considerable expense, then these overextended wells would start yielding good water again, thus undermining the need for the elaborate irrigation scheme just constructed. Furthermore, PR had contradictory interests in the area, since he was looking to sell electricity for the well pumps at a cost low enough to warrant switching from oil-driven engines. But at the same time, the irrigation scheme required that the farmers abandon the use of pumps entirely in favor of the irrigation ("Mr. Rutenberg's schemes embrace two activities which are, in a manner, antagonistic," the report notes). In addition, in the high-lying Jewish areas, where citriculture predominated, the farmers were interested in neither PR's water nor his electricity. By all accounts, they were happy with their wells and oil-driven pumps. For the Arabs, on the other hand, the problem was different. Most Arab land was low-lying and unsuitable for citriculture. In addition, most Arab farmers grew winter cereal and practiced dry farming in the summer. They did not grow the kinds of cash crops that would bring in sufficient revenue. Finally, irrigating Arab-owned land, which Zionists were ultimately looking to acquire, would raise its value.

88. PR to US/SC, July 29, 1937, TNA CO 733/337/2.

89. "Fate of Concessions," *Filastin*, March 19, 1932, CAOG 14/108.

90. Roza el-Eini, *Mandated Landscape: British Imperial Rule in Palestine, 1929–1948* (London: Routledge, 2006), 161–72.

91. Samuel's visit generated a large amount of publicity: "Why *Are* We in Palestine? Huge Electric Monopoly Handed to a Russian: Work for Germany: While the British Taxpayers Find More Millions," *Daily Sketch*, May 25, 1922, TNA CO 733/40; "Mr. Rutenberg: His Palestine Concessions: Kerensky's Chief of Police: Big Contract Given to Germany," *Daily Mail*, May 25, 1922, TNA CAOG 14/109; "Palestine Power Monopoly: Mystery of Order to Germany: Mr. Rutenberg Describes His Schemes: British Business Doubts," *Daily Mail*, May 26, 1922, TNA CO 733/40; "Mare's Nesting in Palestine: Facts on Rutenberg Concessions: A Foolish Outcry," *Daily Chronicle*, May 30, 1922, TNA CO 733/40: article attacked the critics of the PR scheme; the first round of criticism came already in November 1921, when the Opposition MP, Sir W. Joynson-Hicks, asked the S/SC about the matter. November 7, 1921, TNA CO 733/12.

92. July 4, 1922, TNA CO 733/35.

93. July 4, 1922, TNA CO 733/35.

94. "Mr. Churchill's Reply," *The Times*, July 5, 1922, TNA CAOG 14/109.

95. June 1922, TNA CO 733/39; July 4, 1922, TNA CO 733/35.

96. PR to Brandeis, April 23, 1922, IECA 2381-5-60.

97. PR to Brandeis, June 16, 1922, IECA 2381-5-34.

98. *Commercial Bulletin Vol II*, 1922, ISA M-5313/10.

99. For negotiations with Haifa, see ISA M-9/10: report from Treasurer, March 21, 1923, and minutes from a meeting between HC Samuel and the mayor and municipal council of Haifa, May 7, 1923. This agreement for Haifa mirrored the one for Jaffa: Haifa would invest £P15,000, borrowed against a government guarantee, from the Anglo-Egyptian Bank. PR visited Haifa several times in April 1923 to negotiate with the municipality.

100. On April 13, 1923, HC forwarded the draft agreements between PR and the Jaffa and Haifa municipalities. HC to CO, April 28, 1923, ISA M-10/2.

101. Despite the name, the APC was a bank. In 1925, it officially changed its name to the Anglo-Palestine Bank. It still exists today in Israel under the name Bank Leumi. See Nadav Halevi, *Banker to an Emerging Nation: The History of Bank Leumi le-Israel* (Haifa: Shikmona, 1981). The municipal investments were in the amounts of £E25,000 for Jaffa, £E15,000 for Haifa, and £E12,000 for Tel Aviv.

102. Paul Singer to Vernon, May 2, 1923, TNA CO 733/44; proposed Heads of Agreement between the Jaffa Electric Company and Jaffa, sent from the company to CS, April 5, 1923, ISA M-9/10.

103. According to the power company's calculations, Tel Aviv would save £E1,145 a year and Jaffa about £E842. TNA CO 733/44.

104. For Tel Aviv, IECA 0357-197, 0363-329: "'Supply' Agreement for the Supply of Electrical Energy to the Township of Tel-Aviv," dated May 10, 1923; for Haifa, see letter from Vernon to Paul Singer, containing a draft, dated May 4, 1923, of the

Haifa contract and Proposed head of agreement with the municipalities, July 12, 1923, TNA CO 733/61.

105. PR to the Tel Aviv Municipal Council, May 10, 1923, IEC 0363-329.

106. Tel Aviv Council to Assistant DG, Jaffa, April 23, 1923. The reason for the increased sum was the expansion of Tel Aviv. Tel Aviv–Jaffa Municipal Historical Archives (hereafter TAHA) 534/34G.

107. The MCAs were the first organizations of Palestinian nationalism. The Jaffa chapter was the oldest, founded in 1918. Porath, *The Emergence*, 32–33.

108. IECA 0349-42-9: Minutes of meeting on October 15, 1921, at the offices of the Governor of Jaffa; TNA CO 733/62.

109. "Correspondence between Omar Beitar, the Chairman of the Jaffa branch of the Moslem Christian League, and the Mayors of Jerusalem and Jaffa," *Filastin*, June 16 and June 23, 1922.

110. For detailed records of negotiations carried out between PR and the Jerusalem municipality from 1920 to 1922, see IECA 2384-7.

111. Not to mention the fact that Omar Beitar himself had been present at these negotiations. Indeed, his was one of the names (he was the only one who insisted on signing his name in Arabic) on the November 16, 1921, document pledging Jaffa's commitment to PR's scheme.

112. "Correspondence between Omar Beitar, the Chairman of the Jaffa branch of the Moslem Christian League, and the Mayors of Jerusalem and Jaffa, published in Philistine, 16th and 23rd June, 1922," IECA 0349-42.

113. *Filastin*, May 15, 1923.

114. Letter to Governor of Jaffa, May 28, 1923, ISA M-9-10.

115. "Blow to Rutenberg: Jaffa Refuses His Electricity," *Daily Mail*, May 28, 1923, TNA CAOG 14/108.

116. Letter from the committee to the assistant governor of Jaffa, August 28, 1923, ISA M-756/4. The committee consisted of Abdallah Shafik El Dejani, Omar Eff. Bitar, Mohamed Raghed El Iman, Youceph Taleb, Antoine Gelat, and D. Petridis. Also available at ISA M-9/10.

117. Reports from a meeting on May 24, 1923, IECA 0349-42.

118. Undated letter from the municipal committee appointed to study PR's concession, ISA M-756/4.

119. "Fate of Concessions," *Filastin*, March 19, 1932, CAOG 14/108.

120. "Blow to Rutenberg: Jaffa Refuses His Electricity," *Daily Mail,* May 28, 1923, CAOG 14/108.

121. DG, North, to Director of the Department of Commerce and Industry, April 24, 1923, ISA M-9/10.

122. "Beware of Rutenberg's Scheme!!! Declaration for the honorable nation." Pamphlet dated May 8, 1923, put out by the First Arabic Economic Council, established in Jerusalem, and signed by its Chairman Hafez Tuqan, IECA 0349-42-27.

123. *Filastin*, July 17, 1923.

124. "Rutenberg Scheme," *Filastin*, April 24, 1923.

125. The meeting took place on May 31, 1923. *Filastin*, June 1, 1923.

126. *Filastin*, June 5, 1923, IECA 0349-42.

127. 'Abd al-Wahhāb Kayyālī, *Wathā'iq al-muqāwamah al-Filasṭīnīyah al-'Arabīyah ḍidda al-iḥtilāl al-Barīṭānī wa-al-ṣihyūnīyah, 1918–1939* (Bayrūt: Mu'assasat al-Dirāsāt al-Filasṭīnīyah, 1968), 73.

128. MCA, Nazareth to Samuel, November 15, 1923, ISA M-10/2.

129. Minutes from meeting between PR and CS, May 1923 (date illegible), ISA M-756/4.

130. "Inflammatory Speeches Land Two Arab Preachers in Jail," *Jewish Telegraphic Agency*, November 21, 1923.

131. PEC to DPW, July 8, 1923, ISA M-10/2.

132. Statement sent from Jamal Husayni, Secretary of the Executive Committee, 6th Palestine Arab congress, to the national political league, TNA CO 733/59, June 24, 1923. Also ISA M-4/21. The Sixth Arab Palestine Congress was held in Jaffa from June 15 to 20.

133. Parliamentary session on July 16, 1923, TNA CO 733/57.

134. Letter from merchants of Jaffa, July 11, 1923, TAHA 8387/31-0073.

135. Letter from the Jaffa Chamber of Commerce, August 7, 1923, TAHA 8387/31-0073.

136. Samuel to the CO, April 26, 1923, ISA GL-1664/5.

137. PR to Jaffa Office, July 31, 1923, IECA 0357-184.

138. Baharaw to PR, June 25, 1923, IECA 0036-14-1.

139. General Grant to PR, June 5, 1923. IECA 0036-14-1.

140. Later this became a matter of dispute because the moment Tiberias had had its intended effect of breaking the oppositional front against PR, the British were less keen to spend money on a water supply scheme there. As a result, even though the powerhouse was completed and had undergone all the necessary tests by June 10, 1925, it sat idle in anticipation of the water supply plant that would ensure the plant's operating at a profit. On September 4, the company met with representatives of Tiberias and the Palestine Government and drafted an agreement. Before it was officially signed, however, the company commenced supply in November 1925 and, from January 1926, ran continuous street lighting. The supply was informally regulated by an exchange of letters between the town and the PEC. The formal contract, including water supply, was not concluded until December 30, 1927. In the same year, the water supply plant was built, and at which point the British, the Tiberias Municipal Council, and the PEC had negotiated more than thirteen drafts of the supply contract. See Notes on conference at CO: Vernon, Clauson, Holmes, PR, Bradlaw, January 30, 1924, IECA A348-31; June 10, 1925, IECA 2372-2; Third Annual General Meeting of the PEC held on June 29, 1926, IECA 0475-09; IECA 0168-22, 12th draft, dated May 3, 1926; Baharaw to CS, March 19, 1926; Baharaw to PR, March 18, 1927; Baharaw to CS, April 13, 1927; Baharaw to Abraham PR, December 8, 1927; dispatch from HC to S/SC with a revised draft of the agreement between PEC and Tiberias Municipality, August 25, 1926, CO 733/116. See also CO 733/151/7, CS to PEC, November 10, 1926; PEC to CS, April 13, 1927; CS to Colonial Secretary, January 24, 1928.

141. PR's report to the board of directors, December 31, 1923, IECA 0361-275; PR to CS, November 18, 1923, IECA 2372-20.

142. Negotiations began on October 12 between PR and the mayor of Tiberias, Zaki el-Hudeif, in the presence of the assistant DG and a representative of DPW. PR followed up with a letter containing a formal agreement on October 15 and meetings with the municipal council on October 17, 18, and 21, IECA 0361-275-6-23, 0361-275-6-24. First annual report of the board of directors, i.e. report on state of things as of September 30, 1924, IECA 0358-210.

143. Letter from PR to Col. Frederick H. Kisch, Member of the Palestine Executive of the Zionist Organization, November 11, 1923, IECA 0361-275-30.

144. Report to the board of directors of the PEC, December 31, 1923, IECA 0361-275-4.

145. Report to board of directors, March 31, 1927, p. 13, IECA 2377-10.

146. PR to Sir Alfred Mond, November 2, 1923, IECA 2377-7-265.

147. "Decision of the Jaffa Municipality, No. 526," IECA 0361-275-5-48.

148. Report to the board of directors of the PEC, December 31, 1923, IECA 0361-275-4.

149. Baharaw to PR at the Continental-Savoy in Cairo, November 27, 1923, IECA A348-19.

150. *Filastin*, January 18, 1924.

151. Fall 1923, IECA 0352-111.

152. Report to the board of directors of the PEC, December 31, 1923, IECA 0361-275-4.

153. Unsigned note by Palestine Government official, November 17, 1923, ISA M-10/2.

154. Report to the board of directors of the PEC, December 31, 1923, IECA 0361-275-4.

155. Letter dated December 23, 1922, ISA M-9/10.

156. IECA 0349-42-7, "hal al-thulma afdhal min al-nur" [Is darkness preferable to light?], *al-Nafir*, April 14, 1923. It should be noted that *al-Nafir* seems to have been the occasional beneficiary of Zionist funding, and its editor in chief, Iliya Zakka, was generally considered "close to the Jews," and by some the newspaper was even described as "the hired newspaper." See Rashid Khalidi, *Palestinian Identity: The Construction of Modern National Consciousness* (New York: Columbia University Press, 1997), 228n77. According to the Israeli historian Jacob Yehoshua, *al-Nafir* "published articles praising Jewish colonization in the country when it received payment for them, but launched attacks on it any time the payments were interrupted." See Jacob Yehoshua, *Ta'rīkh al-sihāfah al-àrabīyah fī Filastīn fī al-'ahd al-Ūthmāni, 1908–1918* (Jerusalem: Matba at al–Ma arif, 1974), 52–54, also quoted in Khalidi, *Palestinian Identity*, 58.

157. S. O. Richardson, lawyer for the Municipal Council of Jaffa, to AG, December 3, 1924; Richardson to AG, December 29, 1924; AG to Richardson, December 31, 1924, ISA M-756/4.

158. "Arabs Want Destruction of Electricity Plant in Jaffa," *Palestine Bulletin*, October 28, 1929; "Back to Pre-electricity Days," *Palestine Bulletin*, October 31,

1929; "Arab Boycott Causes Skirmich [sic]," November 15, 1929; "Jaffa Municipality Votes for 'LUX,'" *Palestine Bulletin*, November 4, 1929; "Where Are Those Lux Lamps?," *al-Jami'a al-Islamiya*, February 22, 1930. The government came out explicitly against the move in May 1930. "Government Opposes Jaffa Lux Lighting," *Palestine Bulletin*, May 14, 1930.

159. "Week-End Scores," *Palestine Bulletin*, September 29, 1930.

160. PR to Mond, December 19, 1923, IECA 2377-7-297.

161. The third annual general meeting held at the Head Office of the Corporation at the Power House, Jaffa, June 29, 1926, IECA 0475-09-65. For confirmation of profitability of Haifa and Tiberias plants, see Report to fifth annual meeting, for the year ending on December 31, 1927, IECA 0475-09-87.

162. Report to the board of directors of the PEC, December 31, 1923, IECA 0361-275.

163. PR to Baron Rothschild, October 24, 1921, IECA 2369-4-330.

164. Report to the board of directors of the PEC, December 31, 1923. The "deal" was a Heads of Agreement presented by PR to the Mayor of Haifa on March 7, 1923, IECA 0361-275. See Letter from Singer to Halpern, 9 April 1923, IECA 2375-8-32.

165. PR made the formal offer to build a power station in Haifa, March 8, 1922, TNA CO 733/29. The consulting engineers issued a report on Haifa railway workshops, recommending to accept PR's offer, March 29, 1922, TNA CO 733/29; Letters between PR and the CO, dated July 24, July 26, 1922, CO 733/29, CAOG 14/109.

166. Hughes, *Networks of Power*, 18, 201–23, 250–53.

167. Report to the board of directors of the PEC, December 31, 1923, IECA 0361-275-4.

168. First draft of agreement between PEC and Mayor of Haifa, July 26, 1925, IECA 0036-7. *Heskem bein iriya ve-hevrat ha-heshmal, Davar*, November 24, 1925, IECA 0347-12; letter from DC to PR, as secretary of the PEC, formally accepting the lighting scheme on behalf of the municipality, November 30, 1925, IECA 0347-12; IECA 0036-22: letters from January 1, January 11, June 14, December 16, 1926.

169. Report to board of directors, March 31, 1927, IECA 2377-10. The first year of operation generated a surplus of revenue over expenditure of £635.

170. Report to the board of directors of the PEC, December 31, 1923, IECA 0361-265-4; PR asks for an indefinite postponement of Auja hydroplant obligation, August 1, 1924, CO 733/68.

171. Report to the board of directors of the PEC, December 31, 1923, IECA 0361-275; PR to Shuckburgh, February 18, 1924, ISA M-10-2-232; February 18, 1924, TNA CO 733/68; PR to HC CO 733/84; January 25, 1925, CO 733/101,103.

172. March 19, 1925, TNA CO 733/101,103.

173. PR to Pal Gov, June 24, 1925, TNA CO 733/96; PEC Report on the progress of Works, submitted to the US/SC, July 27, 1925, CO 733/101,103.

174. Report to the board of directors of the PEC, December 31, 1923, IECA 0361-275. In fact, however, the first dividends were not paid until the second financial year, October 1925. See IECA 0475-09-48: Report prepared for the second ordinary general meeting held at the powerhouse in Jaffa on Sunday, June 14, 1925:

states the intention to pay dividends. And IECA 0475-09-83: "Report of the Directors to be submitted on the fourth ordinary general meeting of the Corporation to be held on June 9, 1927, London: This confirms that company paid dividends in 1925.

175. The Second Ordinary General Meeting, held at the Power House in Jaffa, June 14, 1925, IECA 0475-09-48.

176. Analysis of kilowatt-hours sold, consumers, etc., June 1924 to February 1925, February 1925, IECA 2377-7-225/226.

177. Memo: "Profits from operations of the Haifa and Tiberias power houses," June 1925, IECA 2377-12, 2377-7: both the Tiberias and the Haifa powerhouses began operations in June: Haifa in June 10; Tiberias in June 25. The period June–December 1925 showed a deficit of £2,620. In 1926, they ran at a surplus of £700, in 1927 of £7,000, and in 1928 of £11,000. The first seven months of 1929 show a surplus (after expenses, which is the figure for all) of £8,499.

178. ISA P-985/17.

179. Shapira, *Israel*, 112.

180. Fifth annual report and balance sheet, December 31, 1927, IECA 2377-12, 0475-09.

181. October 1929, IECA 2377-12; this would rise to £P29,000 in 1929 and £P40,000 in 1930.

182. TNA CO 821/1: *Blue Book on Palestine*, 1926–1927, Section XIII. INDUSTRY AND PRODUCTION, WAGES AND COST OF LIVING. INDUSTRY AND PRODUCTION, p. 51.

183. Fifth annual report and balance sheet, December 31, 1927, IECA 2377-12, 0475-09; December 1930, IECA 2377-13.

184. October 1929, IECA 2377-12; December 1930, IECA 2377-13; Fall 1928, IECA 0036-4; February 23, 1931, IECA 0351-83: answers to Questionnaire from US Gov, sent to PEC via consulate on February 6, 1931; Baharaw to US Consul, December 31, 1932, IECA 0351-83.

185. Report to board of directors, March 31, 1927, IECA 2377-10.

186. Report to board of directors, October 1929, IECA 2377-12.

187. Session of Technical Committee, May 18, 1928, IECA 0477-22.

188. The Postmaster Gen approves the plans for the Kineret-Tiberias-Migdal, May 23, 1930, IECA 0477-15: 22 kV transmission line (thus it can be assumed that this is built soon after). The Postmaster Gen approves the plans for the low-tension grid in Kfar Nahman, June 6, 1930, June 12, 1930, IECA 0477-15.

189. Report to board of directors, June 1931, IECA 2375-11, IECA 2377-13.

190. First Census of Industries, Taken in 1928 by the Trade Section of the Department of Customs, Excise and Trade, Jerusalem 1929, 12, table H. Of the 815 engines in use, 404 were electrical, constituting 49.6 percent.

191. PR to Hoofien, October 19, 1926, IECA 2379-6; Report to board of directors, March 31, 1927, p. 10, IECA 2377-10.

192. Report to board of directors, March 31, 1927, pp. 32–33, IECA 2377-10.

193. For example, in one article, titled "Basta!," Jabotinsky claimed that the Histadrut, the labor Zionist union, and the Zionist Executive were stifling private

initiative and entrepreneurship through its invasive financial activities, channeling money to collective farms and propping up the labor union, making it debilitatingly powerful. Hillel Halkin, *Jabotinsky: A Life* (New Haven, CT: Yale University Press, 2014), 160–61.

194. Indeed, the denial of social influences on the scientific method lies at the heart of the modern condition, according to Bruno Latour. Bruno Latour, *We Have Never Been Modern* (Cambridge, MA: Harvard University Press, 1993), esp. 10–12.

195. Neil Caplan, *Futile Diplomacy*, vol. 1, *Early Arab-Zionist Negotiation Attempts, 1913–1931* (London: Cass, 1983); Neil Caplan, *Futile Diplomacy*, vol. 3, *Arab-Zionist Negotiations and the End of the Mandate* (London: Cass, 1986); Avi Shlaim, *Collusion across the Jordan: King Abdullah, the Zionist Movement, and the Partition of Palestine* (New York: Columbia University Press, 1988), 55, 59–61; IECA, 2375-13, file contains personal correspondence between PR and Abdullah, including birthday cards; Norman Bentwich, *Mandate Memories, 1918–1948* (London: Schocken, 1965), 121.

196. Halkin, *Jabotinsky*, 192–96.

197. Anderson, *Imagined Communities*.

198. Nick Roberts, "Rethinking the Status Quo: The British and Islam in Mandate Palestine, 1917–1929" (PhD diss., New York University, 2010).

199. "Rutenberg's Zionist Current," Sa'd al-Din, *al-Jami'a al-Islamiya*, December 20, 1934 (emphasis added).

200. Shamir, *Current Flow*, 27.

201. Mitchell, *Carbon Democracy*.

202. See Max Weber, *The Protestant Ethic and the "Spirit" of Capitalism and Other Writings* (New York: Penguin Books, 2002); Michel Foucault. *Power/Knowledge: Selected Interviews and Other Writings, 1972–1977* (New York: Pantheon Books, 1980); Michel Foucault, *Discipline and Punish: The Birth of the Prison* (New York: Vintage Books, 1979); James Ferguson. *The Anti-politics Machine: "Development," Depoliticization, and Bureaucratic Power in Lesotho* (Cambridge: Cambridge University Press, 1994).

203. IECA 2371-9.

CHAPTER FOUR

1. Yosef Baratz, "His dismissal" [*Pitorav*], *Davar*, September 28, 1928.

2. White, *The Organic Machine*.

3. The Jordan discharged twenty-five cubic meters per second, and the Yarmuk twenty.

4. DPW to CS, October 31, 1924, CO 733/88, IECA 2372-2; Addison E. Southard, American Consul at Jerusalem, "Palestine: Its commercial resources with particular reference to American trade," US Department of Commerce booklet, October 12, 1922, CO 733/46; Statistical Abstract of Palestine 1937–8, 2, ISA M-4513; Naval Staff, Intelligence Department, *Handbook of Syria (including Palestine)*, June

1919, 648–49, ISA M-4409. The figures given in these sources are occasionally contradictory. The figures given in the text are the ones most consistently cited.

5. Walker describes the devastating effects that attended Japanese industrialization, which meant that certain segments of the population first "knew the state through pain." Walker, *Toxic Archipelago*, 44, 129.

6. Benedict Anderson, "Census, Map, Museum," in *Imagined Communities*, esp. 175, 184.

7. Cf. Sami Zubaida, *Islam, the People, and the State: Political Ideas and Movements in the Middle East* (New York: I. B. Taurus, 1989), 150; see also Owen, *State, Power and Politics in the Making of the Modern Middle East*, 3–4, and his discussion of the colonial state, 7–17.

8. March 29, 1923, IECA 2377-7, 0475-09; April 1923, TNA CO 733/45.

9. CO 733/48; 733/61; August 10, 1923, CAOG 14/109, IECA 2371-17; "First and Second Hydro-Electric Jordan Power Houses" and 13 drawings and plans, 11 diagrams. PEC to Palestine Development Council, January 8, 1924, containing report: "Summarised Description of Works and Estimates of the First Hydro-Electric Installation in Palestine," IECA, 2375-11; see also Consulting Engineers' report, "First and Second Hydro-Electric Jordan Power Houses," May 2, 1924, CO 733/77, CAOG 14/109.

10. In comparison to Western standards, this was relatively low. In Germany and the United States, transmission at voltages of 220,000 was quite common, and Britain's National Grid, inaugurated in the 1930s, transmitted at 132,000 volts. See Hughes, *Networks of Power*, 357.

11. Hughes, *Networks of Power*, 193.

12. "Summarised Description of Works and Estimates of the First Hydro-Electric Installation in Palestine," January 8, 1924, IECA 2375-11.

13. Hughes, *Networks of Power*; Hausman, Hertner, and Wilkins, *Global Electrification*.

14. Consulting Engineers' report, "First and Second Hydro-Electric Jordan Power Houses," May 2, 1924, CO 733/77, CAOG 14/109.

15. DPW to CS, October 31, 1924, CO 733/88, IECA 2372-2.

16. DPW to CS, October 31, 1924, CO 733/88, IECA 2372-2.

17. May 1924, CO 733/82.

18. Official report, issued following Raglan's question, May 28, 1924, TNA CO 733/82.

19. "Report on Mr. Rutenberg's Scheme for Hydroelectric Development in Palestine," IECA 2379-8; 2379-7-207.

20. In Bombay, with a population of just under 1 million, about 34 million units were sold per annum; in Calcutta, a population of 1.5 million consumed 31 million units. Economic Board for Palestine 1922, IECA 0352-107-24.

21. The Economic Board for Palestine orders a report on PR's scheme, June 28, 1922, IECA 2379-8; "Report on Rutenberg Scheme for Palestine," August 1922, prepared by Merz and McLellan, Consulting Engineers, p. 10, IECA 2379-8.

22. Hughes, *Networks of Power*, 249.

23. Hughes, *Networks of Power*, 249–56.
24. Hughes, *Networks of Power*, 319–20, 352–61.
25. For the American organization, then named the Palestine Economic Corporation, the perhaps more important reason was its objection to the preference clause for English manufacturers and machinery then being discussed in negotiations with the British. Late January 1923, IECA 2381-4.
26. PEC to Palestine Development Council, January 8, 1924, IECA 2375-11-213.
27. Louis C. Loewenstein (consulting engineer) to Flexner, May 9, 1924, IECA 2375-11-174.
28. The waterpower station at Paderno in Milan, for instance, was fitted with seven 2,000-hp turbines. For information on the Milan Edison Company's hydropower installation, see Hughes, *Networks of Power*, 211.
29. PR to Flexner, June 12, 1924, IECA 2375-11-161. In a follow-up letter dated September 16, 1924, PR wrote directly to Lowenstein, explaining his vision for future consumption growth, citing the connections about to be made to the Sarafand Air Base and the Silicate Works. He also pointed out that two-thirds of the residents of Tel Aviv and Jaffa were not yet connected to the grid. IECA 2375-11-147.
30. Julian Mack to PR, dated September 19, 1925, IECA 2375-11-123. Once the concession had been signed and funding secured, PR wrote a letter to Flexner, September 27, 1925, equal parts gloating and admonishing, informing him that he had secured all necessary funding for the Jordan powerhouse without the American Zionists. The letter concluded: "I am certain that you and your friends appreciate that your attitude towards this, the most vital undertaking in Palestine has not been the right one." And: "I sincerely hope that in our future relations such waste of time, energy and money on both sides will not take place." See IECA 2381-3.
31. DG North to CS, February 24, 1922, ISA M-9/10.
32. PR to Judge Mack, February 14, 1922, IECA 2381-5.
33. PR to Justice Brandeis, April 23, 1922, IECA 2381-5.
34. DG North to HC, December 23, 1922, ISA M-9/10.
35. DG North to CS, February 24, 1922, ISA M-9/10. This continued to hold through the period. When the Haifa municipality proposed its own lighting scheme for the city in 1923, the treasurer and the director of the Department of Commerce and Industry both determined that PR's plan was the better deal. See Report from the Treasurer of Palestine, March 21, 1923, and minute from Department of Commerce and Industry to CS, March 9, 1923, in ISA M-9/10. The conclusion of the latter was that "I consider that every facility should be given to Mr. Rutenberg to install a Power House at Haifa."
36. PR submits detailed proposal for Jordan scheme, "First and Second Hydro-Electric Jordan Power House," September 1923, TNA CO 733/77; November 1923, TNA CAOG 14/109.
37. TNA CO 733/9; Shaltiel, *Pinhas Rutenberg*, 41.
38. Barry, *Political Machines*, chap. 2, "Technological Zones."
39. "Notes for Tourists," *Palestine Bulletin*, March 27, 1930.

40. Mary C. Wilson, *King Abdullah, Britain, and the Making of Jordan* (Cambridge: Cambridge University Press, 1987), 53.

41. Robins, *History of Jordan*, 13; Gil-Har, "British Commitments," 691.

42. Robins, *History of Jordan*, 19–20; for more on borders, see Gil-Har, "British Commitments," 690.

43. Note by Major Young, April 8, 1925, TNA, CO 733/92.

44. Note by Major Young, April 1924, TNA 733/84; Gil-Har, "British Commitments," 693.

45. This should probably not surprise us, given the improvised and often incompetent administration of the Palestine Mandate in the first years. See, for instance, Krämer, *History of Palestine*, 151–55; Bunton, *Colonial Land Policies*.

46. Letter to *The Times*, June 15, 1922, TNA CO 733/86; dispatch from Philby to Samuel, July 21, 1922; confidential dispatch from Philby to the CO, July 21, 1922. The post of "chief British representative" was changed in 1927 to "British resident" to mark a further separation between Palestine and Transjordan and to bring the post into alignment with British colonial practice; see Wilson, *King Abdullah*, 91.

47. Note by Major Young, April 8, 1925, TNA CO 733/92.

48. Note by Major Young, April 1924, TNA CO 733/84.

49. Report by Bentwich, "Application of the Rutenberg Concession to T.J.," May 1924, TNA, 733/86.

50. Note by Major Young, April 8, 1925, TNA CO 733/92.

51. C. H. F. Cox, Chief British Representative, Amman: Confidential report on the PR concession, February 15, 1925, TNA CO 733/90.

52. Note by Major Young, April 1924, TNA CO 733/84.

53. Note by Major Young, April 1924, TNA CO 733/84; letter from the S/SC to Samuel, April 17, 1924, TNA CO 733/84; handwritten notes by Major Young, May 29, 1924, and R. V. Vernon, June 2, 1924; letter from Shuckburgh to S/SC, June 5, 1924; note by Shuckburgh, January 8, 1925; letter from Samuel to Shuckburgh, May 9, 1924; dispatch from S/SC to Prime Minister, January 15, 1925; memorandum by Bentwich, "Palestine Electric Corporation's Jordan Concession," May 29, 1924, TNA CO 733/86; telegram from Samuel to S/SC, October 10, 1925; note from G. L. M. Clauson, November 4, 1925; letter from Shuckburgh to Nathan, November 23, 1925, TNA, CO 733/98; Abla Mohamed Amawi, "State and Class in Transjordan: A Study of State Autonomy" (PhD diss., Georgetown University, 1993), 298.

54. PR to S/SC, August 22, 1923, IECA 2371-15-145.

55. HC to S/SC, undated, IECA 2370-14-47.

56. Political Report by CS for August 1921, TNA CO 733/6.

57. Tyler, *State Lands*, 31.

58. TNA CO 733/18; *Official Gazette*, October 26, 1922. The actual agreement between the cultivators of the Beisan lands and the government of Palestine is set out in Lewish French, "First Report on Agricultural Development and Land Settlement in Palestine," December 23, 1931, CO 733/214/5, appendix IIIB, 153; Bunton, *Colonial Land Policies*, 72–73; Tyler, *State Lands*, 22–24.

59. Bunton, *Colonial Land Policies*, 76.
60. Weizmann, *Trial and Error*, 274.
61. Porath, *The Emergence*, p. 135.
62. For an authoritative account, see Bunton, *Colonial Land Policies*, esp. chapter 2.
63. Forty-six-page report by PR, March 31, 1927, IECA 2377-10-55.
64. PR to HC, December 10, 1929, IECA 2371-15-71, IECA 2382-2-72. In fact, many of the inhabitants sold their land before they paid the government for it. This led to a lawsuit by the government in 1924. IECA 2370-13.
65. Lease agreement between PICA and Degania Kvutza A and B, December 6, 1927, IECA 2370-8-29.
66. PR to PEC Head Office, November 12, 1925, IECA 2370-12-164.
67. Report by PR, March 31, 1927, IECA 2377-10-55.
68. Sacher to PR, May 12, 1926, IECA 2370-12-120.
69. Report of the Commission on the Palestine Disturbances of August 1929, Cmd. 3530, London: HMSO, March, 1930, 22.
70. Sacher to PR, May 12, 1926, IECA 2370-12-120.
71. IECA 2370-9-44, 2370-12-81.
72. Internal PEC note marked "strictly confidential," February 2, 1927, IECA 2370-9-35.
73. Letter from Cox to HC Plumer, July 27, 1928, TNA CO 733/165/3.
74. Memorandum of agreement made between the Chief Minister Hassan Pasha Abu el Huda and the PEC, March 5, 1927, IECA 2371-15-130.
75. *Alif Ba,* IECA 2370-9-24: Hebrew translation of article.
76. IECA, 2375-13; Bentwich, *Mandate Memories*, 121.
77. Report by PR, March 31, 1927, IECA 2377-10-55.
78. Registration certificate from the Transjordanian Ministry of Justice, IECA, 0428-1128-47; PR to Lord Reading, January 1, 1929, IECA, 2377-12-390; "Note on the Influence on the Corporation's Concession of the Proposed Independence of Transjordan," February 26, 1946; *Official Gazette*, January 23, 1928, IECA, 2383-8-46. See also Amawi, "State and Class in Transjordan," 292.
79. PR to Henri Frank of PICA, December 20, 1927, IECA 1177-10-62.
80. PR to Frank, September 21, 1928, IECA 1177-10-48.
81. David Ben-Gurion's opening address to the fourth convention of the General Federation of Hebrew Workers in Palestine (Histadrut), reprinted in *Davar*, the official newspaper of the Histadrut, February 12, 1933.
82. For the specifics of this, see chapter 3.
83. PR to HC, February 8, 1925, IECA 2372-2.
84. Eleven-page letter from HC to S/SC Amery March 13, 1925, CO 733/90.
85. Diplomatic Correspondent, "The Rutenberg Concession: Court Finds a Right Was Wrongly Granted," *Daily Mail*, March 30, 1925, CAGO 14/108.
86. J. M. N. Jeffries, "Paying for Zionism: More British Money to Back the Illegal Concessions," *Daily Mail*, April 27, 1925, CAOG 14/108.
87. HC to S/SC Amercy, March 13, 1925, CO 733/90.

88. Sacher to CS, October 12, 1925, IECA 0358-205.
89. PR reminds Col. Symes, September 30, 1925, IECA 0358-205.
90. J. Hathorn Hall, September 18, 1925, CO 733/110.
91. February 1926, IECA 0358-205.
92. "Concession for the utilization of the waters of the rivers Jordan and Yarmuk and their effluents for generating and supplying electrical energy," March 5, 1926, CAOG 14/109.
93. Hayyim Hisin, *A Palestine Diary: Memoirs of a Bilu Pioneer, 1882–1887* (New York: Herzl Press, 1976), 10.
94. "Notes for Tourists," *Palestine Bulletin*, March 27, 1930.
95. Y. Bar Zvi, "The Work of Rutenberg," *Davar*, September 13, 1927.
96. Theodor Herzl, *Altneuland: Old-New Land* (Haifa: Haifa Publishing, 1960).
97. "Gesher—Nahalim," *Hapoel Hatsair*, November 25, 1927, IECA 2377-4.
98. Yosef Baratz, "His Dismissal" [*Pitorav*], *Hapoel Hatsair*, September 28, 1928, IECA 2377-4. A century earlier, Thomas Carlyle wrote: "We can remove mountains, and make seas our smooth highway; nothing can resist us. We war with rude Nature; and, by our resistless engines, come off always victorious, and loaded with spoils." Carlyle, *Critical and Miscellaneous Essays* (New York, 1986), vol. 2, 60, quoted in Adas, *Machines as the Measure of Men*, 213.
99. The concept of making legible draws most obviously on Scott, *Seeing Like a State*. But it also has in mind Peter J. Taylor, "Technocratic Optimism, H. T. Odum, and the Partial Transformation of Ecological Metaphor after World War II," *Journal of the History of Biology* 21 (1988): 213–44. For an earlier groundbreaking work in a similar vein, see William Leiss, *The Domination of Nature* (Montreal: McGill-Queen's University Press, 1972).
100. Nye, *America as Second Creation*, 25.
101. March 1926, IECA 1514-1.
102. March/April 1926, IECA 1514-1. See letters from April 13, May 7, and June 4, 1926.
103. December 31, 1926, IECA 0475.
104. Report by PR, March 31, 1927, IECA 2377-10.
105. "The Jordan Station," *Hapoel Hatsair*, May 18, 1927, IECA 2377-4.
106. "Harnessing the Jordan: Electric Power for Palestine," *The Times*, February 25, 1929. This calls to mind Sandra Sufian's account of Zionists' attempts at national regeneration by means of antimalarial campaigns. By healing the land, they sought to heal the nation; thus they succeeded in an inward and outward projection of their healing values, while representing the Arabs as part of the disease. Sandra M. Sufian, *Healing the Land and the Nation: Malaria and the Zionist Project in Palestine, 1920–1947* (Chicago: University of Chicago Press, 2007), esp. 145–46.
107. David Ekbladh, *The Great American Mission: Modernization and the Construction of an American World Order* (Princeton, NJ: Princeton University Press, 2010).
108. Andersen, *British Engineers and Africa*, 13.

109. PR to Henri Frank of PICA, June 29, 1926, IECA 2381-3-7.

110. Report of the Directors to be submitted to the PEC's fourth ordinary general meeting to be held on June 9, 1927, December 31, 1926, IECA 0475. The final agreement was signed between the PEC and the Prudential Assurance Co. on March 24, 1927, Fifth Annual Report & Balance Sheet for the Year Ended December 31, 1927, IECA 0357-188-31. The Trade Facilities Acts of 1921, 1922, 1924, and 1926 provide a perfect example of how international finance congealed with national ambition in the United Kingdom: "Political considerations were linked with national economic development aspirations, from both a home and host nation standpoint. European governments wanted to support exports; and if foreign investment (outward or inward) accomplished that, it was to be favored. Thus, in the 1920s the British passed Trade Facilities Acts (1921, 1922, 1924, and 1926). The British Treasury would guarantee loans to projects, including a number of foreign nongovernmental electrical projects, on the condition that the loaned money be used to purchase British goods." Under these acts, it lent money to Greek, Hungarian, Japanese, and Yugoslavian (last one, approved but not realized) electrical ventures. See Hausman, Hertner, and Wilkins, *Global Electrification*.

111. As of February 1, 1928, IECA 2377-12.

112. Precis of PR's report, November 27, 1928, IECA 2377-12; Precis of PR's Report, January 2, 1929, IECA 2377-12; PR to Reading January 11, 1929, IECA 2377-12.

113. Owen Teedy, "Cheaper Electricity for Palestine: Harnessing the Jordan: Ambitious Plan Almost Completed," *Financial Times*, May 19, 1930.

114. The PEC carried out a successful test of the generators at the Jordan station in late February 1930, and in early March it successfully put the 66,000-volt Haifa-Jordan high-tension line under tension. IECA 0477-15.

115. Report, January 1931, IECA 2377-13.

116. PR to Bradlaw, February 14, 1931; Bradlaw to Charnaud, February 15, 1931; PR to Reading, February 19, 1931, IECA 2377-13; "Power Stations Damage: Palestine Electric," newspaper unknown, TNA CAOG 14/109; "Floods Damage Electric Works," *Palestine Bulletin*, February 16, 1931.

117. PR to Reading, February 19, 1931, IECA 2377-13.

118. Reading to PR, February 27, 1931, IECA 2377-13.

119. See Chandra Mukerji, "The Territorial State as a Figured World of Power: Strategics, Logistics, and Impersonal Rule," *Sociological Theory* 28, no. 4 (December 2010): 414.

120. "Naharayim Works Officially Inaugurated," *Davar*, June 11, 1932.

121. "Wealth from the Dead Sea," *Popular Mechanics*, November 1930, 798.

122. The speech was reprinted in full in several newspapers in both English and Hebrew. For English: "Jordan Electric Scheme Inaugurated," *Palestine Bulletin*, June 12, 1932; for Hebrew, see "The Speeches at the Inauguration of Naharayim," *Davar*, June 14, 1932.

123. This, of course, was not unique to Palestine. India is the most immediate cognate: Goswami, *Producing India*, 37.

124. Blue Book for Palestine, 1932, 169–70, TNA CO 821/7: A list of wages for different professions. The only workers drawing a larger salary than the electricians were skilled masons. The electricians made 400 to 500 mils per day, and the masons 500 to 600 mils per day.

125. "Wealth from the Dead Sea," 798.

126. Norris, *Land of Progress*, 10–11.

127. HC to S/SC, April 20, 1929, ISA Archives Blog on the construction of Haifa harbor, accessed February 10, 2015, http://israelidocuments.blogspot .com/2013/10/blog-post_8865.html?utm_source=feedburner&utm_medium =email&utm_campaign=Feed%3A+blogspot%2FCXXXrP+%28ישראל+-+הסיפור+המתועד%29.

128. Robert Vitalis, *America's Kingdom: Mythmaking on the Saudi Oil Frontier* (Stanford, CA: Stanford University Press, 2007), 100; Myrna I. Santiago, *The Ecology of Oil: Environment, Labor, and the Mexican Revolution, 1900–1938* (New York: Cambridge University Press, 2006), 164–66; Jacob Norris, "Ideologies of Development and the British Mandate in Palestine" (PhD diss., University of Cambridge, 2010), 30–32.

129. Palestine was not unique in this regard. Such a metonymic logic seems integral to how colonial states count. See, for instance, Goswami, *Producing India*, 74–75; Arjun Appadurai, "Number in the Colonial Imagination," in *Modernity at Large: Cultural Dimensions of Globalization* (Minneapolis: University of Minnesota Press, 1996); Bernard Cohn, *Colonialism and Its Forms of Knowledge* (Princeton, NJ: Princeton University Press, 1996).

130. IECA 2377-5.

131. "Jordan Works Officially Inaugurated," *Davar*, June 11, 1932 (emphasis added).

132. *Davar*, February 12, 1933, Ben-Gurion's speech reprinted in full.

133. "There Was Radiance from Naharayim," *Davar*, March 22, 1932; Z. David, "Shorts: There was Radiance from Naharayim, *Davar*, March 25, 1932.

134. Norris, *Land of Progress*, chaps. 3 and 4. The phrase "Zionist industrial complex" is his (120).

135. Hughes, *Networks of Power*, 313–14.

136. I follow Zachary Lockman in my use of the term *labor Zionism* as an umbrella for all worker-oriented organizations and parties in Palestine. See Lockman, *Comrades and Enemies*, 17.

137. Zeev Sternhell, *The Founding Myths of Israel* (Princeton, NJ: Princeton University Press, 1998).

138. Lockman, *Comrades and Enemies*, 49–50.

139. Sternhell, *Founding Myths*, 16.

140. Lockman, *Comrades and Enemies*, 53.

141. Lockman, *Comrades and Enemies*, 54–5.

142. Original Electrification Proposal, December 1920, 61, TNA CO 733/9.

143. The notions of "conquest of labor" (*kibbush ha'avoda*) and "Hebrew labor" (*'avoda 'ivrit*) were well-known tropes of labor Zionism. Lockman, *Comrades and Enemies*, 48.

144. "Farewell Party at Naharayim," *Davar*, October 1, 1930. For an in-depth exposition of working conditions, see "A Day at the Great Jordan Works," *Davar*, July 5, 1929.

145. Metzer, *Divided Economy*, 133; Lockman, *Comrades and Enemies*.

146. *Davar*, February 12, 1933, Ben-Gurion's speech reprinted in full.

147. "Farewell Party at Naharayim," *Davar*, October 1, 1930.

148. "Farewell Party at Naharayim," *Davar*, October 1, 1930. Insisting on singing only songs originating in the Yishuv seems to have been a habit of Hanna Kipnis. Nina S. Spiegel, *Embodying Hebrew Culture: Aesthetics, Athletics, and Dance in the Jewish Community of Mandate Palestine* (Detroit, MI: Wayne State University Press, 2013), 67.

149. "Festivity at Rutenberg Works," *Palestine Bulletin*, October 1, 1930.

150. Penslar, *Zionism and Technocracy*, 154.

151. "A Day at the Great Jordan Works," *Davar*, July 5, 1929.

152. Advertisement, *Palestine Post*, December 15, 1932. Official visits were common from the late 1920s. In December 1928, Naharayim received visitors on three consecutive days: first from the mayor of Haifa and municipal council member Ibrahim Sahyoun and a municipal engineer, then from HC John Chancellor, and then from Amir Abdullah of Transjordan. "Visits at Tel Or," *Davar*, December 28, 1928. In March 1930, the Cultural Commission of the "Poal" went up to Tel Hai, visiting Naharayim on the way. "Ascension to Tel Hai," *Davar*, March 5, 1930. This then became an annual excursion: see "Ascension to Tel Hai," *Davar*, March 21, 1932. Colonel Frederick Kitsch also visited Naharayim in March 1930. *Davar*, March 28, 1930. In September 1930, the British Labour MP E. R. Denman visited Palestine and was taken around by several people. He went to the Jordan Valley, where he visited the central hospital, Ein Harod, Kumi, Degania, and Naharayim. He went on to visit Nazareth and finally Haifa, where he toured the factories; see "Mr. Danman's Visit," *Davar*, September 2, 1930; "Guests in the Valley," *Davar*, April 5, 1932. In January 1933, the Zionist leader Nahum Sokolow visited the works. "Social and Personal," *Palestine Post*, January 26, 1933.

153. Clauson from Burchells, March 9, 1925, TNA CO 733/101.

154. TNA CO 733/72; 733/77; 733/84; 733/86; 733/88; 733/101; 733/103; 733/90; 733/92; 733/93, 733/108; CAOG 14/109; and IECA A348-31; 2371-17; 0358-205; 2371-17; 2372-2.

155. Stuart Elden, *The Birth of Territory* (Chicago: University of Chicago Press, 2013), 17.

156. Henri Lefebvre, *The Production of Space* (Cambridge, MA: Blackwell, 1991), 33.

157. "Power and Irrigation Development in Palestine," *Engineering News-Record* 88, no. 25, June 22, 1922, 1042–6.

CHAPTER FIVE

1. William Robert Wellesley Peel, *Palestine Royal Commission Report* (London: H. M. Stationery Office, 1937), chap. 5, "The Present Situation."

2. Hughes, *Networks of Power*, 140–41, 172.

3. The economy shrunk if Arab citriculture, in which a small number of growers made very large profits, is excluded. If citriculture is included, the Arab economy was stagnant. Amos Nadan, *The Palestinian Peasant Economy under the Mandate: A Story of Colonial Bungling* (Cambridge, MA: Harvard University Press, 2006), 139.

4. Telephone lines were sabotaged 700 times, and the railway and roads, 340. The oil pipeline was damaged twice a week on average. Edward Keith-Roach, *Pasha of Jerusalem: Memoirs of a District Commissioner under the British Mandate* (London: Radcliffe Press, 1994), 191; Matthew Kraig Kelly, *The Crime of Nationalism: Britain, Palestine, and Nation-Building on the Fringe of Empire* (Oakland: University of California Press, 2017).

5. *Statistical Abstract of Palestine*, 1929, Keren Hayesod, Jerusalem, 1930, ISA M-5316.

6. In 1925, PR envisioned selling 800,000 kWh of current for private use, 4,4000,000 kWh for water supply and irrigation, and 9,775,000 kWh for industry. While charging 3 piasters per kilowatt-hour for private use, water would cost 1 piaster, and industry would pay only 0.62 piaster per kilowatt-hour. Report prepared by Messrs. Howard, Howes & Co. for Sir Alfred Mond, "Palestine Electric Corporation Ltd. and Rutenberg Electricity Concessions," July 10, 1925, IECA A348-30.

7. *Blue Book for Palestine* 1928, 412, CO 821/3.

8. *First Census of Industries 1928*, Trade Section of the Department of Customs, Excise and Trade, Jerusalem 1929, p. 17, table N: "Geographical Distribution of Industries in Palestine," and p. 48, table 5: "Geographical Distribution of Industry by Districts and Towns."

9. Metzer, *Divided Economy*, 78.

10. Negotiations between Shell and the PEC for supply of electricity to its facilities in Haifa Port. Letter from Shell to PEC November 18, 1928; Shell to PEC, July 27, 1929, IECA 0036-1; TNA CO 733/150/8.

11. Memo, April 22, 1931. The rates had been reduced from fifteen mils per kilowatt-hour, which was the maximum allowed under the concession, to seven. IECA 2377-13.

12. "Palestine Electric Corporation: Success of Extension Policy: Many New Centres Connected: Lord Reading's Address," *The Times*, September 6, 1935, IECA 0475-15; August/September 1935, IECA 2377-5; March 31, 1931, IECA 2377-13.

13. PR to Reading, December 12, 1928, IECA 2377-12.

14. Memorandum of Association and Articles of Association for Jordan Estates Ltd., IECA 2375-7, IECA 0168-15.

15. The Jordan Investment Ltd. was registered in June 1935. PR to Board, May 21, 1934, IECA 2377-5; "Cold Storage," *Palestine Bulletin*, May 15, 1931.

16. Confidential report prepared for PR, "Possibility of Introducing Jews into the Port Work," October 1929, IECA 2372-18.

17. Indeed, PR and Thompson discussed various means of increasing the share of Jewish workers quite openly. PR also facilitated discussions between Thompson

and Jewish construction interests. See Thompson to D. Hacohen, October 21, 1929; Thompson to PR, October 25, 1929; Thompson to Hacohen, November 19, 1929; Thompson to PR, February 11, 1931, IECA 2372-18.

18. Norris, *Land of Progress*, 120.

19. IECA 0351-83.

20. PR to DPW, December 31, 1938; PR to DPW, July 9, 1929; DPW to PEC, July 22, 1929, IECA 1177-2; Baharaw to DC, north, January 10, 1931, IECA 1177-6; PEC to CS, April 29, 1929, IECA 0352-98.

21. "Report on compensation for damages and payment for the right of fixing poles on private property," June 10, 1929, IECA 0352-98. Kusa is a type of squash particular to Palestine.

22. IECA 0352-98-315.

23. List dated October 27, 1931, IECA 0352-98; Report on Qalkilia, October 30, 1929, IECA 0352-98.

24. PEC to CS, April 29, 1929, IECA 0352-98.

25. Complaint from the *mukhtar* of Kafr Masr in the Beisan to DC, May 16, 1930; PEC to DC, May 18, 1930, IECA 0352-98.

26. PEC to Acting DC, Haifa, November 26, 1930, IECA 0352-98.

27. See correspondence from the fall of 1930, IECA 0352-98.

28. PEC to DO, Beisan, June 17, 1930, IECA 0352-98.

29. This was based on an estimation of the land's value per dunam at £P30 to £P50, which was very high, meaning the compensation was generous. Baharaw to Franci Khayat, November 14, 1929, IECA 0352-98. The PEC paid less on the Jordan-Haifa line than on the Haifa-Hadera line. They paid Dabbourieh, Nazareth District, 200 mils per plot. Same in Maader, Tiberias District, and Massar, Beisan District. In Majdal, Nazareth district, they paid 300 mils. PEC to Acting DC, Haifa, November 26, 1930.

30. IECS 0352-102 contains numerous such contracts.

31. Report on disruptions of work at Qalkilya, November 14, 1929, IECA 0352-98; Baharaw to District Officer Tulkarem, October 1932, IECA 2383-2.

32. Baharaw to DC, north, January 10, 1931; Baharaw to DC, north, May 14, 1935: A list of villages with request for DC to give "Right of Entry" to install 260 high-tension-line poles. Baharaw to DC, north, April 27, 1936: More requests for permissions to enter private lands for the high-tension line from Karkur to Bakhariw. Baharaw to DC, north, September 7, 1936, IECA 1177-6. On September 5 one of the PEC's poles of the 22-kV line to Hadera was set on fire near the village of Ayn Ghazal. DC to PEC, January 2, 1929; letter from the *mukhtars* and elders of Ayn Ghazal village, December 2, 1928; DC to PEC, forwarding the claim from Ayn Ghazal, January 8, 1929; DC forwards the letter along with offer to mediate in the negotiations; Baharaw to DC, January 18, 1929, IECA 0352-98.

33. Internal report, Baharaw, September 19, 1932, IECA 1177-6.

34. "Fate of Concessions," *Filastin*, March 19, 1932.

35. Editorial, unsigned, "A Serious Question Raised by the Jewish Nationalist Policy," *al-Jami'a al-Islamiya*, June 1, 1932. See also "Not Enough for Rutenberg,"

al-Sirat al-Mustaqim, May 22, 1932; *al-Jami'a al-Islamiya*, September 7, 1932; "PR and the River Auja—New Conflict," *Filastin*, March 9, 1932; "Rutenberg and His Undertaking," *Filastin*, June 2, 1932.

36. IECA 0352-109: Damages claims 1930–37. The file contains numerous negotiations and compensation claims.

37. As an indication of the strong British support, in 1928, when the company was about to construct its first high-tension line between Haifa and Hadera, along the coast, the DC in the north sent a letter in Arabic to the *mukhtars* of all the villages in the area (including Tireh, Ayn Hod, Athlit, Ayn Ghazal, Sarafand), which stated that the lines would pass over some of their village lands, adding, "The Corporation is carrying out this work with the sanction of the Government." Letter by DC, November 3, 1928, IECA 0352-98.

38. Kacenelenbogen to DC, Nazareth, April 25, 1930; IECA 0352-98.

39. DC, Nazareth, to Mukhar of Dabbourieh Village, April 28, 1930, IECA 0352-98.

40. It is hard to determine whether the landowners' demands were justified; the land down on the plain where Qalkilya was located, and which they owned and farmed, was certainly more fertile, and thus valuable, than much of the land to the north and east. PEC to DC, north, November 3, 1929; Baharaw to Tawfiq Hamra, November 12, 1929; Baharaw to Francis Khayat, November 14, 1929; Report on disruptions of work in Qalkilya, November 14, 1929, IECA 0352-98.

41. Report on disruptions of work in Qalkilya, November 14, 1929, IECA 0352-98.

42. PEC to Franci Khayat, November 26, 1929, IECA 0352-98.

43. Municipal Council of Binyamina to PEC, January 7, 1929; Municipal Council of Athlit to PEC, February 24, 1929, IECA 0352-98; Municipal Council of Sarona to PEC, in German, June 21, 1930, IECA 0358-98. Another good illustration of the Jewish settlers' way of negotiating directly—and imperiously—with the power company is found in the IECA case related to the construction of a low-tension network in Hadera. IECA 0036-23.

44. Metzer, *Divided Economy*, 22.

45. Eleventh Annual report and Balance Sheet for the year ended December 31, 1934, IECA 0357-188-23.

46. PR to Board, January 7, 1934, IECA 2377-5; Baharaw to US Consul, January 30, 1934, IECA 0351-83; PR to Franck (PICA), February 5, 1934, IECA 1177-10.

47. Meeting between the HC and Dizengoff (mayor of Tel Aviv), V. Konn (PICA), M. Novemeysky (Pal Potash Ltd.), Polak (Nesher), and PR, May 15, 1934, IECA 2377-5; IECA 2372-1.

48. PR to Board, May 21, 1934, IECA 2377-5.

49. PR to Board, January 7, 1934, IECA 2377-5; Baharaw to US Consul, Jerusalem, January 30, 1934, IECA 0351-83; September 2, 1934, IECA 2377-5; Sir Harry Luke and Edward Keith-Roach, eds., *The Handbook of Palestine and Trans-Jordan*, 3rd ed. (London, 1934).

50. PR to Board, May 21, 1934, IECA 2377-5; IECA 2372-1; Baharaw to US Consul, Jerusalem, February 5, 1935, IECA 0351-83; April 5, 1935, IECA 2377-5, IECA 2372-1; report prepared for the thirteenth annual meeting, for the year 1936, IECA 0357-188-13.

51. Under the concession, the company had to submit its plans for approval to the Department of Public Works, but any objection had to be registered within twenty-eight days of submitting the plans, or the right of objection was forfeited. Assessing the extension plans was a complicated procedure; DPW had to forward the grid extension plans to the district offices for review. By 1933–34, the grid expanded at such a rapid clip that these procedures could not keep up. Director of DPW to DC, northern district, January 3, 1934, ISA M-701/23.

52. "Palestine Electric Corporation: Success of Extension Policy: Many New Centres Connected: Lord Reading's Address," *The Times*, September 6, 1935, IECA 0475-15; PR's report to the Board of Directors, February 7, 1935, IECA 2372-1.

53. February 19, 1935, PR to Board. Already by March 1935, two thousand new connections a month were made, IECA 2372-1; March 1935, IECA 2377-5.

54. Annual Report of the Department of Agriculture and Forests for the year ended 31st March, 1936, by Mr. M. T., Dir of Agriculture and Forests, Jerusalem.

55. March 1935, IECA 2377-5.

56. PR's report to the Board, May 12, 1935, IECA 2372-1; "Palestine Electric Corporation: Success of Extension Policy: Many New Centres Connected: Lord Reading's Address," *The Times*, September 6, 1935, IECA 0475-15; Baharaw to US Consul, March 10, 1938, IECA 0351-83.

57. Shamir, *Current Flow*, 27.

58. Bruno Latour, *Science in Action: How to Follow Scientists and Engineers through Society* (Cambridge, MA: Harvard University Press, 1987); Sheila Jasanoff, *States of Knowledge: The Co-production of Science and Social Order* (London: Routledge, 2004).

59. Development officer to Baharaw, December 5, 1935; Baharaw's reply, December 19, 1935, IECA 0478-6.

60. Tulkarm: PEC to DPW, August 5, 1929, IECA 1177-2; Internal report, May 21, 1940, IECA 2372-1.

61. March 10, 1935, IECA 2377-5.

62. *Filastin*, August 25, 1932.

63. Letter from merchants of Jaffa, July 11, 1923, TAHA 8387/31-0073.

64. PR to DWP, May 20, 1923, IECA 0036-14.

65. Dr. W. D. Bathgate to PR, August 18, 1923; H. F. Lechmere Taylor to PR, September 19, 1923; Mayor of Nazareth to the PEC, June 12, 1929; Mayor to PEC, October 6, 1930; Mayor to PEC, October 24, 1930; Acting DC of the north to PEC, December 3, 1930; minutes of meeting between Ittin, PEC's "fixer," and the mayor, November 25, 1930; Mayor to PEC, January 16, 1931; PEC to Mayor, January 23, 1931; DC to PEC, April 14, 1931; Mayor to PEC, June 18, 1931; DC to PEC, January 22, 1932; Mayor to PEC, March 3, 1932; Baharaw to PR, April 26, 1932, IECA 0036-14.

66. Contract between the bank, Nazareth, and the PEC, July 4, 1932; Heads of Agreement between Nazareth and PEC, August 2, 1932, IECA 0036-14.

67. Contract, August 17, 1932, IECA 0036-14.

68. Minutes from meeting in Nazareth between representatives from the PEC, Nazareth, and the Palestine Government, August 26, 1932, IECA 0036-14.

69. Acting CS to DC, November 17, 1932, IECA 0036-14.

70. Report of directors for year ending December 31, 1934, presented on September 5, 1935, IECA 0475-15.

71. *Filastin*, August 19, 1932; *Filastin*, September 28, 1932; *Filastin*, November 1, 1932; *al-Jami'a al-Islamiya*, October 21, 1932; *al-Karmel*, September 21, 1932; *al-Jami'a al-Islamiya*, September 7, 1932; *al-Jami'a al-Islamiya*, September 5, 1932.

72. *Filastin*, August 19, 1932.

73. Report by Baharaw to PR, November 2, 1932, IECA 0036-14.

74. Baharaw to DPW, August 23, 1932, IECA 0036-14.

75. Letter from "the Arab Association" in Nablus to the PEC, February 26, 1940.

76. *al-Jami'a al-Islamiya*, September 7, 1932.

77. Nour El Din, January 27, 1925; Sulayman Tuqan, September 6, 1926, Nablus Municipal Library Archives (hereafter NMLA) 5/7.

78. Baharaw to PR, May 20, 1927, IECA 0429-1135.

79. Tuqan to Nathmi al-Tamimi and Nathmi Kamal, March 18, 1936, NMLA 3/7. In 1931, the company was approached by Spiro Spyrindon, and in 1933 the power company was approached by the Nabulsi Yusef al-Hijawi. Shapiro to Head Office, November 19, 1931; PR to Shapiro, November 25, 1931; Shapiro to Spyrindon, November 27, 1931; Spyrindon to Shapiro, November 18, 1933; Shapiro to Head Office, November 20, 1933, IECA 0429-1135; Yusef al-Hijawi to PEC, February 15, 1933; Baharaw to al-Hijawi, February 20, 1933; Hijawi to PEC, May 13, 1933; Baharaw to Hijawi, June 19, 1933, IECA 0429-1135.

80. Muhammad Saber Shanaar to PEC, December 18, 1933, IECA 0429-1135; Nur match company to PEC, December 24, 1933; Baharaw to Nur, January 10, 1934, IECA 0429-1135.

81. Minutes of conversation between Shapira and the deputy mayor of Nablus, Afif Ashour, and the Nablus notable Hilmi al-Fotiani on December 2, 1934; report dated December 31, 1934, IECA 0429-1135. See also letter from A. Koch, May 31, 1934, and the PEC's reply, June 4, 1934, IECA 0429-1135.

82. Baharaw to PR, May 20, 1927, IECA 0429-1135.

83. Unsigned and undated internal PEC memo, IECA 0429-1135.

84. "Arab Nablus," *Filastin*, June 24, 1933.

85. M. Jakobson to PEC, February 18, 1934, IECA 0429-1135; "Attempt to Introduce Rutenberg in Nablus," *al-Difa'*, December 17, 1934; editorial, *Mirat al-Sharq*, December 19, 1934; "Attempts Destined to Fail," *al-Difa'*, December 21, 1934.

86. Nablus mayor Tuqan was part of the opposition faction and was in favor of bringing electricity to Nablus. So were all the other members of the municipal council who supported the idea: Ahmad Shika', Abd al-Rahim al-Nabulsi, Taher Masri,

Nimr al-Nabulsi, and Rashid Masri. M. Jakobson to PEC, February 18, 1934, IECA 0429-1135; Imil al-Ghuri, "The Crime in Nablus," *al-Wahda al-Arabiya*, January 1, 1935; *Filastin*, whose editor 'Isa al-'Isa was also a member of the Opposition, published a reply to Ghuri, rejecting his accusations: *Filastin*, January 12, 1935.

87. "Rutenberg's Electricity in Nablus," *Filastin*, January 4, 1934; "Extension of Rutenberg Electricity to Nablus," *al-Jami'a al-Islamiya*, January 31, 1934; "Rutenberg's Electricity," *Filastin*, February 6, 1934; "Rutenberg's Electricity," *Filastin*, June 19, 1934; "Who Will Light Nablus with Electricity?," *al-Difa'*, June 18, 1934; "Nablus Municipality and Rutenberg's Electricity," *Filastin*, June 21, 1934; "The Honorable Mayor of Nablus Denies the Matter of an Agreement with Rutenberg," *al-Difa'*, June 21, 1934; *al-Karmel*, December 9, 1934; "Nablus Uses Motors Instead of Rutenberg," *al-Jami'a al-Islamiya*, December 11, 1934; *Mirat al-Sharq*, December 12, 1934; *Mirat al-Sharq*, December 13, 1934; *Filastin*, December 13, 1934; *al-Jami'a al-Islamiya*, December 16, 1934; "Attempt to Introduce Rutenberg in Nablus," *al-Difa'*, December 17, 1934; editorial, *Mirat al-Sharq*, December 19, 1934; "Nablus and the Rutenberg Electric Company," *al-Jami'a al-Arabiya*, December 12, 1934; "Rutenberg's Zionist Current," *al-Jami'a al-Islamiya*, December 20, 1934; "Attempts Destined to Fail," *al-Difa'*, December 21, 1934; Imil al-Ghuri, "The Crime in Nablus," *al-Wahda al-Arabiya*, January 1, 1935.

88. "Nablus and the Rutenberg Electric Company," *al-Jami'a al-Arabiya*, December 12, 1934.

89. Editorial, *Mirat al-Sharq*, December 19, 1934.

90. "Rutenberg's Zionist Current," *al-Jami'a al-Islamiya*, December 20, 1934.

91. "Nablus Uses Motors Instead of Rutenberg," *al-Jami'a al-Islamiya*, December 11, 1934; special correspondent, Nablus, "The Lighting of Nablus by Motors Instead of Rutenberg," *Filastin*, August 26, 1947.

92. "Attempt to Introduce Rutenberg in Nablus," *al-Difa'*, December 17, 1934.

93. *al-Karmel*, December 9, 1934; for other examples of this usage, see "Rutenberg's Zionist Current," *al-Jami'a al-Islamiya*, December 20, 1934; "Attempts Destined to Fail," *al-Difa'*, December 21, 1934; Imil al-Ghuri, "The Crime in Nablus," *al-Wahda al-Arabiya*, January 1, 1935.

94. Imil al-Ghuri, "The Crime in Nablus," *al-Wahda al-Arabiya*, January 1, 1935.

95. General Manager of the Arab Bank in Nablus to Tuqan, March 31, 1936, 5/8 NMLA; *al-Liwa*, February 27, 1936. See also "The Rutenberg Company Will Light Nablus," *Filastin*, March 17, 1936; "The Zionists and the National Stronghold," *al-Jami'a al-Islamiya*, March 16, 1936; "Jews Begin Work in Arab Nablus," *al-Liwa*, March 16, 1936.

96. T. S. Boutagy & Sons to PEC, July 11, 1935; PEC's reply, July 18, 1935; Boutagy to PEC, March 17, 1936; the PEC's reply, March 24, 1936, IECA 0429-1135.

97. "Palestine Electric Corporation: Success of Extension Policy: Many New Centres Connected: Lord Reading's Address," *The Times*, September 6, 1935, IECA 0475-15.

98. PR to Rothschild, April 25, 1935, IECA 2369-4.

99. Matthew Kraig Kelly, "Crime in the Mandate: British and Zionist Criminological Discourse and Arab Nationalist Agitation in Palestine, 1936–39" (PhD diss., University of California, Los Angeles, 2013).

100. Charles Anderson, "From Petition to Confrontation: Palestinian National Movement and the Rise of Mass Politics, 1929–1939" (PhD diss., New York University, 2013), 591; Weldon C. Matthews, *Confronting an Empire, Constructing a Nation: Arab Nationalists and Popular Politics in Mandate Palestine* (London: I. B. Tauris, 2006), 236, 244–46.

101. Royal Air Force monthly summary of intelligence (February 1935), CZA S/25, 22735, cited in Anderson, "From Petition to Confrontation," 598.

102. Jacob Norris, "Repression and Rebellion: Britain's Response to the Arab Revolt in Palestine of 1936–9," *Journal of Imperial and Commonwealth History* 36, no. 1 (March 2008): 28; Ted Swedenburg, *Memories of Revolt: The 1936–1939 Rebellion and the Palestinian National Past* (Fayetteville: University of Arkansas Press, 2003); Matthews, *Confronting an Empire*, 256–57.

103. Baharaw to Commandant of Police, Northern District, Haifa, September 5, 1932, IECA 2383-2; Kelly, "Crime in the Mandate," 10–11.

104. PR to Board, 1938: M-301/8.

105. Report to board, October 3, 1938, IECA 2372-1. For instances of sabotage: August 2, 1938, IECA 2372-1; March 19, 1938, IECA 2372-1; September 20, 1938, IECA 2372-1.

106. Anderson, "From Petition to Confrontation," 988.

107. PR to Board, 1938: M-301/8; PR to Board, October 3, 1938, IECA 2372-1.

108. The triangle's three points were Jenin, Nablus, and Tulkarm. Kelly, *The Crime of Nationalism*, 45.

109. PR to Board, October 3, 1938, IECA 2372-1.

110. PR to Board, October 3, 1938, IECA 2372-1; Walid Khalidi, ed., *From Haven to Conquest: Readings in Zionism and the Palestine Problem until 1948* (Washington, DC: Institute for Palestine Studies, 1971); David Ben-Gurion, "Our Friend: What Wingate Did for Us," *Jewish Observer and Middle East Review*, September 27, 1963, 15–16.

111. September 14, 1936, IECA 1177-6: Report about pole burning.

112. Report and Balance Sheet for the year ending December 31, 1936, IECA 0357-188-14.

113. See fifteenth and sixteenth annual meeting reports for years 1937 and 1938, IECA 0357-188-18 and IECA 0357-188-20.

114. PEC to Tel Aviv Municipality, May 3, 1937, TAHA 4-1098.

115. Report on Palestine and Transjordan to the Council of the League of Nations, December 31, 1937, M-4384.

116. March 19, 1938, IECA 2372-1.

117. The PEC had originally wanted to buy the land, but the government refused. PR to CS, December 22, 1936; CS to PR, principal approval, January 16, 1937, CS to PR; CS to PR, January 27, 1937; PR to CS, February 24, 1937; April 7, 1936, IECA 2379-16; PR to Board, September 17, 1937, IECA 2372-1.

118. PR to Board, September 17, 1937, IECA 2372-1; August 2, 1938, IECA 2372-1.
119. PR to Board, October 3, 1938, IECA 2372-1.
120. November 1, 1937, IECA 2372-1.
121. PR to Board, 1938, M-301/8; PR to Board, October 3, 1938, IECA 2372-1.
122. PR to Board, September 17, 1937, IECA 2372-1.
123. November 1, 1937, IECA 2372-1; Report: "New settlements in the Beisan District," by A. Karzfeld, September 8, 1937, IECA 1177-12.
124. Report on Palestine and Transjordan to the Council of the League of Nations, December 16, 1937, M-4384.
125. March 19, 1938, IECA 2372-1.
126. May 23, 1938, M-4384.
127. From £P145,289 to £P1,333,008.
128. *Palnews: Economic Annual of Palestine* 1937, p. 20, ISA M-5331/5.
129. Matthews, *Confronting an Empire*, 258.
130. Nadan, *Palestinian Peasant Economy*, 25.
131. *Palnews*, p. 22, ISA M-5331/5.
132. Prices in the Jewish market were 21 percent higher on average. Nadan, *Palestinian Peasant Economy*, 24.
133. *Palnews*, pp. 20–22, ISA M-5331/5.
134. *Statistical Abstract of Palestine 1940*, M-4512.
135. Jewish Agency for Palestine Economic Research Institute, *Statistical Handbook of Middle Eastern Countries, Palestine, Cyprus, Egypt, Iraq, The Lebanon, Syria, Transjordan, Turkey* (Jerusalem, 1945). Turkey produced 357,605,000 kWh in 1939 and 414,393,000 kWh in 1941.
136. Jewish Agency for Palestine Economic Research Institute, *Statistical Handbook of Middle Eastern Countries*.
137. This is why David E. Nye refers to electricity as an "enabling technology," while noting that its significance is often overlooked in historical accounts focusing on the activities that the availability of electricity enabled. Nye, *Electrifying America*, 26.
138. *Statistical Abstract of Palestine, 1943*, 7th ed., Government Department of Statistics.
139. "Jordan Valley Authority: A Plan for Irrigation and Hydro-Electric Development in Palestine," Memorandum submitted by Abel Wolman, Chairman, Engineering Consulting Board Commission on Palestine Surveys, 65–66.
140. Letter signed "Arab Association" in Nablus to PEC, February 26, 1940, IECA 0429-1135.
141. Newton to Shapiro, January 10, 1940; Shapiro to Newton, January 31, 1940, IECA 0429-1135.
142. PR's report to the Board, May 21, 1940, IECA 2372-1.
143. *al-Sirat al-Mustaqim*, March 7, 1940, IECA 0429-1135.
144. Open letter from Nimr al-Nabulsi, *al-Sirat al-Mustaqim*, March 19, 1940.
145. Assistant DC, Nablus, to PEC, March 21, 1940, IECA 0429-1135.
146. IECA 0429-1135-71.

147. Minutes of a meeting, May 5, 1940, IECA 0429-1135.

148. Minutes of meeting at the Haifa head office, February 11, 1945, IECA 0429-1135.

149. *Nida' al-Ard*, August 11, 1947: article giving a critical account of the Nablus municipality's negotiations with the PEC; two days later, in a letter to the editor, Mahmud Nabih al-Baytar, Nablus town clerk, defended the negotiations and the agreement that was about to be made between Nablus and the PEC, *Nida' al-Ard*, August 19, 1947; special correspondent, Nablus, "The Lighting of Nablus by Motors Instead of Rutenberg," *Filastin*, August 26, 1947: this article describe the group of "hidden hands" that are working against the negotiations with the PEC; special correspondent, Nablus, "Electricity in Nablus and the Government's Loan to the Municipality," *Filastin*, August 27, 1947: contains, among other things, an interview with the mayor of Nablus, defending the negotiations.

150. Letter signed by the Arab League Committee to Boycott Zionist Goods to the PEC, July 8, 1946, IECA 0429-1132.

151. Proclamation of the Nablus Municipal Council (Authorisation of Loan) Order, 1947, *Palestine Gazette* 1617, suppl. 2, p. 1491, October 2, 1947, IECA 0429-1132-88. Resolution no. 179 of the Municipal Corporation of Nablus, September 11, 1947; resolution no. 201 of the Municipal Corporation of Nablus, October 5, 1947, IECA 0429-1132-85. Loan Agreement between the Municipal Corporation of Nablus and the PEC, October 20, 1947, witnessed by Nabih Zu'aytir and signed by Nabih Bitar, town clerk, for Nabus, and by George Bradlaw, secretary, and Abraham PR, managing director, for the PEC, IECA 0429-1132-77. Supply Agreement between the PEC and Nablus, February 10, 1948, IECA 0429-1132-37.

152. Actually, this matter was first pursued through Jordanian courts in 1958–61; the suit was settled in Nablus's favor, but no payments were made. PEC to counsel, Lifshitz & Associates, December 28, 1858, IECA 0429-1132-28; S. Horowitz & Co. to Israel Electric Corp., December 14, 1967, IECA 0429-1132-8; unsigned legal opinion, dated January 10, 1968, IECA 0429-1132-7; "Schem tovaat mechev' hacheshmal 60,000 lirot falastinit," *Ma'ariv*, March 25, 1968, IECA 0429-1132.

153. Metzer, *Divided Economy*, 13.

154. Ziman, *Hakalkala ha'eretzyisra'elit bemisparim*. For a pioneering study of the central role of statistics in making commodities, markets, and ultimately "the economy," see Emmanuel Didier, "Do Statistics 'Perform' the Economy?," in *Do Economists Make Markets? On the Performativity of Economics*, ed. Donald A. MacKenzie, Fabian Muniesa, and Lucia Siu (Princeton, NJ: Princeton University Press, 2007), 276–310.

155. Metzer, *Divided Economy*, 155; Troen, *Imagining Zion*, 14; Shamir, *Current Flow*, 132. That the British were relying on Zionist figures was not entirely new. For the 1919 *Handbook of Syria (and Palestine)*, prepared by the Colonial Office, most of the figures were supplied by the leading Zionist and statistician Arthur Ruppin. Naval Staff, Intelligence Department, *Handbook of Syria (and Palestine)*, June 1919, ISA M-4409. In August 1939, the British intensified their

statistical activities in Palestine by publishing the *General Monthly Bulletin of Current Statistics*. See *Monthly Bulletin of Economic Statistics and Special Bulletins*, ISA M-5310/8.

156. E.g., *Special Bulletin No. 13 Wage Census 1943*, "Table VI—All-Groups Index Numbers and Individual Group Index Numbers in the First Five Months of 1942 and of 1943, in Three Arab Markets and in Three Jewish Markets," p. 10, ISA M-5303/27.

157. Jewish Agency for Palestine Economic Research Institute, *Statistical Handbook of Middle Eastern Countries*.

158. *Statistical Abstract of Palestine, 1937–1938*, Government of Palestine, 1938, Preface by S. A. Cudmore.

159. "Jordan Valley Authority: A Plan for Irrigation and Hydro-Electric Development in Palestine," Memorandum submitted by Abel Wolman, Chairman, Engineering Consulting Board Commission on Palestine Surveys, 65–66.

160. Palestine Partition Commission Report, presented by S/SC to Parliament, October 1938 (London, 1938), 93.

161. ISA M-5314/11: Statistical Abstract of Palestine 1944–45.

162. The 1939 White Paper officially announced that Britain had fulfilled its pledges according to the Balfour Declaration.

163. *Statistical Abstract of Palestine 1944–45*, table 4—"General Summary for Each Section of Industry Enumerated for the Year 1939," p. 53; Fuel and Light Index, p. 117, ISA M-5314/11.

164. Timothy Mitchell talks about "the evolution of precision" in *Rule of Experts*, 82.

165. *Statistical Abstract of Palestine, 1937–8*, preface, ISA M-4513.

166. Jewish Agency for Palestine Economic Research Institute, *Statistical Handbook of Middle Eastern Countries*.

167. Herbert Sidebotham, *Great Britain and Palestine* (London: Macmillan, 1937), 107.

168. Sidebotham, *Great Britain and Palestine*, 111.

169. Sidebotham, *Great Britain and Palestine*, 112.

170. Official Records of the Second Session of the General Assembly Supplement No. 11, United Nations Special Committee on Palestine, Report to the General Assembly, Vol. 1, Chapter 2, Paragraph 43 (emphasis added).

171. "Palestine Could Easily Support Five Times Present Population If Modern Methods of Farming Are Introduced," *Jewish Telegraphic Agency*, March 27, 1931.

172. Sidebotham, *Great Britain and Palestine*, 115. This calls to mind Churchill's assurances in 1922 that the Zionists would build a "new Palestine" and thus not rely on "one scrap of what was there before." Parliamentary Debates, House of Commons, Motion on Rutenberg Plan, July 4, 1922.

173. Shapira, *Israel*, 115.

174. Nadan, *Palestinian Peasant Economy*, 7, 139; Metzer, *Divided Economy*, 130–31.

175. Nadan, *Palestinian Peasant Economy*, 200.

176. Nadan, *Palestinian Peasant Economy*, 92.
177. Judith Tendler, *Electric Power in Brazil: Entrepreneurship in the Public Sector* (Cambridge, MA: Harvard University Press, 1968), 20–21.

CHAPTER SIX

1. CZA S8/2260/1.
2. Ad for Café Weintraub, *Palestine Post*, January 5, 1935; "'EVA' Comes to Tel Aviv," *Palestine Post*, January 7, 1935.
3. The Electricity Concession (Jerusalem) Bill, 1930, *Official Gazette of the Government of Palestine*, January 13, 1930.
4. G. L. M. Clauson, November 3, 1922, TNA CO 733/39.
5. S/SC memo, January 7, 1922, TNA CO 733/17b.
6. Shuckburgh to Ormsby-Gore, February 17, 1923, TNA CO 733/39.
7. Smith, *Roots of Separatism in Palestine*, 123–24.
8. ISA P-920/3.
9. Case before the District Court of Jerusalem, Civil Case No. 456/30, June 1931; Testimony of General Manager, Campbell-Brown, Awni 'Abd al-Hadi's papers, ISA P-152/17; DC to US/SC, November 25, 1943, ISA P-21/36.
10. *Official Gazette of the Government of Palestine*, January 13, 1930.
11. "Electricity in Jerusalem," *Palestine Bulletin*, September 8, 1927.
12. "Important Electricity Case This Week," *Palestine Bulletin*, June 1, 1930; "Hassolel Case in Court," *Palestine Bulletin*, June 8, 1930; "The Electricity Case," *Palestine Bulletin*, June 24, 1930; "Darkness in Jerusalem," *Palestine Bulletin*, July 13, 1930.
13. "Jerusalem and Its Electricity," *Palestine Bulletin*, September 30, 1930.
14. DC to US/SC, November 25, 1943, ISA P-21/36 and P-21/37.
15. Testimony of Campbell-Brown, June 6, 1946, ISA P-21/37. See also "Seasonal variation in the consumption of electric power in Palestine, 1939," ISA M-5311/6.
16. Bernard Joseph Trial, testimony of Joseph Wager, inspector; ISA P-21/38.
17. Comparison diagram covering the years 1931–45, compiled from the General Monthly Bulletin of Current Statistics, ISA P-21/38.
18. "May Boycott Jerusalem Electricity," *Sentinel,* April 26, 1929; "The Jerusalem Electric Corporation," *Palestine Bulletin*, June 24, 1929; "World Over," *Bnai Brith Messenger,* July 12, 1929.
19. Sherene Seikaly, "Meatless Days: Consumption and Capitalism in Wartime Palestine, 1939–1948" (PhD diss., New York University, 2007), 36.
20. Khadir to JEPSC, June 24, 1936; JEPSC to Khadir, June 26, 1936; Khadir to JEPSC, June 27, 1936, ISA P-334/19.
21. Khadir to JEPSC, September 4, 1936; JEPSC to Khadir, September 13, 1936; Announcement of rebate, published in English, Hebrew, and Arabic, on November 10, 1936, ISA P-334/19.
22. J. Gordon Boutagy to Campbell-Brown (Arab Chamber of Commerce CC'd), July 22, 1938; on July 26, 1938, Khadir wrote another note to the JEPSC

following up on Boutagy's complaint and urging it to investigate it thoroughly, ISA P-334/19.

23. S. O. Richardson & Co to A. G., July 1, 1940, ISA M-704/28.

24. "Stealing Electricity: Novel Case in Court," *Palestine Bulletin*, October 21, 1930; "Theft of Electricity," *Palestine Bulletin*, November 4, 1930.

25. "Electricity for Nothing," *Palestine Bulletin*, November 20, 1930; "Electricity—For Payment," *Palestine Bulletin*, January 18, 1931.

26. Henri Cattan's papers, ISA P-184/17.

27. Suit filed against Lorenzo by the Attorney General, based on claim by the JEPSC, July 6, 1938; Criminal Case No. 96/38, District Court of Jerusalem, ISA P-225/31.

28. Uri Kupferschmidt, *The Supreme Muslim Council: Islam under the British Mandate for Palestine* (Leiden: E.J. Brill, 1987), 70.

29. JEPSC to AG, February 2, 1938. N. Abcarius Bey, lawyer for the accused, made the formal request to the AG to drop the case. Criminal Case No. 21 of 1938, District Court, Jerusalem, letter dated February 17, 1938, ISA M-704/15.

30. Alan Rose, on behalf of AG to A.J. Kingsley, Deputy Inspector General at the Criminal Investigation Department of the Palestine Police Force, ISA M-704-15.

31. Government Official Communique, December 10, 1946, M-5001/33; Campbell-Brown to A.S. Valentine of Balfour, Beatty & Co., Ltd., June 13, 1943, ISA P-21/36. The prices of coal, wood, and kerosene all rose precipitously during the war. "Jerusalem Electric Explains: Required Machinery Arrived Late," *Palestine Post*, February 28, 1946.

32. According to this report, the drastic increase in sales did not see a corresponding increase in profits, as a result of the increased costs for materials. But, significantly, they also paid out cost-of-living allowances for staff of £P20,274 in 1943, and in 1945 this amount was raised to £P 28,287. ISA P-21/38.

33. Campbell-Brown to A. S. Valentine, June 13, 1943, ISA P-21/36.

34. Campbell-Brown to A. S. Valentine, June 13, 1943, ISA P-21/36; DC to US/SC, November 25, 1943, ISA P-21/36; Government Official Communique, December 10, 1946, ISA M-5001/33.

35. Campbell-Brown to Valentine, October 22, 1943, ISA P-21/36; DC to US/SC, November 25, 1943 ISA P-21/36; Record of proceedings in *Bernard Joseph v. JEPSC*, hearing on March 23, 1946; testimony of Wilson Edwin Blaze, Station Engineer at the main station; testimony of Campbell-Brown, June 6, 1946, ISA P-21/37.

36. Campbell-Brown to Valentine, October 22, 1943; Valentine to Campbell-Brown, November 27, 1943; Brook to CO, March 26, 1945; JEPSC to DPW, June 10, 1945, ISA P-21/36; William Shearer's address for the year under review, 1944, ISA P-21/38.

37. JEPSC to Controller of Heavy Industries, November 30, 1943; JEPSC to CS, December 15, 1943; DPW to JEPSC, December 4, 1943; JEPSC to DPW, December 20, 1943; JEPSC to Secretary of War Supply Board, December 29, 1943: the import license was granted on December 25, 1943; War Supply Board to JEPSC, February

25, 1944, with notification that it has recommended expedited manufacture and shipment; Brook to F. Waugh, CO, March 16, 1944; Waugh to Brook, March 22, 1944; Secretary of War Supply Board to JEPSC, May 16, 1944; War Supply Board to JEPSC, April 10, 1944: the Middle East Supply Centre has recommended giving the license priority, ISA P-21/36.

38. War Supply Board to JEPSC, April 10, 1944, ISA P-21/36; DC to CO, March 26, 1945; DC to CO, May 16, 1945, ISA P-21/36. The manufacturer of the generator the company had ordered, Crossley Premier Engines, Ltd., was deprived of thirty of its workers, who were called in for service in the armed forces.

39. Testimony, Campbell-Brown, June 6, 1946, ISA P-21/38.

40. JEPSC to DPW, June 10, 1945, ISA P-21/36.

41. JEPSC to Valentine, February 19, 1945; Brook to CO, May 16, 1945, ISA P-21/36; Record of proceedings in the case of *Bernard Joseph v. JEPSC*; testimony from the first witness for the defense, Wilson Edwin Blaze, Station Engineer (main station), hearing on March 23, 1946, ISA P-21/37; reprinted from the Times Company Meetings, November 10, 1945; William Shearer's address for the year under review, 1944, ISA P-21/38.

42. Government Official Communiqué, December 10, 1946, ISA M-5001/33; ISA P-21/38; November 6, 1945, London, Reuters, ISA P-21/38.

43. ISA P-285/6.

44. Government Official Communiqué, December 10, 1946, ISA M-5001/33.

45. Matron to DC, October 21, 1946; Tuqan's reply, October 24, 1946; the matron's second letter, October 24, 1946; Tuqan's final reply, November 2, 1946, ISA M-861/3.

46. Harry Gottlieb to JEPSC, October 6, 1946; Gottlieb to DC, October 6, 1946; DC to JEPSC, October 9, 1946; JEPSC to DC, October 11, 1946, ISA M-5001/33.

47. Qubayn to JEPSC, November 1946; Qubayn to JEPSC, December 8, 1946; Qubayn to CS, February 9, 1947, ISA M-5001/33.

48. "LP .20 Damages against J. E. C.: Dr. Joseph Wins His Case," *Palestine Post*, October 8, 1946.

49. Letter signed "Mukhtars of Wadi al-Joz and Aqabat al-Suwana and others" to DC, March 11, 1947; DC to JEPSC, March 14, 1947; Municipal Corporation of Jerusalem to DC, March 25, 1947, ISA M-5001/33.

50. Beth Israel to DC, January 4, 1947; DC to JEPSC, January 17, 1947; JEPSC to DC, January 19, 1947; DC to Beth Israel, January 22, 1947, ISA M-861/3.

51. Otter, *Victorian Eye*, 240.

52. Case before the District Court of Jerusalem, Civil Case No. 456/30, June 1931, ISA P-152/17: Awni Abd al-Hadi's papers.

53. Record of proceedings in the case of *Bernard Joseph v. JEPSC*; testimony from the first witness for the defense, Wilson Edwin Blaze, Station Engineer (main station), hearing on March 23, 1946. See also testimony of Joseph Vager, inspector at the subsidiary station and repairman. According to the Station Log Book, it is clear that almost all the engines were in a bad way, ISA P-21/37.

54. ISA P-21/38. The notice appeared in *Yediot Hayom* (German), *Davar* (Hebrew), November 26; *Palestine Post* (English), *Haaretz* (Hebrew), *Haboker* (Hebrew), November 27 (English); *Hage* (Hebrew), November 28; *al-Difa'* (Arabic), November 29; *Yediot Khadashot* (Hebrew), November 30; *Filastin* (Arabic), December 1, 1945. According to the copy of the bill, the expense for this advertising campaign was £P70.

55. Campbell-Brown's testimony, June 6, 1946, ISA P-21/37.

56. "Jerusalem Electric Explains: Required Machinery Arrived Late," *Palestine Post*, February 28, 1946.

57. *Palestine Post*, December 16, 1945; *Palestine Post*, December 19, 1945.

58. "Sha'rorit hakheshmal beyerushalaim magia le'si khadash" [Electricity scandal in Jerusalem as a result of new peak], *Ha'aretz*, February 17, 1946; see also "Consumer Sues J'lem Electric: Advocates' Office Claims Damages," *Palestine Post*, February 17, 1946.

59. Levitsky to JEPSC, March 15, 1946; Levitsky to JEPSC, March 31, 1946; JEPSC to Levitsky with check for £P222.070; Levitsky to JEPSC, bill for £P152.410, July 7, 1946; Levitsky to JEPSC, bill for £P236.278 [n.d.], ISA P-21/37.

60. *Davar*, May 2, 1946.

61. One notable example of this came after the verdict fell in October. Bernard Joseph immediately wrote to the power company demanding to be paid the full amount of the fine (£P54.335) within five days. If they had not received it, they threatened to take the company to court again. The power company dragged its feet and also appealed to the court for a stay of execution, which was denied on November 4, 1946. When the power company finally paid out the fine, in late February 1947, the court deducted £P1.380 in administrative fees for serving as intermediary. Bernard Joseph then immediately wrote to the power company demanding the outstanding £P1.380 and did not let up until Levitsky sent a check for the amount on March 13, 1947. See Joseph to JEPSC, October 23, 1946; Levitsky to Copland, October 23, 1946; ruling against JEPSC's appeal for a stay of execution, November 4, 1946; meeting protocol, Brook, HG Balfour, EM Berstrom, November 4, 1946; Joseph to JEPSC, February 24, 1947; Levitsky to Joseph, with check, March 13, 1947, ISA 21/37.

62. For a summary of the case, see ISA P-21/36, E. N. Berouti to Campbell-Brown, October 26, 1946.

63. Ruling, October 7, 1946, ISA P-21/37; "LP .20 Damages against J. E. C.: Dr. Joseph Wins His Case," *Palestine Post*, October 8, 1946; "The Jerusalem Electric Company Sentenced to Pay LP. 20," *Haboker*, October 8, 1946; "The Jerusalem Electric Company Has to Pay Compensation to Its Customers," *Ha'aretz*, October 8, 1946. For the comment on locally produced spare parts, see Blaze's testimony in Bernard Joseph trial, March 23, 1946.

64. Record of proceedings in the case of *Bernard Joseph v. JEPSC*; testimony from the first witness for the defense, Wilson Edwin Blaze, Station Engineer (main station), hearing on March 23, 1946; testimony of Campbell-Brown, June 6, 1946, ISA P-21/37; "Jerusalem Electric Explains: Required Machinery Arrived Late," *Palestine Post*, February 28, 1946.

65. JEPSC to Gibson, September 17, 1947; Jerusalem Electric & Public Service Corporation Proposed Bulk Supply Agreement with the Palestine Electric Corporation (1947), December 2, 1947; CS to JEPSC, November 8, 1947, draft of amendments to the Electricity Concession from 1930; JEPSC's amended Articles of Association, validated on January 21, 1948; Note from the Solicitor General, February 27, 1948, ISA M-701/17.

CHAPTER SEVEN

1. Sidebotham, *Great Britain and Palestine*, 222.
2. Shalom Reichman, Yossi Katz, and Yair Paz, "The Absorptive Capacity of Palestine, 1882–1948," *Middle Eastern Studies* 33, no. 2 (April 1997): 338–61. Susan Pedersen's account of the attitudes of the Permanent Mandates Commission of the League of Nations confirms this single-minded focus on economic issues and "economic absorptive capacity." Pedersen, *The Guardians*, 102. "Report of the Commission on the Palestine Disturbances of August 1929," more commonly known as the "Shaw Report," after its chairman, Walter Shaw, Presented by the Secretary of State for the Colonies to Parliament by Command of His Majesty, 1930. "Report on Immigration, Land Settlement and Development," more commonly known as the "Hope Simpson Report," after the chairman, John Hope Simpson, October 1, 1930.
3. Royal Commission Report, 380–386; Biger, *Boundaries*, 200.
4. Biger, *Boundaries*, 200.
5. PR to US/SC, July 29, 1937; see also letter signed by the "directors" of the PEC, PR, Samuel (chairman), Hirst (director), July 2, 1937, TNA CO 733/337/2.
6. Khalidi, *From Haven to Conquest*, 333.
7. Officer Administering the Government to S/SC, September 14, 1937; see also minute, Shuckburgh, July 30, 1937, TNA CO 733/337/2.
8. Biger, *Boundaries*, 206.
9. Palestine Partition Commission Report, presented by S/SC to Parliament, October 1938 (London, 1938), 47.
10. Palestine Partition Commission Report, presented by S/SC to Parliament, October 1938 (London, 1938), 295, Appendix 6.
11. Note by Major Young, April 8, 1925, TNA CO 733/92.
12. Biger, *Boundaries*, 212.
13. Biger, *Boundaries*, 214–18.
14. IECA 1177-6; Khalidi, *All That Remains*.
15. Draft letter, PEC to Mayor Tuqan of Nablus, March 29, 1948, IECA 0429-1131-31.
16. Paul R. Mendes-Flohr and Jehuda Reinharz, eds., *The Jew in the Modern World: A Documentary History* (New York: Oxford University Press, 1980), 630.
17. PEC to Abdullah, August 5, 1948, IECA, 0347-2-30.
18. PEC to Abdullah, April 25, 1949, IECA, 0347-2-18.

19. Undated draft of letter from PEC to Abdullah from 1953; "Final Report of the United Nations Economic Survey Mission for the Middle East," pt. I, p. 4, IECA, 2383-8-10.

20. *Filastin*, February 12 and 13, 1953; Internal PEC memorandum on legal status of concession, May 7, 1953, IECA, 2383-8-7.

21. Undated draft of letter from PEC to Abdullah from 1953, IECA 2383-8-10.

22. Undated draft of letter from PEC to Abdullah from 1953, IECA 2383-8-10.

23. Undated draft of letter from PEC to Abdullah from 1953, IECA 2383-8-10.

24. Copy of Law no. 9 of 1954, IECA 2383-8-4.

25. Twenty-fifth annual general meeting for the year ended 1947, held on September 30, 1948, IECA 0357-188-9.

26. Twenty-eighth annual general meeting of the PEC, held on September 28, 1951, for the year ended December 31, 1950, IECA 0357-188-11.

27. A first such set had been put into commission in December 1949. Twenty-seventh annual general meeting for the year ended 1949, IECA 0357-188-19.

28. "Doing Business in the West Bank & Gaza: 2014 Country Commercial Guide for U. S. Companies," Department of Commerce (accessed March 8, 2015), http://www.buyusainfo.net/docs/x_6914413.pdf.

29. Twenty-Ninth Annual Report and Accounts for the Year Ended December 31, 1951, IECA 0357-188-32; Thirtieth Annual Report and Accounts for the Year Ended December 31, 1952, IECA 0357-188-33. For the early Israeli leaders' efforts at industrialization, especially within the context of the Sharon Plan, see Troen, *Imagining Zion* Kindle Location 2742–47. Under the Sharon Plan, Israel's leaders envisioned a future in which 80 percent of the population would live in towns and cities.

30. Thirty-first annual general meeting of the PEC for the year ended 1953, held on September 30, 1954, IECA 0357-188-1. Samuel took over from Lord Reading upon his death in 1936.

CONCLUSION

1. Alfred Bonne, "The Holy Land Electrified," *B'nai B'rith Magazine*, January 1931, in CZA S90/2124/7.

2. Charles Gide, *Communist and Co-operative Colonies* (London: Harrap, 1930), 182.

3. "Jordan Electric Scheme Inaugurated," *Palestine Bulletin*, June 12, 1932.

4. Thomas P. Hughes, "The Evolution of Large Technological Systems," in *The Social Construction of Technological Systems: New Directions in the Sociology and History of Technology*, ed. Wiebe E. Bijker, Thomas P. Hughes, and Trevor Pinch (Cambridge, MA: MIT Press, 2012), 45; Winner, "Do Artifacts Have Politics?"; David E Nye. *Technology Matters: Questions to Live With* (Cambridge, MA: MIT Press, 2006); Hughes, *Networks of Power*.

5. White, *Railroaded*, 28.

6. Said, "Zionism from the Standpoint of Its Victims," 36.

7. Bruno Latour, "On Some of the Affects of Capitalism" (lecture presented at the Royal Academy, Copenhagen, February 26, 2014).

8. David Edgerton, *The Shock of the Old: Technology and Global History since 1900* (Oxford: Oxford University Press, 2007), 5–6.

9. Said, "Zionism from the Standpoint of its Victims," 49.

BIBLIOGRAPHY

PRESS SOURCES

English

B'nai B'rith Magazine
Bnai Brith Messenger
Daily Chronicle
Daily Herald
Daily Mail
Daily Telegraph
Financial Times
Manchester Guardian
Palestine Bulletin
Palestine Post
Sentinel
The Times of London

Arabic

al-Difa'
Alif Ba
al-Jami'a al-Arabiya
al-Jami'a al-Islamiya
al-Karmel
al-Liwa
al-Sirat al-Mustaqim
al-Wahda al-Arabiya
Filastin
Mirat al-Sharq
Nida' al-Ard

Hebrew

Davar
Ha'aretz
Haboker
Hage
Hapoel Hatsair
Ma'ariv
Yediot Hayom
Yediot Khadashot

Miscellaneous

Krig og Fred (Norwegian)
Teknisk Ukeblad (Norwegian)
Yediot Hayom (German)

ARCHIVES

Central Zionist Archives (CZA), Jerusalem, Israel
Haifa Municipal Archives (HMA), Haifa, Israel
Israel Electric Corporation Archives (IECA), Haifa, Israel
Israel State Archives (ISA), Jerusalem, Israel
Jerusalem Municipal Archives (JMA), Jerusalem, Israel
Nablus Municipal Library Archives (NMLA), Nablus, Palestine
National Archives of the United Kingdom (TNA), London, UK
Tel Aviv–Jaffa Municipal Historical Archives (TAHA), Tel Aviv, Israel
Weizmann House Archives (WHA), Rehovoth, Israel
Weizmann Institute Archives (WIA), Rehovoth, Israel

UNPUBLISHED MANUSCRIPTS

Amawi, Abla Mohamed. "State and Class in Transjordan: A Study of State Autonomy." PhD diss., Georgetown University, 1993.
Anderson, Charles. "From Petition to Confrontation: Palestinian National Movement and the Rise of Mass Politics, 1929–1939." PhD diss., New York University, 2013.
Cookson-Hills, Claire Jean. "Engineering the Nile: Irrigation and the British Empire in Egypt, 1882–1914." Ph.D. diss., Queen's University, 2013.
Coopersmith, Jonathan. "The Electrification of Russia, 1880 to 1925." PhD diss., Oxford University, 1985.

Fakher Eldin, Munir. "Communities of Owners: Land Law, Governance, and Politics in Palestine, 1858–1948." PhD diss., New York University, 2008.
Kelly, Matthew Kraig. "Crime in the Mandate: British and Zionist Criminological Discourse and Arab Nationalist Agitation in Palestine, 1936–39." PhD diss,, University of California, Los Angeles, 2013.
Norris, Jacob. "Ideologies of Development and the British Mandate in Palestine." PhD diss., University of Cambridge, 2010.
Roberts, Nick. "Rethinking the Status Quo: The British and Islam in Mandate Palestine, 1917–1929." PhD diss., New York University, 2010.
Seikaly, Sherene. "Meatless Days: Consumption and Capitalism in Wartime Palestine, 1939–1948." PhD diss., New York University, 2007.

BOOKS AND ARTICLES

Abu El-Haj, Nadia. *Facts on the Ground: Archaeological Practice and Territorial Self-Fashioning in Israeli Society*. Chicago: University of Chicago Press, 2001.
Abu-Lughod, Ibrahim A. *The Transformation of Palestine; Essays on the Origin and Development of the Arab-Israeli Conflict*. Evanston, IL: Northwestern University Press, 1971.
Adas, Michael. *Machines as the Measure of Men: Science, Technology, and Ideologies of Western Dominance*. Ithaca, NY: Cornell University Press, 1989.
Alatout, Samer. "Bringing Abundance into Environmental Politics: Constructing a Zionist Network of Water Abundance, Immigration, and Colonization," *Social Studies of Science* 39, no. 3 (June 2009): 363–94.
Alder, Ken. *Engineering the Revolution: Arms and Enlightenment in France, 1763–1815*. Princeton, NJ: Princeton University Press, 1997.
Amery, Leopold. *My Political Life*. Vol. 2. London: Hutchinson, 1953.
Amsterdamska, Olga. "Surely You Are Joking, Monsieur Latour!" *Science, Technology, and Human Values* 15 (1990): 495–504.
Andersen, Casper. *British Engineers and Africa, 1875–1914*. London: Pickering and Chatto, 2011.
Anderson, Benedict. *Imagined Communities*. New York: Verso, 1991.
Anghie, Antony. "Civilization and Commerce: The Concept of Governance in Historical Perspective." *Villanova Law Review* 45, no. 5 (2000): 887–911.
———. *Imperialism, Sovereignty, and the Making of International Law*. Cambridge: Cambridge University Press, 2005.
Appadurai, Arjun. *Modernity at Large: Cultural Dimensions of Globalization*. Minneapolis: University of Minnesota Press, 1996.
Arato, Andrew, and Eike Gebhardt, eds. *The Essential Frankfurt School Reader*. New York: Continuum, 1982.
Arnold, David. *Science, Technology, and Medicine in Colonial India*. New York: Cambridge University Press, 2000.

Arrighi, Giovanni. *The Long Twentieth Century: Money, Power and the Origins of Our Times.* New York: Verso, 2010.
Baber, Zaheer. *The Science of Empire: Scientific Knowledge, Civilization, and Colonial Rule in India.* Albany: State University of New York Press, 1996.
Barak, On. *On Time: Technology and Temporality in Modern Egypt.* Berkeley: University of California Press, 2013.
Barnes, Barry. *Scientific Knowledge and Sociological Theory.* London: Routledge, 1974.
Baron, Salo. *The Russian Jew under Tsars and Soviets.* New York: Macmillan, 1964.
Barry, Andrew. *Political Machines: Governing a Technological Society.* New York: Athlone Press, 2001.
Beit-Hallahmi, Benjamin. *Original Sins: Reflections on the History of Zionism and Israel.* New York: Olive Branch, 1993.
Ben-Gurion, David, and Yitzhak Ben-Zvi. *Eretz Yisra'el be'avar uvahoveh.* Jerusalem: Hotsa'at Yad Yitsḥak Ben-Tsevi, 1979.
Bentwich, Norman. *Mandate Memories, 1918–1948.* London: Schocken, 1965.
Berman, Bruce, and John Lonsdale. *Unhappy Valley: Conflict in Kenya and Africa.* London: John Currey, 1992.
Biger, Gideon. *The Boundaries of Modern Palestine, 1840–1947.* London: Routledge, 2004.
Bijker, Wiebe E., Thomas P. Hughes, and Trevor Pinch, eds. *The Social Construction of Technical Systems: New Directions in the Sociology and History of Technology.* Anniversary edition. Cambridge, MA: MIT Press, 2012.
Blumberg, Arnold. *Zion before Zionism 1838–1880.* Syracuse, NY: Syracuse University Press, 1985.
Bourdieu, Pierre. *In Other Words: Essays towards a Reflexive Sociology.* Stanford, CA: Stanford University Press, 1990.
Brenner, Neil. *New State Spaces: Urban Governance and the Rescaling of Statehood.* Oxford: Oxford University Press, 2004.
Brett, E. A. *Colonialism and Underdevelopment in East Africa: The Politics of Economic Change 1919–1939.* London: Heinemann, 1973.
Brubaker, Rogers. *Nationalism Reframed: Nationhood and the National Question in the New Europe.* New York: Cambridge University Press, 1996.
Bunton, Martin. *Colonial Land Policies in Palestine, 1917–1936.* Oxford: Oxford University Press, 2007.
Burchell, Graham, Colin Gordon, and Peter Miller, eds. *The Foucault Effect: Studies in Governmentality.* Chicago: University of Chicago Press, 1991.
Cain, P. J., and A. G. Hopkins. "Gentlemanly Capitalism and British Expansion Overseas II: New Imperialism, 1850–1945." *Economic History Review,* n.s., 40, no. 1 (February 1987): 1–26.
Callon, Michel, ed. *The Laws of the Markets.* Oxford: Blackwell, 1998.
Caplan, Neil. *Futile Diplomacy.* Vol. 1, *Early Arab-Zionist Negotiation Attempts, 1913–1931.* London: Cass, 1983.

———. *Futile Diplomacy*. Vol. 3, *Arab-Zionist Negotiations and the End of the Mandate*. London: Cass, 1986.
Casazza, Jack, and Frank Delea. *Understanding Electric Power Systems: An Overview of the Technology and the Marketplace*. Hoboken, NJ: John Wiley and Sons, 2010.
Chatty, Dawn. *Displacement and Dispossession in the Modern Middle East*. Cambridge: Cambridge University Press, 2010.
Clavin, Patricia. *Securing the World Economy: The Reinvention of the League of Nations, 1920–1946*. Oxford: Oxford University Press, 2013.
Cohn, Bernard. *Colonialism and Its Forms of Knowledge*. Princeton, NJ: Princeton University Press, 1996.
Collins, Harry. *Changing Order: Replication and Induction in Scientific Practice*. Beverly Hills, CA: Sage, 1985.
Conklin, Alice L. *A Mission to Civilize: The Republican Idea of Empire in France and West Africa, 1895–1930*. Stanford, CA: Stanford University Press, 1997.
Constantine, Stephen. *The Making of British Colonial Development Policy 1914–1940*. London: F. Cass, 1984.
Coopersmith, Jonathan. *The Electrification of Russia, 1880–1926*. Ithaca, NY: Cornell University Press, 1992.
Cotterell, Paul. *The Railways of Palestine and Israel*. Abingdon: Tourret, 1984.
Cowen, Michael, and Robert W. Shenton. *Doctrines of Development*. London: Routledge, 1996.
Cronon, William. *Nature's Metropolis: Chicago and the Great West*. New York: W. W. Norton, 1991.
Daston, Lorraine, and Peter Galison, *Objectivity*. New York: Zone Books, 2007.
Davis, Diana K. *Resurrecting the Granary of Rome: Environmental History and French Colonial Expansion in North Africa*. Athens: Ohio University Press, 2007.
Davis, Diana K., and Edmund Burke, eds. *Environmental Imaginaries of the Middle East and North Africa*. Athens: Ohio University Press, 2011.
Deakin, Alfred. *Irrigated India: An Australian View of India and Ceylon, Their Irrigation and Agriculture*. London, 1893.
DeGraaf, Leonard. "Corporate Liberalism and Electric Power System Planning in the 1920s." *Business History Review* 64, no. 1 (Spring 1990): 1–31.
Delaney, David, and Helga Leitner. "The Political Construction of Scale." *Political Geography* 16, no. 2 (1997): 93–97.
Denoon, Donald. *Settler Capitalism: The Dynamics of Dependent Development in the Southern Hemisphere*. Oxford: Clarendon Press, 1983.
Deutscher, Isaac. *The Prophet: The Life of Leon Trotsky*. Brooklyn: Verso, 2015.
Dirks, Nicholas B., ed. *Colonialism and Culture*. Ann Arbor: University of Michigan Press, 1992.
Doumani, Beshara. *Rediscovering Palestine: Merchants and Peasants in Jabal Nablus, 1700–1900*. Berkeley: University of California Press, 1995.
Drayton, Richard Harry. *Nature's Government: Science, Imperial Britain, and the "Improvement" of the World*. New Haven, CT: Yale University Press, 2000.

Edgerton, David. *The Shock of the Old: Technology and Global History since 1900.* Oxford: Oxford University Press, 2007.

Ekbladh, David. *The Great American Mission: Modernization and the Construction of an American World Order.* Princeton, NJ: Princeton University Press, 2010.

Elden, Stuart. *The Birth of Territory.* Chicago: University of Chicago Press, 2013.

El-Eini, Roza. *Mandated Landscape: British Imperial Rule in Palestine, 1929–1948.* London: Routledge, 2006.

Elkins, Caroline, and Susan Pedersen. *Settler Colonialism in the Twentieth Century: Projects, Practices, Legacies.* New York: Routledge, 2005.

Engel, David. *Zionism.* Harlow, UK: Pearson/Longman, 2009.

Eshel, Tzadok. *Naharayim: Sipura Shel Tachanat Ko'ach.* Haifa: Israel Electric Corporation, 1990.

Ezrahi, Yaron. *The Descent of Icarus: Science and the Transformation of Contemporary Democracy.* Cambridge, MA: Harvard University Press, 1990.

Fakher Eldin, Munir. "British Framing of the Frontier in Palestine, 1918–1923: Revisiting Colonial Sources on Tribal Insurrection, Land Tenure, and the Arab Intelligentsia." *Jerusalem Quarterly* 60 (2014): 42–58.

Ferguson, James. *The Anti-politics Machine: "Development," Depoliticization, and Bureaucratic Power in Lesotho.* Cambridge: Cambridge University Press, 1994.

Ferguson, Niall. *Empire: The Rise and Demise of the British World Order and the Lessons for Global Power.* New York: Basic Books, 2003.

Fields, Karen E. *Revival and Rebellion in Colonial Central Africa.* Princeton, NJ: Princeton University Press, 1985.

Fink, Carole. *Defending the Rights of Others.* Cambridge: Cambridge University Press, 2006.

Foucault, Michel. *The Archaeology of Knowledge.* New York: Pantheon Books, 1972.

———. *Discipline and Punish: The Birth of the Prison.* New York: Vintage Books, 1979.

———. *Power/Knowledge: Selected Interviews and Other Writings, 1972–1977.* New York: Pantheon Books, 1980.

Fox, Francis. *River, Road, and Rail: Some Engineering Reminiscences.* London: John Murray, 1904.

Friesel, Evyatar. "Herbert Samuel's Reassessment of Zionism in 1921." *Studies in Zionism* 5, no. 2 (Autumn 1984): 213–37.

Fromkin, David. *A Peace to End All Peace: Creating the Modern Middle East, 1914–1922.* New York: H. Holt, 1989.

Gallagher, John, and Ronald Robinson. "The Imperialism of Free Trade." *Economic History Review* 6, no. 1 (1953): 1–15.

Gelvin, James L. *Divided Loyalties: Nationalism and Mass Politics in Syria at the Close of Empire.* Berkeley: University of California Press, 1998.

———. *The Israel-Palestine Conflict: One Hundred Years of War.* New York: Cambridge University Press, 2005.

Gerber, Haim. "'Palestine' and Other Territorial Concepts in the 17th Century." *International Journal of Middle East Studies* 30, no. 4 (November 1998): 563–72.

———. *Remembering and Imagining Palestine: Identity and Nationalism from the Crusades to the Present*. New York: Palgrave Macmillan, 2008.

Gide, Charles, *Communist and Co-operative Colonies*. London: Harrap, 1930.

Gieryn, Thomas F. "Boundary-Work and the Demarcation of Science from Non-science: Strains and Interests in Professional Ideologies of Scientists." *American Sociological Review* 48, no. 6 (1983): 781–95.

———. *Cultural Boundaries of Science: Credibility on the Line*. Chicago: University of Chicago Press, 1999.

Gil-Har, Yitzhak. "Boundaries Delimitation: Palestine and Trans-Jordan." *Middle Eastern Studies* 36, no. 1 (2000): 68–81.

———. "British Commitments to the Arabs and Their Application to the Palestine–Trans-Jordan Boundary: The Issue of the Semakh Triangle." *Middle Eastern Studies* 29, no. 4 (1993): 690–701.

Gilbar, Gad G. *Ottoman Palestine, 1800–1914: Studies in Economic and Social History*. Leiden: Brill, 1990.

Good, Kenneth. "Settler Colonialism: Economic Development and Class Formation." *Journal of Modern African Studies* 14, no. 4 (1976): 597–620.

Gooday, Graeme. *Domesticating Electricity: Technology, Uncertainty and Gender, 1880–1914*. London: Pickering and Chatto, 2008.

Goswami, Manu. *Producing India: From Colonial Economy to National Space*. Chicago: University of Chicago Press, 2004.

Guroff, Gregory. "The Legacy of Pre-revolutionary Economic Education: St. Petersburg Polytechnic Institute." *Russian Review* 31, no. 3 (July 1972): 272–81.

Halevi, Nadav. *Banker to an Emerging Nation: The History of Bank Leumi le-Israel*. Haifa: Shikmona, 1981.

Halkin, Hillel. *Jabotinsky: A Life*. New Haven, CT: Yale University Press, 2014.

Hannah, Leslie. *Electricity before Nationalisation*. Baltimore: Johns Hopkins University Press, 1979.

Haraway, Donna. *Primate Visions: Gender, Race, and Nature in the World of Modern Science*. New York: Routledge, 1989.

Harding, Sandra G. *Is Science Multicultural? Postcolonialisms, Feminisms, and Epistemologies*. Bloomington: Indiana University Press, 1998.

Harris, Leila M., and Samer Alatout. "Negotiating Hydro-Scales, Forging States: Comparison of the Upper Tigris/Euphrates and Jordan River Basins." *Political Geography* 29, no. 3 (2010): 148–56.

Harvey, David. *Social Justice and the City*. Rev. ed. Athens: University of Georgia Press, 2009.

———. *Spaces of Global Capitalism: Toward a Theory of Uneven Geographical Development*. Brooklyn: Verso, 2006.

Hausman, William J., Peter Hertner, and Mira Wilkins. *Global Electrification: Multinational Enterprise and International Finance in the History of Light and Power, 1878–2007*. Cambridge: Cambridge University Press, 2008. Kindle Edition.

Headrick, Daniel. *The Tentacles of Progress: Technology Transfer in the Age of Imperialism, 1850–1940*. Oxford: Oxford University Press, 1988.

———. *The Tools of Empire: Technology and European Imperialism in the Nineteenth Century*. New York: Oxford University Press, 1981.

———. "The Tools of Imperialism: Technology and the Expansion of European Colonial Empires in the Nineteenth Century." *Journal of Modern History* 51 (June 1979): 231–63.

Hecht, Gabrielle, ed. *Entangled Geographies: Empire and Technopolitics in the Global Cold War*. Cambridge, MA: MIT Press, 2011.

Hecht, Gabrielle. *Being Nuclear: Africans and the Global Uranium Trade*. Johannesburg: Wits University Press, 2012.

———. *The Radiance of France: Nuclear Power and National Identity after World War II*. Cambridge, MA: MIT Press, 1998.

Hertzberg, Arthur. *The Zionist Idea: A Historical Analysis and Reader*. Garden City, NY: Doubleday, 1959.

Herzl, Theodor. *Altneuland: Old-New Land*. Haifa: Haifa Publishing, 1960.

Hetherington, Penelope. *British Paternalism and Africa 1920–1940*. Totowa, NJ: Frank Cass, 1978.

Hisin, Hayyim. *A Palestine Diary: Memoirs of a Bilu Pioneer, 1882–1887*. New York: Herzl Press, 1976.

Hodge, Joseph Morgan. *Triumph of the Expert: Agrarian Doctrines of Development and the Legacies of British Colonialism*. Athens: Ohio University Press, 2007.

Huber, Valeska. *Channelling Mobilities: Migration and Globalisation in the Suez Canal Region and Beyond, 1869–1914*. New York: Cambridge University Press, 2013.

Hughes, Thomas P. *Networks of Power: Electrification in Western Society, 1880–1930*. Baltimore: Johns Hopkins University Press, 1983.

———. "The Seamless Web: Technology, Science, Etcetera, Etcetera." *Social Studies of Science* 16, no. 2 (1986): 281–92.

Huler, Scott. *On the Grid: A Plot of Land, an Average Neighborhood, and the Systems That Make Our World Work*. Emmaus, PA: Rodale, 2010.

Huneidi, Sahar. *A Broken Trust: Herbert Samuel, Zionism and the Palestinians, 1920–1925*. London: I. B. Tauris, 2001.

Jasanoff, Sheila. "Science, Politics, and the Renegotiation of Expertise at EPA." *Osiris* 7 (1992): 195–217.

———, ed. *States of Knowledge: The Co-production of Science and Social Order*. London: Routledge, 2004.

Joerges, Bernward. "Do Politics Have Artifacts?" *Social Studies of Science* 29, no. 3 (1999): 411–31.

Jones, Toby Craig. *Desert Kingdom: How Oil and Water Forged Modern Saudi Arabia*. Cambridge, MA: Harvard University Press, 2010.

Kaiser, Robert, and Elena Nikiforova. "The Performativity of Scale: The Social Construction of Scale Effects in Narva, Estonia." *Environment and Planning D: Society and Space* 26, no. 3 (2008): 537–62.

Kamen, Charles. *Little Common Ground: Arab Agriculture and Jewish Settlement in Palestine, 1920–1948*. Pittsburgh: University of Pittsburgh Press, 1991.

Kaufman, Asher. *Contested Frontiers in the Syria-Lebanon-Israel Region: Cartography, Sovereignty, and Conflict.* Baltimore: Johns Hopkins University Press, 2014.

Kayyālī, 'Abd al-Wahhāb. *Wathā'iq al-muqāwamah al-Filasṭīnīyah al-'Arabīyah ḍidda al-iḥtilāl al-Barīṭānī wa-al-ṣihyūnīyah, 1918–1939.* Beirut: Mu'assasat al-Dirāsāt al-Filasṭīnīyah, 1968.

Keith-Roach, Edward. *Pasha of Jerusalem: Memoirs of a District Commissioner under the British Mandate.* London: Radcliffe Press, 1994.

Kelly, Matthew Kraig. *The Crime of Nationalism: Britain, Palestine, and Nation-Building on the Fringe of Empire.* Oakland: University of California Press, 2017.

Khalidi, Rashid. *The Iron Cage: The Story of the Palestinian Struggle for Statehood.* Boston: Beacon Press, 2006.

———. *Palestinian Identity: The Construction of Modern National Consciousness.* New York: Columbia University Press, 1997.

Khalidi, Walid. *All That Remains: The Palestinian Villages Occupied and Depopulated by Israel in 1948.* Washington, DC: Institute for Palestine Studies, 1992.

———. *From Haven to Conquest: Readings in Zionism and the Palestine Problem until 1948.* Washington, DC: Institute for Palestine Studies, 1971.

Kimmerling, Baruch. *Zionism and Territory: The Socio-Territorial Dimensions of Zionist Politics.* Berkeley: University of California Press, 1983.

Kline, Ronald R. *Consumers in the Country: Technology and Social Chance in Rural America.* Baltimore: Johns Hopkins University Press, 2000.

Knorr Cetina, Karin. *Epistemic Cultures: How the Sciences Make Knowledge.* Cambridge, MA: Harvard University Press, 1999.

Kocka, Jörgen. *Capitalism: A Short History.* Princeton, NJ: Princeton University Press, 2016.

Koponen, Juhani. *Development for Exploitation: German Colonial Policies in Mainland Tanzania, 1884–1914.* Helsinki: Finnish Historical Society, 1994.

Krämer, Gudrun. *A History of Palestine: From the Ottoman Conquest to the Founding of the State of Israel.* Princeton, NJ: Princeton University Press, 2008.

Kupferschmidt, Uri. *The Supreme Muslim Council: Islam under the British Mandate for Palestine.* Leiden: E. J. Brill, 1987.

Lapin, G. G. "70 Years of Gidroproekt and Hydroelectric Power in Russia." *Hydrotechnical Construction* 34, nos. 8–9 (August 2000): 374–79.

Laqueur, Walter. *A History of Zionism.* New York: Holt, Rinehart and Winston, 1972.

Lash, Scott, and John Urry. *The End of Organized Capitalism.* Madison: University of Wisconsin Press, 1987.

Latour, Bruno. "On Actor-Network Theory: A Few Clarifications." *Sociale Welt* 47, no. 4 (1996): 369–81.

———. "On Some of the Affects of Capitalism." Lecture presented at the Royal Academy, Copenhagen, February 26, 2014.

———. *Pandora's Hope: Essays on the Reality of Science Studies.* Cambridge, MA: Harvard University Press, 1999.

———. *The Pasteurization of France*. Cambridge, MA: Harvard University Press, 1988.
———. *Reassembling the Social: An Introduction to Actor-Network-Theory*. Oxford: Oxford University Press, 2005.
———. *Science in Action: How to Follow Scientists and Engineers through Society*. Cambridge, MA: Harvard University Press, 1987.
———. *We Have Never Been Modern*. Cambridge, MA: Harvard University Press, 1993.
Law, John. "Notes on the Theory of the Actor-Network: Ordering, Strategy and Heterogeneity." *Systems Practice* 5, no. 4 (August 1992): 379–93.
———, ed. *Power, Action and Belief: A New Sociology of Knowledge?* London: Routledge, 1986.
Law, John, and John Hassard, eds. *Actor Network Theory and After*. Oxford: Blackwell/Sociological Review, 1999.
Lefebvre, Henri. *The Production of Space*. Cambridge, MA: Blackwell, 1991.
———. *The Survival of Capitalism*. New York: St. Martin's Press, 1976.
Leiss, William. *The Domination of Nature*. Montreal: McGill-Queen's University Press, 1972.
Lewis, Bernard. "On the History and Geography of a Name." *International Review* 2, no. 1 (January 1980): 1–12.
Lockman, Zachary. *Comrades and Enemies: Arab and Jewish Workers in Palestine, 1906–1948*. Berkeley: University of California Press, 1996.
Ludden, David. "Patronage and Irrigation in Tamil Nadu: A Long-Term View." *Indian Economic and Social History Review* 16, no. 3 (September 1979): 358–60.
Lugard, Frederick D. *The Dual Mandate in British Tropical Africa*. London: William Blackwood and Sons, 1922.
MacDonald, Robert H. *The Language of Empire: Myths and Metaphors of Popular Imperialism, 1880–1918*. Manchester, UK: Manchester University Press, 1994.
MacKenzie, Donald A., Fabian Muniesa, and Lucia Siu. *Do Economists Make Markets? On the Performativity of Economics*. Princeton, NJ: Princeton University Press, 2007.
MacMillan, Margaret. *Paris 1919: Six Months That Changed the World*. New York: Random House, 2002.
Mandel, Neville J. *The Arabs and Zionism before World War I*. Berkeley: University of California Press, 1976.
Marlowe, John. *Milner: Apostle of Empire*. London: Hamish Hamilton, 1976.
Marx, Charles D. "Social and Economic Aspects of Hydro-Electric Power." *Proceedings of the American Society of Civil Engineers* 48, no. 9 (November 1922): 1781–87.
Marx, Leo. "Technology: The Emergence of a Hazardous Concept." *Technology and Culture* 51, no. 3 (July 2010): 561–77.
Massad, Joseph. *Colonial Effects: The Making of National Identity in Jordan*. New York: Columbia University Press, 2001.
Mathew, William M. "The Balfour Declaration and the Palestine Mandate, 1917–1923: British Imperialist Imperatives." *British Journal of Middle Eastern Studies* 40, no. 3 (2013): 231–50.

———. "War-Time Contingency and the Balfour Declaration of 1917: An Improbable Regression." *Journal of Palestine Studies* 40, no. 2 (Winter 2011): 26–42.
Matthews, Weldon C. *Confronting an Empire, Constructing a Nation: Arab Nationalists and Popular Politics in Mandate Palestine*. London: I. B. Tauris, 2006.
McMullin, Ernan, ed. *The Social Dimensions of Science*. Notre Dame, IN: University of Notre Dame Press, 1992.
Medoff, Rafael, and Chaim I. Waxman. *The A to Z of Zionism*. Lanham, MD: Scarecrow Press, 2009.
Meijers, Anthony, ed. *Handbook of the Philosophy of Science*. Vol. 8, *Philosophy of Technology and Engineering Sciences*. London: Elsevier, 2009.
Meinertzhagen, Richard. *Middle East Diary, 1917–1956*. New York: Yoseloff, 1960.
Melancon, Michael S., and Donald J. Raleigh, eds. *Russia's Century of Revolutions: Parties, People, Places: Studies Presented in Honor of Alexander Rabinowitch*. Bloomington, IN: Slavica, 2012.
Mendes-Flohr, Paul R., and Jehuda Reinharz, eds. *The Jew in the Modern World: A Documentary History*. New York: Oxford University Press, 1980.
Metcalf, Thomas R. *Ideologies of the Raj*. Cambridge: Cambridge University Press, 1994.
Metzer, Jacob. *The Divided Economy of Mandatory Palestine*. Cambridge: Cambridge University Press, 1998.
Mitchell, Timothy. *Carbon Democracy: Political Power in the Age of Oil*. London: Verso, 2011.
———. "The Limits of the State: Beyond Statist Approaches and Their Critics." *American Political Science Review* 85, no. 1 (1991): 77–96.
———. *Rule of Experts: Egypt, Techno-Politics, Modernity*. Berkeley: University of California Press, 2002.
———. "The Work of Economics: How a Discipline Makes Its World." *European Journal of Sociology* 46, no. 2 (2005): 297–320.
Moon, Suzanne. "Empirical Knowledge, Scientific Authority, and Native Development: The Controversy over Sugar/Rice Ecology in the Netherlands East Indies, 1905–1914." *Environment and History* 10, no. 1 (2004): 59–81.
Moore, Jason W. *Capitalism in the Web of Life: Ecology and the Accumulation of Capital*. New York: Verso, 2015.
Morris, Benny. *Righteous Victims: A History of the Zionist-Arab Conflict, 1881–1999*. New York: Alfred A. Knopf, 1999.
Mrázek, Rudolf. *Engineers of Happy Land: Technology and Nationalism in a Colony*. Princeton, NJ: Princeton University Press, 2002.
Mukerji, Chandra. *A Fragile Power: Scientists and the State*. Princeton, NJ: Princeton University Press, 1989.
———. "The Territorial State as a Figured World of Power: Strategics, Logistics, and Impersonal Rule." *Sociological Theory* 28, no. 4 (December 2010): 402–24.
Nadan, Amos. *The Palestinian Peasant Economy under the Mandate: A Story of Colonial Bungling*. Cambridge, MA: Harvard University Press, 2006.
Nash, Linda. "The Agency of Nature or the Nature of Agency?" *Environmental History* 10, no. 1 (January 2005): 67–69.

———. *Inescapable Ecologies: A History of Environment, Disease, and Knowledge.* Berkeley: University of California Press, 2006.

Needham, Andrew. *Power Lines: Phoenix and the Making of the Modern Southwest.* Princeton, NJ: Princeton University Press, 2014.

Nelson, Cary, and Lawrence Grossberg, eds. *Marxism and the Interpretation of Culture.* Urbana: University of Illinois Press, 1988.

Netanyahu, Binyamin. *A Durable Peace: Israel and Its Place among the Nations.* New York: Warner Books, 2000.

Norris, Jacob. *Land of Progress: Palestine in the Age of Colonial Development, 1905–1948.* Oxford: Oxford University Press, 2013.

———. "Repression and Rebellion: Britain's Response to the Arab Revolt in Palestine of 1936–9." *Journal of Imperial and Commonwealth History* 36, no. 1 (March 2008): 25–45.

Nye, David E. *America as Second Creation: Technology and Narratives of New Beginnings.* Cambridge, MA: MIT Press, 2003.

———. *Electrifying America: Social Meanings of a New Technology.* Cambridge, MA: MIT Press, 1990.

———. *Technology Matters: Questions to Live With.* Cambridge, MA: MIT Press, 2006.

Oren, Michael B. "Ha-Masa o-Matan lehasgat Zikayion LeKhevrat Ha-Kheshmal" [The negotiation to obtain a concession to electrify Palestine]. *Cathedra* 26 (5743; 1982): 133–76.

Orwell, George. *All Art Is Propaganda: Critical Essays.* New York: Mariner Books/Houghton Mifflin Harcourt, 2009.

Otter, Chris. *The Victorian Eye: A Political History of Light and Vision in Britain, 1800–1910.* Chicago: University of Chicago Press, 2008.

Owen, Roger. *State, Power and Politics in the Making of the Modern Middle East.* New York: Routledge, 2004.

———, ed. *Studies in the Economic and Social History of Palestine in the Nineteenth and Twentieth Centuries.* London: Macmillan, 1982.

Pappe, Ilan. *The Rise and Fall of a Palestinian Dynasty: The Husaynis, 1700–1948.* Berkeley: University of California Press, 2010.

Peckham, Robert. "Economies of Contagion: Financial Crisis and Pandemic." *Economy and Society* 42, no. 2 (May 2013): 226–48.

Pedersen, Susan. *The Guardians: The League of Nations and the Crisis of Empire.* New York: Oxford University Press, 2015.

Penslar, Derek. *Zionism and Technocracy: The Engineering of Jewish Settlement in Palestine, 1870–1918.* Bloomington: Indiana University Press, 1991.

Pickering, Andrew, ed. *Science as Practice and Culture.* Chicago: University of Chicago Press, 1992.

Pitts, Jennifer. *A Turn to Empire: The Rise of Imperial Liberalism in Britain and France.* Princeton, NJ: Princeton University Press, 2005.

Porath, Yehoshua. *The Emergence of the Palestinian-Arab National Movement, 1918–1929.* London: Cass, 1974.

Provence, Michael. *The Great Syrian Revolt and the Rise of Arab Nationalism*. Austin: University of Texas Press, 2005.

Rabinow, Paul, ed. *The Foucault Reader*. New York: Pantheon Books, 1984.

Reichman, Shalom, Yossi Katz, and Yair Paz. "The Absorptive Capacity of Palestine, 1882–1948." *Middle Eastern Studies* 33, no. 2 (April 1997): 338–61.

Renton, James. *The Zionist Masquerade: The Birth of the Anglo-Zionist Alliance, 1914–1918*. New York: Palgrave Macmillan, 2007.

Robins, Philip. *A History of Jordan*. Cambridge: Cambridge University Press, 2004.

Rogan, Eugene L. *Frontiers of the State in the Late Ottoman Empire: Transjordan, 1850–1921*. Cambridge: Cambridge University Press, 1999.

Said, Edward. "Zionism from the Standpoint of Its Victims." *Social Text*, no. 1 (Winter 1979): 7–58.

Santiago, Myrna I. *The Ecology of Oil: Environment, Labor, and the Mexican Revolution, 1900–1938*. New York: Cambridge University Press, 2006.

Saul, S. B. "The Economic Significance of 'Constructive Imperialism.'" *Journal of Economic History* 17, no. 2 (June, 1957): 173–92.

Sayes, Edwin. "Actor-Network Theory and Methodology: Just What Does It Mean to Say That Nonhumans Have Agency?" *Social Studies of Science* 44, no. 1 (2014): 134–49.

Schmorak, E. *Palestine's Industrial Future*. Jerusalem: Rubin Mass, 1946.

Schneer, Jonathan. *The Balfour Declaration: The Origins of the Arab-Israeli Conflict*. New York: Random House, 2010.

Scott, James C. *Seeing Like a State: How Certain Schemes to Improve the Human Condition Have Failed*. New Haven, CT: Yale University Press, 1998.

Segev, Tom. *One Palestine, Complete: Jews and Arabs under the Mandate*. New York: Metropolitan Books, 2000.

Seikaly, May. *Haifa: Transformation of an Arab Society 1918–1939*. London: I. B. Tauris, 2002.

Seikaly, Sherene. *Men of Capital: Scarcity and Economy in Mandate Palestine*. Stanford, CA: Stanford University Press, 2015.

Shafiee, Katayoun. *Machineries of Oil: An Infrastructural History of BP in Iran*. Cambridge, MA: MIT Press, 2018.

———. "A Petro-Formula and Its World: Calculating Profits, Labour and Production in the Assembling of Anglo-Iranian Oil." *Economy and Society* 41, no. 4 (November 2012): 585–614.

Shaltiel, Eli. *Pinhas Rutenberg: 'Aliyato unefilato shel "ish chazak" beeretz yisra'el, 1879–1942*. Tel Aviv: Am Over, 1990.

Shamir, Ronen. *Current Flow: The Electrification of Palestine*. Stanford, CA: Stanford University Press, 2013.

Shapin, Steven. *The Scientific Revolution*. Chicago: University of Chicago Press, 1996.

———. *A Social History of Truth: Civility and Science in Seventeenth-Century England*. Chicago: University of Chicago Press, 1994.

Shapin, Steven, and Simon Schaffer. *Leviathan and the Air-Pump: Hobbes, Boyle, and the Experimental Life.* Princeton, NJ: Princeton University Press, 2011.
Shapira, Anita. *Israel: A History.* Waltham, MA: Brandeis University Press, 2012.
Shlaim, Avi. *Collusion across the Jordan: King Abdullah, the Zionist Movement, and the Partition of Palestine.* New York: Columbia University Press, 1988.
———. *The Iron Wall: Israel and the Arab World.* New York: W.W. Norton, 2000.
Sidebotham, Herbert. *Great Britain and Palestine.* London: Macmillan, 1937.
Slyomovics, Susan. *The Object of Memory: Arab and Jew Narrate the Palestinian Village.* Philadelphia: University of Pennsylvania Press, 1998.
Smith, Barbara J. *The Roots of Separatism in Palestine: British Economic Policy, 1920–1929.* Syracuse, NY: Syracuse University Press, 1993.
Smith, Neil. *Uneven Development: Nature, Capital, and the Production of Space.* Athens: University of Georgia Press, [1984] 2008.
Smuts, Jan C. "African Settlement." In *Africa and Some World Problems*, 109–31. Oxford: Oxford Clarendon Press, 1930.
Snow, A. "The First National Grid." *Engineering Science and Education Journal* 2, no. 5 (October 1993): 215–24.
Spiegel, Nina S. *Embodying Hebrew Culture: Aesthetics, Athletics, and Dance in the Jewish Community of Mandate Palestine.* Detroit, MI: Wayne State University Press, 2013.
Sternhell, Zeev. *The Founding Myths of Israel.* Princeton, NJ: Princeton University Press, 1998.
Sufian, Sandra M. *Healing the Land and the Nation: Malaria and the Zionist Project in Palestine, 1920–1947.* Chicago: University of Chicago Press, 2007.
Swedenburg, Ted. *Memories of Revolt: The 1936–1939 Rebellion and the Palestinian National Past.* Fayetteville: University of Arkansas Press, 2003.
Taylor, Peter J. "Geographical Scales within the World Economy Approach." *Review* 5, no. 1 (1981): 3–11.
———. "A Materialist Framework for Political Geography." *Transactions of the Institute of British Geographers* 7, no. 1 (1982): 15–34.
———. "Technocratic Optimism, H. T. Odum, and the Partial Transformation of Ecological Metaphor after World War II." *Journal of the History of Biology* 21 (1988): 213–44.
Tendler, Judith. *Electric Power in Brazil: Entrepreneurship in the Public Sector.* Cambridge, MA: Harvard University Press, 1968.
Tibawi, Abdul Latif. *Anglo-Arab Relations and the Question of Palestine, 1914–1921.* London: Luzac, 1977.
Tilley, Helen. *Africa as a Living Laboratory: Empire, Development, and the Problem of Scientific Knowledge, 1870–1950.* Chicago: University of Chicago Press, 2011.
Troen, S. Ilan, *Imagining Zion: Dreams, Designs, and Realities in a Century of Jewish Settlement.* New Haven, CT: Yale University Press, 2003. Kindle Edition.
Tyler, Warwick P. N. *State Lands and Rural Development in Mandatory Palestine, 1920–1948.* Brighton: Sussex Academic Press, 2001.

Van Beusekom, Monica M. *Negotiating Development: African Farmers and Colonial Experts the Office Du Niger, 1920–1960*. Portsmouth, NH: Heinemann, 2002.

Vilbush, Nahum. *Haroshet ha-ma'aseh: kovets letoldot ha-ta'asiyah baaretz: lezekher po'olo shel Nahhum Vilbush, haluts hata'asiyah ha'ivrit hahhadishah baaretz*. Tel Aviv: Hotsa'at Milo, 1974.

Vitalis, Robert. *America's Kingdom: Mythmaking on the Saudi Oil Frontier*. Stanford, CA: Stanford University Press, 2007.

———. *When Capitalists Collide Business Conflict and the End of Empire in Egypt*. Berkeley: University of California Press, 1995.

Walker, Brett L. *Toxic Archipelago: A History of Industrial Disease in Japan*. Seattle: University of Washington Press, 2010.

Wasserstein, Bernard. *The British in Palestine: The Mandatory Government and the Arab-Jewish Conflict, 1917–1929*. London: Royal Historical Society, 1978.

Weber, Max. *The Protestant Ethic and the "Spirit" of Capitalism and Other Writings*. New York: Penguin Books, 2002.

Weisbord, Robert G. *African Zion: The Attempt to Establish a Jewish Colony in the East Africa Protectorate, 1903–1905*. Philadelphia: Jewish Publication Society of America, 1968.

Weizmann, Chaim. *Trial and Error: The Autobiography of Chaim Weizmann*. New York: Harper, 1949.

White, Richard. *The Organic Machine*. New York: Hill and Wang, 1995.

———. *Railroaded: The Transcontinentals and the Making of Modern America*. New York: W. W. Norton, 2012.

Whitley, R. D. "Black Boxism and the Sociology of Science: A Discussion of the Major Developments in the Field." *Sociology of Science, Sociological Review Monograph*, no. 18 (1972): 61–92.

Wilson, Mary C. *King Abdullah, Britain, and the Making of Jordan*. Cambridge: Cambridge University Press, 1987.

Winner, Langdon. "Do Artifacts Have Politics?" *Daedalus* 109, no. 1 (Winter 1980): 121–36.

Wise, M. Norton, ed. *The Values of Precision*. Princeton, NJ: Princeton University Press, 1997.

Wright, Quincy. *Mandates under the League of Nations*. Chicago: University of Chicago Press, 1930.

Yehoshua, Jacob. *Ta'rīkh al-sihāfah al-àrabīyah fī Filastīn fī al-'ahd al-Ùthmāni, 1908–1918*. Jerusalem: Matba at al-Ma arif, 1974.

Young, Robert, ed. *Untying the Text: A Post-structuralist Reader*. Boston: Routledge, 1981.

Zammito, John H. *A Nice Derangement of Epistemes: Post-positivism in the Study of Science from Quine to Latour*. Chicago: University of Chicago Press, 2004.

Zerubavel, Yael. "The Politics of Interpretation: Tel Hai in Israeli Collective Memory." *Association for Jewish Studies Review* 16 (1991): 133–60.

———. *Recovered Roots: Collective Memory and the Making of Israeli National Tradition*. Chicago: University of Chicago Press, 1995.

Zhuravlev, Sergei. "'Little People' and 'Big History': Foreigners at the Moscow Electric Factory and Soviet Society, 1920s–1930s." *Russian Studies in History* 44, no. 1 (Summer 2005): 10–86.

Ziman, Yehoshua. *Hakalkala ha'eretzyisra'elit bemisparim* [The economy of the land of Israel in numbers]. Tel Aviv: Davar, 1929.

Zimmerman, Andrew. "'What Do You Really Want in German East Africa, *Herr Professor*?': Counterinsurgency and the Science Effect in Colonial Tanzania." *Comparative Studies in Society and History* 48, no. 2 (April 2006): 419–61.

Zubaida, Sami. *Islam, the People, and the State: Political Ideas and Movements in the Middle East*. New York: I. B. Taurus, 1989.

INDEX

Aaronsohn, 47
Abadieh, 26, 120, 123
Abdullah, Emir, 73, 114, 126, 130, 132, 138, 139, 141, 214, 215
Absorptive capacity, 40–41, 209
Acre, 109, 161, 166
Actor-Network Theory, 19, 162–163
AEG, 61, 93
Ahdut HaAvoda, 81
Aliyah: first, 45; fifth, 160; second, 45
Allenby, Edmund, 142
American Jewish Congress, 25
Anderson, Benedict, 10, 115, 119
Anglo-French border negotiations, 70, 72
Anglo-Palestine Company, 113, 253n101
Arab Executive, The, 7, 79, 82, 83
Auja basin, 64–65, 81, 95, 100, 158, 252n87
Auja River, 27, 63–67, 81
Auja scheme, 63–67

Baharaw, Yakutiel, 105, 132, 166, 167, 178
Balfour, Arthur, 37
Balfour Declaration, 12–13, 35–38, 43, 72, 183, 219
al-Baytar, 'Umar, 98–99
Beisan, 26, 54, 55, 118, 129, 157, 161, 174, 221
Ben-Ami, Pinhas, 25
Ben-Gurion, David, 3, 46, 48, 134, 144, 214
Bentwich, Norman, 127
Ben-Zvi, Yitzhak, 46, 48
Bible, The, 13, 21, 43, 47, 49, 70
Bilad al-Sham. *See* Syria
Bishara, Salim, 165

Blackouts, 200–207
Bols, Louis J., 27
Bolsheviks, 23, 26, 32, 33, 49, 50
Borochov, Ber, 25
Boundary-work, 11, 108, 114–116, 193, 213, 220; oppositional, 80, 169
Brandeis, Louis, 96
Burchells & Co., 67, 148
al-Bustani, Wadi', 101

Cairo Conference, 126
Cecil, Lord, 38, 236n90
Centralization, 7, 27, 28, 32–33, 41, 121–122, 169, 220, 222
Clapp Report, 214
Clauson, G. L. M., 68, 69, 136
Chamberlain, Joseph, 29, 30, 32, 37, 38
Churchill, Winston, 21, 26, 40, 60, 66, 70
Civilizing mission, 35, 38, 77
Colonial development, 14, 28–33, 39, 80, 149, 220
Common sense, 2, 19–20, 30, 31, 160, 221
Constructive imperialism, 29–30, 37, 40, 47
Consulting engineers, 28, 61, 63, 64, 66–67, 76, 84, 120, 121
Council of People's Commissars, 23
Cox, Henry, 130

al-Dajani, Kamal, 101
Dawson, Philip, 75
de Bunsen report, 142
Degania, 26, 129, 210

Development. *See* colonial development
Disenchantment, 136–137
Division of labor, 16–17, 141–143, 148–149
Dual mandate, 33–34, 36, 59

Economies of scale, 16–17, 122
Eder, David, 93
Egypt, 29, 31, 32
Electricity theft, 195–196
Electrocution, 202–203

Foucault, Michel, 15–16, 116
Free market. *See* free trade
Free trade, 9, 38, 88–89, 222, 228n26

Gapon, Father, 25
General Electric, 48, 61, 75, 121, 139
General Organization of Workers in the Land of Israel, The. *See* histadrut
Gide, Charles, 218
GOELRO Plan. *See* State Commission for the Electrification of Russia
Governmentality, 14, 16, 221
Great Arab Revolt, The, 12, 41, 151–152, 170–176
Greater Syria. *See* Syria

Hadera, 112, 155, 171, 172, 173
al-Hadi, Awni abd, 203
Haifa, 2, 49; economic and population growth, 154; electrification, 46, 54, 57, 62, 81, 101, 107–109, 110, 112, 123–124, 143, 161–162; port, 48, 142–143; protests, 88
Herzl, Theodor, 25, 30, 36, 137
Heterogeneous engineering, 3–4, 19, 24, 77–78, 150, 155, 164, 208
Hirst, Hugo, 48, 61, 75, 139
Histadrut, 134, 146
Hoofien, Siegfried, 113
al-Hudayf, Zaki, 105
Husayn-McMahon correspondence, 43, 72, 125, 238–239n121
al-Husayni, Amin, 168, 169, 171, 196
al-Husayni, Fakhri, 196
al-Husayni, Jamal, 103–104
Hybrid causation, 55

India, 31, 32, 33, 121

Infrastructural state, 6
Insull, Samuel, 90
Integration: horizontal, 16; technological, 24, 139, 159, 215; vertical, 16–17
Iraq, 35, 59, 76, 114, 126, 214,
Irrigation, 27, 40, 46, 49, 53, 65, 67, 81, 95, 100, 133, 176, 179, 215, 252n85, 252n87
Island of Peace, 215
Israel Electric Corporation, 10

Jabotinsky, Vladimir, 25, 113, 258–259n193
Jaffa: electrification, 46, 54, 57, 61, 63–65, 67, 81, 84, 97, 103, 108, protests, 2, 4, 11, 12, 27, 98–102, 104; railway, 62–63; 1921 riots, 84–85; telegraph lines, 44
Jaffa municipal council, 84–89, 99–101, 106
Jenin, 163
Jerusalem: electrification, 27, 98; protests, 102, water supply, 95
Jerusalem District Electric Company, 217
Jerusalem Electric & Public Service Corporation, Ltd., 190
Jewish Legion, 25
Jewish National Council. *See* Vaad Leumi
Jezreel Valley, 161
Jisr al-Mujami', 56, 129, 130, 137
Jordan River, 40, 72, 117–118, 119, 125–127, 210
Joynson-Hicks, William, 95–96, 127, 253n91

Kerensky, Alexander, 25
Khalidi, Rashid, 13
al-Khalidi, Yusuf Diya', 37
Kidd, Benjamin, 39, 233n52
King-Crane commission, 43

Land sales: Auja basin, 91–2, 94–95; Beisan, 128–130; Haifa, 251n68; Jaffa, 91–92; Jordan Valley, 133; Tel Aviv, 93; Transjordan, 130–133
Large Technological Systems, 16–20, 23–24, 27–33, 54, 150–153
Latour, Bruno, 19. *See also* Actor-Network Theory
League of Nations, 34–35, 46, 73, 125. *See also* mandates system, permanent mandates commission

Legibility, 6, 27, 137, 138, 149, 264n99. *See also* quantification
Lenin, Vladimir, 23, 28
Lowenstein, M. L. C., 121
Lugard, Frederick, 34, 35, 40
Lydda, 110, 163

Mandates system of the League of Nations, 34–35
Marx, Charles, 30, 55
Mavrommatis, Euripides, 44, 46, 189, 190
McLellan, William, 121–122
Mendeleev, Dimitri, 23
Merz, Charles H., 90, 121–122
Mesopotamia. *See* Iraq
Midian Ltd., 44
Migdal, 112
Mikveh Israel, 110
Mill, John Stuart, 29
Miller, Oskar von, 90
Milner, Alfred, 34, 38, 46
Momentum. *See* systemic momentum, technological momentum
Mond, Alfred, 28, 52
Morrison-Grady plan, 211
Mount Scopus, 191
Muslim-Christian Association (MCA), 98, 99, 102, 254n107

Nablus, 2; electrification, 46, 135, 166–170, 176–182
al-Nabulsi, Nimr, 177
Naharayim, 119, 134, 142–149, 155, 160, 191, 210, 213–215, 216
Nashashibi, Ragheb, 98–99
National space, 7, 48, 118, 125, 137, 149, 152, 219
Nazareth, 27, 102, 105, 161, 164–165
Negev, 174
Nesher, 162
Nes Ziona, 113
Netanya, 112,
New Deal, The, 28
New Economic Policy, The, 216
New imperialism. *See* constructive imperialism
Nigeria, 32
1948 War, 2–3, 5–7, 13, 14, 213–216

Novomeysky, Moshe, 144, 213

Ormsby-Gore, William, 35, 38, 104, 189, 210, 241n146
Ottoman Empire, 34, 41–42, 44, 45, 218
Orwell, George, 4

Palestine mandate charter, 36, 39, 126
Palestine Royal Commission of 1937, 12, 39, 174, 209–212
Paris Peace Conference, 26
Peel Commission. *See* Palestine Royal Commission of 1937
Permanent Mandates Commission of the League of Nations, 13, 36, 91, 111
Philby, Harry St. John, 127
Pinsker, Leo, 25
Policy of detail, 26
Politics of non-politics, 20, 89–90, 95, 114, 116, 189, 196, 214
Preece, Cardew, and Rider, 64, 84
Putilov Plant, 25

Quantification, 5–6, 18, 26–27, 32, 56, 77, 120, 138, 140–141, 148–149, 153–154, 182–184, 221–222

Ra'anana, 112
Ramleh, 109, 110, 206
Randall, Palmer, and Triton, 142
Rationalization, 6, 18, 221. *See also* politics of non-politics
Rehovot, 110, 112
Rishon Letzion, 110, 112, 113, 136, 174
Round Table, The, 34, 46, 47
Rothschild, James de, 7, 170
Russian Revolution of 1905, 25
Russian Revolution of 1917, 25–26
Rural Electrification Administration, 28
Rutenberg, Abraham, 204–205
Rutenberg, Pinhas, 21–22; British attitudes toward, 7, 28, 33, 40, 59, 121; and borders, 69–73; as chairman of the Vaad Leumi, 145; education, 22–23; friendship with Emir Abdullah, 132; and fundraising in America, 96–97; and the Jordan Plan, 53–58; negotiations with the British, 52–54; opposition to,

Rutenberg *(continued)*
  2, 11, 20, 79–80, 95–96, 115, 135; and the Palestine Electric Corporation, 5; at the Paris Peace conference, 26; relations with PICA, 133; and the Russian revolutions, 25; survey work, 26, 51; as systems entrepreneur, 24, 50, 109–110; and technocapitalism, 48; and Zionism, 25, 60

Sabotage, 11, 80–83, 87, 88, 89, 93, 94, 102, 107, 115, 116, 152, 166, 168, 171–178, 268n4
Sacher, Harry, 92, 93, 119
Said, Edward, 26, 42, 222, 231n23
al-Sa'id, 'Issam, 84, 99, 101, 108
al-Said, Nuri, 114
Samakh, 55, 56, 73, 126–128, 158, 210, 211
Samuel, Herbert, 38, 51, 59, 69, 81–82, 217
San Remo conference, 35–36
Sarafand, 110
Sarona, 110
Schumpeter, Joseph, 6
Scramble for Africa, 31
Sea of Galilee, 26, 53, 56, 58, 72, 112, 117–120, 124, 126, 133, 148, 158, 210
Segev, Tom, 13
Settler colonialism, 36, 40
Shamir, Ronen, xiii, 10–11, 66, 115, 162–164
Shapira, Anita, 13
Shaykh Muwannis (alt. Sheikh Munes), 85–86, 91
Shemen, 112
Shuckburgh, John, 51, 59, 60, 68
Sixth Arab Palestinian Congress, 2, 101–102, 109
Smuts, Jan, 38, 39, 236n90
Snell, John, 58, 62, 76
Social determinism, 17
Social-Revolutionary Party, 24–25
Sovereignty, 6, 14
Standardization, 157–160
State building, 4, 41, 45, 152, 223
State Commission for the Electrification of Russia (GOELRO) Plan, 23–24
Statistics, 23, 153, 170, 182–184. *See also* quantification

Stuart Hall: "the horizon of the taken-for-granted," 19, 31. *See also* common sense
St. Petersburg Polytechnic Institute, 22–23
Supreme Muslim Council, 115, 196
Syria, 11, 35, 55, 215
Sykes-Picot Agreement, 43, 72, 125, 238–239n121
Syro-Ottoman Agricultural Company, 44
Systemic momentum, 159–160, 185
Systems entrepreneurs, 24, 50, 90, 109, 122, 148

Tabari, Abd al-Salam, 105
Technocapitalism, 13–14, 19–20, 41–49, 76, 105, 143, 144, 183, 219–222
Technological determinism, 17
Technological momentum, 150–151
Technological utopianism, 28, 32, 49, 53, 67
Technopolitics, 41, 65, 77, 107
Tel Aviv: electricity before Rutenberg, 27; grid, 54, 57, 64, 79, 81, 98, 102, 110, 143, 161; powerhouse, 94, 97, 173–174, 216
Tel Or, 132, 137–140, 146, 148
Tennessee Valley Authority, 139
Tiberias, 44, 56, 86, 105–106, 111, 124
Tolkowski, Shmuel, 47
Trade Facilities Act, 139
Transjordan, 35, 73, 114, 125–134
"Triangle of terror" (Nablus, Qalqilya, Tulkarm), 172, 173
Triumph of the expert, 29
Tocqueville, Alexis de, 29
Tulkarm, 1, 46, 101, 163, 171, 173, 178, 214
Tuqan, Sulayman, 168, 177, 179, 200

Uganda Offer, 30
Uneven development, 3–5, 6, 151–152, 160, 162–164, 176
United Nations Special Committee on Palestine (UNSCOP), 211–212

Vaad Leumi (Jewish National Council), 145
Vernon, R. V., 69, 87, 89, 244n47
Vilbush, Nahum, 45, 54

Weber, Max, 6, 19, 116
Weizmann, Chaim, 21, 27, 37, 41, 49, 72, 129
World War I, 1–2, 26, 34

Yarkon River. *See* Auja River
Yarmuk River, 53–56, 67, 72–73, 76, 118–120, 125, 139, 140, 141
Yarmuk-Jordan Valley Project, 214
Young, Hubert, 59, 73, 127

Zionist Commission, 27, 45, 93
Zionist Organization, 21, 30, 36, 38; and commercial concessions 27, 45–46, 49; at the Paris Peace conference, 47; and Rutenberg, 52, 68, 90, 121